Reinforced plastics durability

Edited by

Geoffrey Pritchard

CRC Press
Boca Raton Boston New York Washington, DC

WOODHEAD PUBLISHING LIMITED
Cambridge England

Published by Woodhead Publishing Limited, Abington Hall, Abington
Cambridge CB1 6AH, England

Published in North and South America by CRC Press LLC, 2000 Corporate Blvd,
NW Boca Raton FL 33431, USA

First published 1999, Woodhead Publishing Ltd and CRC Press LLC
© 1999, Woodhead Publishing Ltd
The authors have asserted their moral rights.

British Library Cataloguing in Publication Data
A catalogue record for this book is available from the British Library.

Library of Congress Cataloging in Publication Data
A catalog record for this book is available from the Library of Congress.

Woodhead Publishing ISBN 1 85573 320 X
CRC Press ISBN 0-8493-0547-0
CRC Press order number: WP0547

Cover design by The ColourStudio
Typset by Best-set Typesetter Ltd., Hong Kong
Printed by St Edmundsbury Press, Suffolk, England

Contents

Preface

Primitive reinforced plastics products were known in the 1920s and 1930s, but the more advanced fibre reinforced materials we know today only became significant commercially as structural materials in the 1950s, and even then, the more recent reinforcing fibres such as carbon/graphite, aramid (e.g. Kevlar®, Twaron®) and polyethylene (e.g. Dyneema®) fibres were all still completely unknown. The great majority of reinforced plastics articles we use today have been manufactured since 1975.

Thus, we have a very limited number of case histories of structural applications with which to prove the durability of fibrous composites beyond dispute, although the evidence we have is very encouraging. This promising start is not surprising because, except in a very few special cases, they resist microbial organisms and are unaffected by electrolytic corrosion.

As we reach the end of the present millenium, we notice a great expansion of reinforced plastics into new fields such as load bearing parts of buildings, bridges over highways and various infrastructure pipelines. A mere 25 years of useful life will not suffice for these applications. In many cases the target is 60, 75 or even 100 years. This is a longer period than the entire previous history of reinforced plastics. The contribution that theory can make to predicting the achievable lifetimes of reinforced plastics is growing, but it is too early yet to base decisions entirely on conclusions that are unsupported by any practical or experimental confirmation. Therefore we have to use an amalgam of medium term laboratory research, case histories and a knowledge of the theoretical principles of the main degradation processes.

Major studies are currently being undertaken into composites durability. One of these programmes, the Composites Durability Study, is being carried out in the USA by the Civil Engineering Research Foundation and the Society of the Plastics Industry's Composites Institute Market Development Alliance, for the benefit of the design and construction industry and the civil engineering profession. It is said that particular concerns include performance degradation in severe operating environments, fatigue life, creep, fire resistance, weatherability and maintenance.

The problem with many technical books is that although they contain valuable information, it is not always made sufficiently accessible to those who most need it. This is often because of the failure to explain basic terminology to readers from different backgrounds and specialties. The editor has therefore made a considerable effort to ensure that at least the early chapters of this book are readily understandable by people from discipline areas other than composites science. It has not been possible to eliminate altogether the chemical and mathematical equations which the outsider often finds alienating, but they have been reduced to a minimum. Those who wish that the subject matter had been treated more rigorously – and no doubt there will be many – will find numerous references at the end of each chapter, indicating further literature with a more specialized orientation.

This book is therefore targeted at those concerned for the first time with using reinforced plastics in building, highway engineering, offshore engineering, civil and chemical engineering, marine and electrical areas. It is an entry level book for engineers, architects, entrepreneurs, managers, designers, and graduate students who have already completed courses in engineering or science and who now need to be able to converse better with consultants and specialists in the reinforced plastics world. If research chemists, resin technologists or plastics process engineers should also find something of interest, that will be a bonus.

The editor takes the blame for the frequent use of the phrase 'reinforced plastics' rather than the increasingly favoured, up-market term 'advanced composites'. He recognizes that glass fibre reinforced composites are customarily excluded from the advanced composite category in aerospace circles, chiefly because of their relatively low modulus. But when their technical qualities as a whole are considered, including their durability in the broad sense, and when their cost effectiveness is also taken into account, reinforced plastics scarcely need rebadging. They are among the best materials in the world in the context of durability.

List of contributors

Manufacturing Centre, School of Manufacturing, Materials and Mechanical Engineering, Plymouth University, PL4 8AA, England

Chapter 8 Survey of long term durability of fiberglass-reinforced plastics tanks and pipes
Ben Bogner, Amoco Chemicals, 150 West Warrenville Road D-7, Naperville, Illinois 60563–8460, USA

Chapter 9 Epoxy vinyl ester and other resins in chemical process equipment
Paul Kelly, Dow Deutschland Inc., Postfach 20, Industriestraße 1, D-77836 Rheinmünster, Germany

Chapter 10 Repairs using fibre reinforced plastics
Wing Kong Chiu and Rhys Jones, Department of Mechanical Engineering, Monash University, Wellington Road, Clayton Campus, Victoria 3168, Australia

Chapter 11 Fatigue performance: The role of the interphase
Nikhil E Verghese and John J Lesko, Materials Response Group, Department of Engineering Science and Mechanics, Virginia Polytechnic Institute and State University, Blacksburg, VA 24061-0219, USA

Chapter 12 Computer models for predicting durability
Samit Roy, Department of Mechanical Engineering, University of Missouri-Rolla, 1870 Miner Circle, Rolla, MO 65401, USA

1

An introduction to plastics for non-specialists

GEOFFREY PRITCHARD

1.1   Durability

Most materials have a finite life. Metals corrode; they can also suffer from fatigue. Wood rots, and concrete cracks or suffers from various chemical degradation processes. Natural rubber can perish as a result of ozone attack. All these materials have been around for long enough for us to know and make allowance for their weaknesses.

Our concern here is with identifying and assessing the weaknesses, if any, of reinforced plastics. The experience of museums is that plastics artifacts, even when kept on display in glass cases and preserved from rough handling, nevertheless eventually show signs of deterioration. We cannot assume that the same will apply to reinforced plastics, because the resins used in reinforced plastics are different. Of all the resins commonly used in reinforced plastics, phenolics are the oldest, having been known for over nine decades. They date commercially from about 1909, being used first as wood lacquers rather than in composites, whereas polyesters have been used structurally in reinforced plastics since the 1940s. All these resins have evolved considerably since then, because new improved varieties of resin come out every seven years or thereabouts, so it is arguable that we cannot know by experience what the lifetime of any of the modern equivalents on sale today is likely to be. Reinforced plastics are now being specified for applications designed to last for 30, 40 or even 60 years without loss of functional effectiveness. The accelerating trend towards using reinforced plastics in bridges and buildings means a further extension of the required lifetime, possibly to almost a century – longer than the entire history of the reinforced plastics industry.

This book is designed to bridge the gap between composite materials specialists and end-users. It will discuss in broad terms the ways in which deterioration in reinforced plastics occurs, and the ways to delay it. We shall concentrate mainly, but not exclusively, on time-dependent processes, such as weathering, corrosion, wear, fatigue and heat aging, rather than on sudden destruction by impact or fire, although the latter dangers are mentioned too.

Inevitably, by its nature, the book highlights potential *problems*, rather than opportunities, for the user of reinforced plastics. The resulting impression could easily be too negative, as when a layman anxiously consults a medical textbook about his health – there seem to be too many things to go wrong! Such a response would be unduly pessimistic. We should recall that reinforced plastics have surpassed early expectations by their durable and reliable performance in outdoor use. Some applications have been very demanding, notably in the marine sector, the offshore oil industry, aerospace, and perhaps most obviously, in the chemical process equipment field.

Moreover, many of the early successes were achieved in the early days of the new material, in the 1940s, 1950s and 1960s, without the benefit of any substantial theoretical basis or long term performance database. The approach then was understandably one of cautious overdesign. Later, designers moved towards more cost effective and lightweight structures, without sacrificing too much durability. It was not sufficient to build a boat that survived, it had to win races. Aircraft had to lose as much weight as possible. Actual performance continued to exceed the pessimistic expectations produced by an obsession with the literature on failure mechanisms. Whether this happy state of affairs will continue, with ever more demanding applications and expectations, remains to be seen, but it now at least seems possible to identify the application areas where reinforced plastics can safely be used.

1.2   Cost effectiveness and product lifetimes

A product is at the end of its useful life when it no longer fulfils its technical function. But technical performance alone is not enough. To be useful, the product must continue to do its job *in a cost-effective way*. The criteria for cost effectiveness depend on the application and on the financial situation of the organization concerned, including its available investment capital and its perceptions of likely future return on capital. Such concepts are clearly outside our present terms of reference, but it is worthwhile to remind ourselves that cost effectiveness is inseparable from technical considerations, especially as energy saving features and low maintenance costs, rather than initial outlay, are so often mentioned as reasons for using reinforced plastics products.

1.3   When does a fibre reinforced plastics product have to be replaced?

When can we say that a product's life has finally ended? Bridges must not fall down. They must remain safe to use, more specifically, despite the stresses and the external weathering they experience over decades. They must not become too expensive to maintain. Fibre reinforced plastics (FRP) boats are exposed to several hostile forces – water, weather, repetitive wave action and occasional impact. The chief concern is again structural integrity, but most small boats also have cosmetic appeal to their owners, and the cost of maintaining pristine appearance, free from worrying defects, must also be considered.

The FRP cladding panels of buildings do not have to support large stresses, because a steel framework can do this, but in the same way as with boat hulls, their aesthetic appearance and moisture barrier qualities must not be allowed to deteriorate. Often they have been subject to colour changes caused by the action of ultraviolet light on flame retardant additives in the resin.

Translucent roofing over an indoor swimming pool must remain translucent and failure to do so cannot be compensated by structural virtues such as excellent retention of flexural strength and modulus. In contrast, the appearance of a chemical storage tank is of minor importance provided that it does not leak. Pipes and tanks need to maintain both their load carrying and chemical resistance qualities.

In most of these applications, the visible signs of deterioration appear gradually and can include one or more of the following: cracks of various kinds, surface pitting, blisters, swelling, delamination, softening and occasionally, discoloration. The possibility of repair is an attractive feature of reinforced plastics. They can have an extension of their useful life because they are more easily repaired than some other materials and can even be used to extend the life of structures originally made from something else, such as concrete or metals.

1.4   Health and safety

Health and safety considerations need to be taken into account. Legislation, even though not yet enacted, could require a product to be replaced while it is still functionally effective and cost effective. One of the best fibrous reinforcements in the FRP world used to be asbestos. Its strength, stiffness and especially its heat resistance were outstanding, but since the mid-1970s, removing asbestos-based materials from factories, schools and other buildings has been an industry in itself. It is therefore prudent to consider the long term health and safety status of all materials proposed for use in durable structures.

Although this author claims no special expertise in the toxicology of materials, it seems fair to say that the weight of opinion at present is reassuring about fibres in existing use. Glass, aramid and polyethylene fibres are all much safer than asbestos. In common with traditional materials such as wood and cotton, they must always be handled with care, especially if they are finely divided and therefore in respirable forms, i.e. small (<3 μm) diameter short fibres or fine dust.

Typical reinforcing glass fibres have much larger diameters (8–16 μm) and unlike asbestos, the filaments do not readily split along their axes to form smaller diameter fibres. Once incorporated in a resin matrix, they are no longer respirable except when the composite is machined.

Health-related issues have been discussed in more detail by Braddock [1].

Another safety consideration is the flammability of reinforced plastics, together with their capacity for smoke evolution during a fire. This matter is therefore discussed in Chapter 4. It should be safe to assume that the standards required by the relevant authorities will become progressively more demanding over the next few decades. The actual lifetimes of reinforced plastics structures being designed today could well be determined by future fire legislation as much as by weathering performance or fatigue resistance. This fact may account for the increasing popularity of phenolic resin matrix composites, which have good records in this respect. There is a widespread residual prejudice against synthetic materials and it should be remembered that timber, wool and cotton are also flammable.

Among thermoplastics, PEEK (polyether ether ketone) is also very promising but at the time of writing, much too expensive for most purposes outside medicine and aerospace. The way to improve the fire and smoke performance of an ordinary resin is conventionally by using grades containing appropriate additives [2].

1.5   What causes deterioration in fibre reinforced plastics?

We should not assume automatically that there will be any deterioration! But if a reinforced plastics product was competently designed and soundly fabricated, material deterioration usually begins through one or more of the following four influences:

1   mechanical stress, including static loading, fatigue, repeated minor impact, erosion (including water erosion) and abrasion
2   chemicals (water, solvents, fuels, oils, acids, cleaning liquids, atmospheric oxygen, oxidizing agents, caustic alkalis, etc.)

3   radiation (including sunlight)
4   heat, including high temperatures, and large and rapid fluctuations in temperature.

Note that outdoor weathering can involve all four factors simultaneously. The subject of weathering is considered in Chapter 6. Chemicals are considered in Chapter 3, and heat in Chapter 4. Mechanical stress involves several aspects; fatigue is considered in Chapters 5 and 11. Chapter 12 links mechanical and environmental effects.

Materials can often survive individual threats such as ultraviolet light or a specific solvent, but they can still succumb to a combination of influences. The combination of mechanical stress and chemical action is well known and will be discussed in Chapter 3 (see environmental stress cracking in Section 3.7.3). A combination of heat and oxygen is harmful to carbon fibres, whereas neither influence need cause concern in the absence of the other.

1.6   Biodegradation

Biodegradation through microorganisms has very little importance in the degradation of most plastics, whether reinforced or not. The few exceptions to this generalization, none of which are likely to be of great interest in the structural composites area, include vulnerable resins such as cellulosics, and any that contain susceptible additives, for example, certain plasticizers, starch and wood fillers. Most reinforced plastics products can safely be buried underground for decades without rotting. Specific examples of this are given in Chapter 8.

Rodents are not regarded as a widespread problem for fibre reinforced plastics.

1.7   Terms

The middle sections of this chapter are intended for readers without much previous specialized knowledge of either plastics or composites. Those coming to the subject from other fields will no doubt rely on specialists for expert advice but the purpose of this chapter is to help in the communication process by providing key concepts and terminology.

Those who already possess a good knowledge of plastics, resins and composite materials are advised to proceed to Chapter 2 straight away. Others can continue with this chapter and then any remaining gaps can be filled by reference, where necessary, to the standard textbook by Birley *et al.* [3] or by McCrum *et al.* [4] (for engineering and physics aspects). Engineering design with composites has been discussed by Powell [5], who incidentally mentions some of the factors affecting durability, and more recently by Holloway [6].

When illustrating the characteristic mechanical behaviour of unreinforced plastics, we shall rely mainly on schematic diagrams rather than actual data, for clarity. It follows that the diagrams are often simplifications of real behaviour, although they represent real trends.

1.7.1 The matrix

Strictly, the word *matrix* refers to the continuous phase in a composite material containing a discontinuous, dispersed phase such as fibres or filler particles (see Fig. 1.1) and the matrix could in principle be a resin, cement, glass, carbon, metal or ceramic. Here, it will always mean an organic resin. We shall refer to several types of resin, but the two important categories are the thermosetting and thermoplastic varieties (the difference is explained in Section 1.8). Among thermosets, the best known are the epoxy resins and the unsaturated polyesters. The latter are familiar from their use in repair kits and in fibreglass boats. Types of thermoplastics are more numerous and we do not need to list all in order to make progress, but we can start by mentioning three kinds: polypropylene, polyamides (nylons) and polyethylene terephthalate (PET).

1.7.2 Composites, interfaces and the interphase

We have already mentioned the word 'composites' without explaining it. The combination of dispersed fibres and a continuous resin phase is a classic example of a *composite*, that is, a material made up of at least two different phases, distinguishable by examination under a powerful microscope. Naturally occurring composites include wood, bone and teeth. All these materials are surprisingly tough, except for the bones in sufferers from osteoporosis.

When the fibres are all aligned in one direction, we say that a fibrous composite is *unidirectional*, to distinguish it from bidirectional materials and those with randomly aligned fibres. The word *interface* refers to the boundary region where the resin and fibres (or resin and filler particles) are in immediate contact and where they must adhere to each other if load transfer is to take place. Without adhesion, there is no reinforcement, other than that provided by frictional forces. Adhesion between fibres and matrix is sometimes easy to achieve, and sometimes it needs special effort, such as by surface treating the fibre surfaces.

The more recently established term *interphase* is applicable to a wider region, embracing both the interface defined above and the immediate neighbouring parts of each contiguous material (Fig. 1.1). The interphase

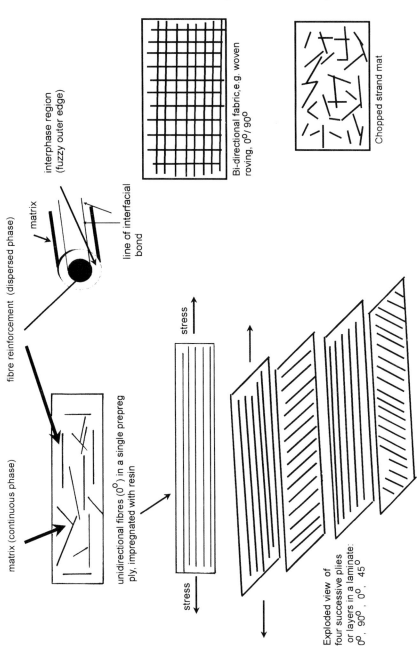

matrix (continuous phase)

fibre reinforcement (dispersed phase)

matrix

interphase region
(fuzzy outer edge)

line of interfacial
bond

unidirectional fibres (0°) in a single prepreg
ply, impregnated with resin

stress

stress

Exploded view of
four successive plies
or layers in a laminate:
0°, 90°, 0°, 45°

Bi-directional fabric, e.g. woven
roving, 0°/90°

Chopped strand mat

Figure 1.1 Reinforced plastics terminology

concept arises from the fact that the part of the matrix which lies closest to the fibres can be different from the bulk matrix, for various reasons, such as that the matrix molecules are unusually restricted in their movement. The structure and properties of the interphase are not uniform but graded. It is not easy to say exactly where the interphase ends and the matrix proper begins.

The interphase is mentioned again briefly at the beginning of Chapter 3, and the subject is treated more extensively in relation to dynamic loading in Chapter 11. Many writers still refer only to the interface and ignore the gradation of structure between fibre and matrix.

A layered structure such as a stack of glass fabric layers or paper sheets, impregnated with resin, is known as a *laminate*. Some boat hulls (not all) and kitchen worktops are therefore laminates as well as composites.

1.7.3 Composite 'systems'

System is a term used in the technical literature to describe a material made from a specific resin combined with its recommended hardener, together with specific reinforcing fibres. Sometimes the constituents are identified simply by manufacturers' reference numbers, for example, a carbon fibre reinforced epoxy resin prepreg (see Section 2.2) supplied by Toray and identified by the user as prepreg number P2212-15 was described in more detail as 'T800H/#3631' in one recent paper. The carbon fibre was denoted by T800H and the epoxy resin by #3631. The fibre reference number may identify the fibre's surface treatment process, if any, as well as the nature of the fibres themselves. This approach to nomenclature gives no unnecessary information to anyone and can be particularly frustrating to chemists, because they are denied the opportunity to hazard a guess at the resin's main characteristics from a knowledge of its systematic name and structural category, but it has advantages for end-users in that they know they have the precise product.

1.7.4 Plastics

The word 'plastics' means a material (usually synthetic) which at some stage in its processing history has shown *plastic* behaviour, that is, irreversible deformation under stress. This state is most commonly achieved on heating during fabrication operations. It should not be assumed that plastic behaviour is a property of the same material on cooling down again. The letter 's' in plastics distinguishes the noun from the adjective.

The common tendency to confine the word 'plastics' to the cheaper commodity materials such as polystyrene has no logical basis. There are many expensive, exotic plastics such as PTFE (or 'polytetrafluoroethene'),

PEEK and the polyimides. Such materials are used because of their technical advantages, and/or because of their cost effectiveness over the whole
product lifetime rather than because of initial material costs.

1.7.5 Polymers

In practice, the terms 'plastics', 'polymer', 'resin' and 'matrix' mean very
similar things and are used almost interchangeably in reinforced plastics
circles without much confusion, although there are minor differences of
interest mainly to the specialist.

The word polymer is the odd one. It simply means a substance composed
of very large molecules, which in turn are built up repetitively from small
chemical building blocks called 'monomers' or 'mers'. Examples of polymers are polypropylene, nylons (properly called polyamides), epoxy resins,
polyimides, etc.

The overwhelming majority of commercial polymers are organic, that is,
they are based on carbon chemistry, and they therefore possess poor heat
resistance, so they can be ruled out from true high temperature applications, other than sacrificial ones such as ablative shields. But there is no
reason why polymers should always be organic in future (the family of
silicone rubbers provide an example of inorganic polymers).

If new inorganic plastics are ever successfully commercialized, they will
be completely different from today's epoxy resins, sulfones and nylons, etc.
They will probably lose the great advantage associated with today's organic
polymers – easy processing at fairly low temperatures. They may not even
be called plastics, because that word has a poor image, dating from early
failures in the 1920s, and from the brittle, ephemeral nature of cheap toys,
etc. An attempt has already been made to escape from this by using the
term advanced composites, even when it is plain that the material under
discussion is really reinforced plastics.

1.7.6 Additives

Polymers in their raw state are usually technically unsatisfactory in one
respect or another, such as their stability to light or heat, or their
processability, or flammability, or colour, or opacity, or antistatic characteristics, etc., and they simply could not be used in commercial applications
successfully without the incorporation of one or more additives [2] to
modify behaviour. The additives are often present at very low concentrations (0.1–3 parts per hundred of resin, by weight) and are called stabilizers,
UV absorbers, viscosity modifiers, lubricants, fire retardants, pigments, etc.
Fillers may be present at 50 or even 150 parts per hundred of resin, by
weight. Thermosetting resins tend to have fewer additives of the kind

mentioned above but they still need flame retardants, viscosity modifiers, exotherm modifiers and/or pigments.

Adding something to the polymer often creates some new problem, so we end up with further additions and the final complex recipe is called a *formulation*. The substance becomes a mixture, rendering the term polymer, with its implication of a single constituent, an inaccurate way to describe it. The mixture is instead called a *resin*, which is a gratifyingly vague term, originally applied only to viscous or solid substances occurring in nature, for example in the barks of trees, and so on. It is now applied to a wide range of mainly synthetic compositions containing a polymer, and is often used indiscriminately, whether any additives are present or not. Thus an epoxy resin is called a resin, even at the stage before any fillers and so on are added, and very few people bother to call it an epoxy polymer. 'Resin' is a useful and flexible term.

The mechanical properties are often virtually unchanged by additives, although a few (fillers, impact modifiers) are expressly designed to have that effect. In some cases it can be difficult to tell the difference visually between the base polymer before and after addition of the additives, unless of course they include a pigment. Nevertheless trace substances can have tremendous effects on the ultimate durability and the selection of the correct grade or mixture for an application is essential. The advice of the materials suppliers must be considered when selecting a resin for durability.

Particulate fillers are additives that are used at much higher concentrations, often more than 100 parts per 100 of resin by weight. They increase the modulus of the resin, but they also alter its processing characteristics and frequently embrittle it, resulting in lower strength and lower ultimate elongation at break. The precise effects depend on the filler used. Significant quantities of rubbers are sometimes added to polymers, to produce toughened or 'high impact' grades.

1.7.7 Grades

Suppliers of a given type of plastics, such as polyethersulfones or epoxy resins, usually offer a whole range of grades. It is impossible to tell from company literature whether two resin grades are fundamentally different, or similar in all respects but for one detail, such as their ease of processing.

The differences between grades are of two kinds:

1 The base polymers themselves can be different, for example in chain length (molecular weight). The distinction will probably only be meaningful to chemists, but it can have important effects on specific characteristics such as processing.

2   The base polymers could be exactly the same and any differences could be caused only by the additives. A single large batch of resin could be sold for quite different applications, after being divided into smaller batches and each batch rendered suitable by various additives for use in different end-user products. A UV stabilizer or antibacterial agent can greatly increase the outdoor durability of a resin matrix. Very few additives actually reduce the lifetime, but the possibility cannot be ruled out. For example a water-soluble additive will be partially leached out during exposure to wet environments and this could leave porosity.

Occasionally, one additive will interfere with the effectiveness of another. Therefore suppliers should ideally provide laboratory or case-history evidence of the durability of the precise grade proposed for future use, with all the ingredients, and not simply rely on data from earlier grades. In practice this is a counsel of perfection, because users cannot reasonably expect the latest advances in resins, light stabilizers, antioxidants, fillers, and so on and also 20 years' test data!

1.8   Thermoplastics and thermosetting resins

Nearly all plastics fall into two broad categories: thermoplastics and thermosets. The processing and fabrication are very different, but in both cases have a major effect on the durability of the final products.

1.8.1   Thermoplastics

Thermoplastics [7, 8] are capable of being repeatedly reshaped by heating and deforming while hot. From the engineer's standpoint, the big advantage that they have over thermosetting resins is that the shaping or forming processes do not involve chemical reactions and do not require the addition of catalysts or accelerators. There is no danger of the resin permanently solidifying in the wrong part of the processing machinery as a result of chemical hardening reactions. Indefinite storage prior to use does not involve any risk of premature solidification, although other very slow changes can sometimes occur (see, for example, Section 1.17 on Physical ageing, later in this chapter, and again in Sections 4.5 and 12.4). Scrap, off-cuts and so on can be recycled. (Thermosetting resins have only recently been taken seriously as recyclable materials).

Unreinforced thermoplastics are ideal for mass production processing technology and together they constitute the great majority of plastics usage worldwide, although more than half of the output consists of commodity plastics, not intended for durable products. Although thermoplastics can be reinforced with fibres, they have until very recently not been well represented among load-bearing, structural reinforced plastics applications. This situation is now changing, because of improvements in fabrication techniques, as discussed in Chapter 2, Section 2.8. The problem has been that incorporating fibre reinforcement into thermoplastics is difficult compared with thermosets, because of their extremely high melt viscosity. Most reinforced thermoplastics simply contained randomly oriented, very *short* fibres in small quantities (10–40% by weight). The achievement of good strength and stiffness, retained at elevated temperatures, requires high concentrations of *long* fibres, with good control over their fibre orientation and distribution. The processing techniques originally developed for unreinforced thermoplastics, such as injection moulding (i.e. heating the material and forcing it while liquid into a relatively cool mould cavity) could be adapted to the moulding of short fibre reinforced plastics, provided that the total amount of fibres was not very great, but these processes resulted in further fibre length reduction as well as machine wear. Recent developments have addressed these problems and several ways of reinforcing thermoplastics with long or continuous fibres are now becoming available (see Section 2.8).

The molecules of thermoplastic polymers are linear or string-like and remain essentially unchanged throughout the various processing operations. This is where they differ from thermosetting resins, the molecules of which are small at first but later are chemically reacted to form molecular networks. The linear structures of thermoplastics result in lower softening temperatures, other factors being equal, and this has kept thermoplastics from being used much in load-bearing, high temperature applications. New and improved thermoplastics are now being developed with superior resistance to high temperatures, and with other virtues such as toughness and flame resistance. Unfortunately history shows the need for a long development period between the invention and the widespread commercial exploitation of load-bearing plastics. Typically this has been about 30 years and these advanced thermoplastics only really started on their development journey in the early 1980s.

1.8.2 Thermosetting resins

Thermosetting resin products [9] are usually made in two stages: first the production of a liquid, or a low-melting solid, consisting of chain-like or other intermediate size molecules, and second the chemical linking together or 'crosslinking' (by the fabricator, not the resin supplier) of these intermediates by means of a substance known as a hardener or catalyst, to form a hard, often brittle solid. The formation of the final moulded product is simultaneous with the chemical reaction that forms the crosslinked resin, and the hardening stage is known as the *cure*. It usually, but not invariably requires heating. It is also an exothermic reaction, often generating enough heat to produce cracks unless there is a heat sink available to remove heat.

The curing reaction tends to be incomplete (that is, there are still some unreacted chemical groups which have not been used and the physical properties have not yet reached constant values). The extent to which the curing reaction approaches completion is called the *degree of cure*, and it can be further progressed by raising the temperature in steps over a lengthy timescale. This process is known as *postcure*. It is a fine judgment whether postcure is worthwhile, as not only does it cost money, but brittleness or impact damage resistance can increase, although chemical and moisture resistance tend to be improved, along with many other properties. Failure to postcure boat hulls is probably responsible for their absorbing excessive water in their first 15 years in the water, and their needing much more maintenance than would otherwise be the case, in order to achieve a 30 or 40 year lifespan.

Thermosetting resins, once cured or hardened, can never be reshaped, nor do they melt. Heat is resisted well until the temperature rises towards the point of onset of irreversible chemical decomposition.

Because of the ease with which fibres can be incorporated at the first (low viscosity) stage, thermosetting resins have until recently dominated the market for load-bearing reinforced plastics, and have proved especially suitable for large, low volume applications such as boat hulls, chemical storage tanks, pipework, cladding for buildings, parts of aircraft, helicopter rotor blades and playground equipment. Thermosetting processes and moulds with few exceptions need less capital outlay than mass production thermoplastics, but in the last five years the moves towards automation of thermosetting processes for the vehicle body part industry has made great strides.

Some examples of both types of plastics material are given in Table 1.1. Note that polyimides are unusual since the ones made by the addition polymerization process are thermosetting resins, but others are formally thermoplastics; the difference depends on their detailed synthesis and structure. Very high temperature thermoplastics are not easily processed by conventional thermoplastics methods.

1.9   Blends

As technology advances, materials have been developed which evade the above classification. Mixtures of thermoplastics and thermosetting resins have been developed in which the thermoplastics material acts as a toughening agent for the thermosetting resin. Alloys or *blends* of two or more thermoplastics, or of plastics and rubbers, are becoming commonplace.

Table 1.1 Categories of resin

Category	Examples
Commodity thermoplastics	Polyethylene; polypropylene; polystyrene; PVC (polyvinyl chloride)
Engineering thermoplastics	Nylon 6; nylon 6, 6; polycarbonate; polyethylene terephthalate; polybutylene terephthalate; acetal; ABS; polyphenylene oxide
High performance thermoplastics	PEEK; polyetherimides; certain polyimides; polyamide-imide; polysulfone; polyethersulfone
Thermosetting resins	Unsaturated polyesters; phenol formaldehyde; urea formaldehyde; melamine formaldehyde; epoxies; vinyl ester resins; cyanate ester resins; bismaleimides; certain polyimides

ABS = acrylonitrile-butadiene-styrene.

A small quantity of a third polymer is usually required, to render the two main constituents compatible. The third polymer is then known as a *compatibilizer*.

1.10   Commodity, engineering and high performance plastics

The various resins on the market can usefully be divided into three categories: 'commodity', 'engineering' and 'high performance' plastics. The first two terms are applied chiefly to thermoplastics. Ordinary, low cost, *commodity* plastics such as polyethylene and polystyrene are not really engineering materials. Sales depend on price rather than performance, but it is worth noting the virtues that have propelled them into prominent positions in the manufacturing world. The most obvious is their ease of fabrication (partly a consequence of their low softening temperatures. This advantage is almost lost with high performance plastics).

Other virtues include light weight, good electrical and thermal insulation, and in some cases, corrosion resistance, especially where dilute acids and alkalis are concerned. Their low modulus is useful for packaging film applications. Polyethylene comes in several varieties and some of them are quite tough.

The mid-range group, the *engineering* plastics [7,8] (see Table 1.1) such as polycarbonate, acetal and the polyamides (nylons), have slightly better moduli and strength than the commodity plastics and their toughness is commendable but they still need fibre reinforcement before they are suitable for anything other than very light engineering applications. Few are claimed to be especially heat resistant, although the polyamides such as nylon 6 are much better than the commodity plastics.

The third category is the *high performance* plastics (Table 1.1). In most cases the polymers are so classified, not because they have outstanding mechanical properties – they are actually worse in some ways – but mainly because of their superior tolerance of high temperatures.

For reasons explained in Chapter 4, laboratory data often show impressive heat resistance for short periods under favourable conditions, for example, 1000h in an inert dry atmosphere, without any applied stress, but their continuous use temperatures in air or air/steam mixtures are still surprisingly low. As a sweeping generalization, we can only say that most high performance resins can be used in air under significant stresses for very long periods above 160°C and a few can be used for long periods at temperatures well in excess of 250°C without distortion, oxidation or decomposition.

1.11   Crystallinity in thermoplastics

Polymers are completely different from metals and other crystallizable solids. Their molecular structure is based on very long chains of atoms and so their opportunities to form genuine crystals are limited. Commercial, crosslinked thermosets do not crystallize at all. Certain thermoplastics can achieve ordered structures, that is, partial crystallinity, if cooled slowly enough from the liquid state. This happens during the fabrication operation and different processes have different cooling rates, so the degree of crystallinity varies. This translates into differences in solvent resistance and therefore different durability in chemical environments. There always remains some amorphous material alongside the crystalline regions of commercial polymers, although polymer single crystals do exist.

The significance of crystallization in thermoplastics therefore is that it greatly increases their resistance to solvent attack, and in addition it usually improves the toughness and enables the modulus to hold up at temperatures in excess of the *glass transition temperature*, when it would normally fall sharply (see Section 1.13). Often the transparency is reduced. Thermosetting resins cannot crystallize. Nor can some thermoplastics with irregular structures that cannot be close packed for reasons of geometry, charge repulsion, chain rigidity, and so on. Ordinary polystyrene and PMMA (polymethyl methacrylate) are examples of this.

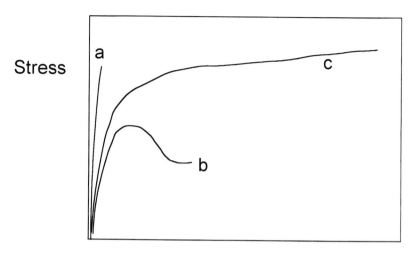

Strain

Figure 1.2 Typical schematic tensile stress–strain curves for polymers: (a) brittle, amorphous thermoplastic, (b) same polymer with toughening additive, (c) intrinsically tough, semi-crystalline thermoplastic. The curves should be taken only to convey trends and not relative breaking stresses, which vary with the precise materials

1.12   Stress–strain behaviour of unreinforced plastics

Plastics do not obey Hooke's Law very closely. The increase in length caused by a tensile stress is not directly proportional to the magnitude of the stress applied and there is no straight line plot of load against extension in a tensile experiment, even for small loads.

Typical stress–strain curves for plastics are shown in Fig. 1.2. The figure shows qualitatively (a) the steep, but non-linear curve associated with amorphous, brittle thermoplastics such as unmodified polystyrene, (b) the equivalent graph for a similar brittle thermoplastics material to which rubber has been added to produce a 'high impact' grade and (c) an intrinsically tough thermoplastics material, such as a nylon (polyamide).

1.12.1 Modulus

It follows that the modulus, or stiffness of polymers cannot always be computed directly from the initial slope of the stress–strain graph, because of its previously mentioned non-linearity. It can be derived instead in various other ways. The *initial tangent modulus* is obtained using the slope of the tangent to the stress–strain curves at the origin, and similarly we

Shear modulus

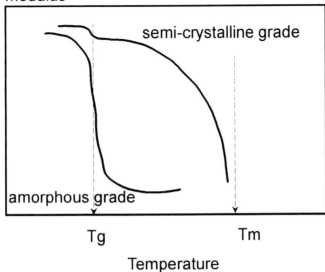

Figure 1.3 Effect of crystallinity on modulus–temperature curves. The two curves refer to the same generic polymer, differing only in rate of cooling from the melt. Both samples show a T_g

could use the slope of the tangent to the curve at some other arbitrarily chosen point, or we could use the slope of a secant.

The moduli of polymers change considerably with temperature. Figure 1.3 shows the effect of temperature on the shear modulus of an amorphous but potentially crystallizable polymer such as poly(ethylene terephthalate) (PET). There is a very steep decline in modulus at a given characteristic temperature, see Section 1.13.

If the same polymer had been in the crystalline form, the second curve in Fig. 1.3 would apply, indicating a more gradual decline in modulus above T_g, the glass transition temperature, but a rapid fall as the melting temperature, T_m, is approached.

The Young's modulus of a hard thermosetting resin such as a crosslinked polyester or epoxy usually lies in the range 2.5–4 GPa at room temperature. This compares with 75 GPa for the Young's modulus of E-glass fibres and with 230–960 GPa for carbon fibres (See Table 1.2; [10–17]). As already mentioned, the moduli of organic polymers are reduced even further by heating and can reach the very low levels associated with rubbers. Resin stiffness can be virtually neglected when calculating the stiffness of a unidirectionally reinforced plastics material in the direction parallel to the fibre orientation.

Table 1.2 Tensile properties of plastics, metals, wood and reinforcing fibres [10–17]

Material	Form	Reference	Property	Units	Value
E-glass	Virgin filaments	[10]	Modulus	GPa	71
E-glass	Virgin filaments	[10]	Strength	MPa	3.4
E-glass	Virgin filaments	[11]	Elongation	%	5.0–5.6
Boron	Filaments	[12]	Strength	MPa	3600
Boron	Filaments	[12]	Modulus	GPa	400
Boron	Filaments	[11]	Elongation	%	0.7
Aramid	Kevlar 29 dry yarn, twisted	[10]	Modulus	GPa	59
Aramid	Kevlar 29 dry yarn, twisted	[10]	Strength	MPa	3.45
Aramid	Kevlar 29 dry yarn, twisted	[10]	Elongation	%	4.0
Aramid	Kevlar 49 dry yarn, twisted	[10]	Modulus	GPa	124
Aramid	Kevlar® 49 dry yarn, twisted	[10]	Strength	MPa	3.62
Aramid	Kevlar® 49 dry yarn, twisted	[10]	Elongation	%	2.5
Carbon	high modulus PAN-based[d]	[11]	Modulus	GPa	294–588
Carbon	high modulus PAN-based	[11]	Strength	MPa	2500–3900
Carbon	high strength PAN-based	[11]	Modulus	GPa	230–294
Carbon	high strength PAN-based	[11]	Strength	MPa	3500–7100
Carbon	high modulus pitch-based	[11]	Modulus	GPa	520–960
Carbon	high modulus pitch-based	[11]	Strength	MPa	2100–2200
Asbestos	chrysotile	[13]	Modulus	GPa	162
Asbestos	chrysotile	[13]	Strength	MPa	3040
UHMWPE[e]	Dyneema® SK60	[16]	Modulus	GPa	87
UHMWPE	Dyneema® SK60	[11]	Strength	MPa	2700
UHMWPE	Dyneema® SK60	[16]	Poisson's ratio	—	0.29

Material	Description	Ref	Property	Units	Value
Epoxy	unreinforced	[14]	Impact strength	Jm^{-1}	0.7[c]
Polyester	unreinforced, unsaturated	[15]	Modulus	GPa	3.5
Polyester	unreinforced, unsaturated	[15]	Strength	MPa	60
Epoxy	unreinforced	[16]	Modulus	GPa	3.4
Epoxy	unreinforced	[16]	Poisson ratio	—	0.37
Epoxy	unmodified, DGEBA[f]	[17]	Yield stress	MPa	83–112
Polypropylene	semi-crystalline	[8]	Modulus	GPa	1.14–1.55
Polypropylene	semi-crystalline	[8]	Strength	MPa	31–41
Polypropylene	semi-crystalline	[8]	Elongation	%	100–600
Nylon 6,6	crystalline	[8]	Modulus	GPa	1.6–3.8
Nylon 6,6	crystalline	[8]	Strength	MPa	76–95
Mild steel		[15]	Modulus	GPa	210
Mild steel		[15]	Strength	MPa	240 yield
Aluminium	N8 alloy	[15]	Modulus	GPa	70
Aluminium	N8 alloy	[15]	Strength	MPa	140 yield
Oak	12% moisture	[15]	Modulus	GPa	9.5/1.5[a]
Oak	12% moisture	[15]	Strength	MPa	90[b]

[a] Longitudinal/transverse direction relative to grain.
[b] Parallel to grain.
[c] Izod; Joules per metre of notch.
[d] Polyacrylonitrile.
[e] Ultra high molecular weight polyethylene.
[f] Diglycidyl ether of bisphenol A.

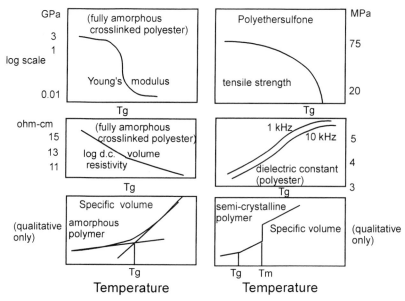

Figure 1.4 Changes in the properties of amorphous or partly amorphous polymers at the glass transition temperature

1.13   Glass transition temperature

The glass transition temperature, T_g, is one of the most important of all properties of polymers. It is the temperature region where not only the modulus but almost all the physical and mechanical properties change rapidly during heating or cooling. It is the temperature at which the plastics material changes to become almost rubber-like.

One fact reduces the impact of the T_g on plastics behaviour. It is a phenomenon associated only with the amorphous, that is, non-crystalline or disordered phase. Many physical and mechanical properties change at or around T_g. For example, Fig. 1.4 shows typical changes in modulus, tensile strength, electrical resistivity, dielectric constant and specific volume (reciprocal density). Note that the graph of specific volume, V, versus temperature, T, shows a change in dV/dT, but there is no step change, that is, no ΔV at T_g. There is no latent heat of fusion at the melting temperature, T_m, nor are there the sharp changes in X-ray and infrared spectra of the kind usually associated with a melting point.

In physical terms, T_g is the temperature at which, on heating, the molecular chain structure ceases to be 'frozen' and suddenly becomes much more mobile, because of first, an increase in the thermal energy required to impart motion and second, increasing availability of sufficient space for these movements.

In practice, it is necessary to avoid using amorphous resins in structural applications above, or even near to, the T_g. (This restriction applies to all thermosets, because they are invariably amorphous, but it does not apply to crystalline thermoplastics, otherwise polypropylene chairs, with a T_g of around $-15°C$, could only be used safely in Alaska and in Siberian winters). A margin of safety of $20°C$ is required for reasonable durability, and a much larger one in humid conditions. Traces of water, or any other solvent, lower the T_g to below its usual value, so the maximum safe working temperature of a load-bearing amorphous resin is lower in humid conditions than in a dry environment. There is a widely quoted rule of thumb that each 1% moisture absorbed by the matrix lowers the T_g by $20°C$. This comes from epoxy resin data in wet atmospheres and different figures apply to other resins and liquids.

Typical values for the T_g of polymers in their amorphous state are given in Table 1.3. The properties of polymers that are normally crystalline are not controlled as much by the T_g.

The T_g of thermosetting resins is a much less sudden transition than it is with thermoplastics. The properties change considerably, but only slowly, that is, over a very wide temperature range, (typically more than $30°C$); see Fig. 1.5.

A mixture of two incompatible polymers will have two separate transitions, as shown in Fig. 1.5(c). Each transition approximates to that of the constituent polymer. Much smaller changes in properties, termed minor or secondary transitions, often occur as well at lower temperatures than T_g. The impact resistance of the resin is usually superior when the temperature is above that of the secondary transitions.

1.14   Stress and time: viscoelastic behaviour

Plastics are intermediate in character between solids and liquids. Consequently, their mechanical behaviour is akin to that of exceedingly viscous liquids, and as with liquids, there is a time-dependent aspect. Solids respond instantly to an applied stress, as with springs, whereas unreinforced plastics respond and deform rather more slowly. They are said to exhibit a *viscoelastic* mechanical response. This shows itself under sustained loading through the phenomenon of *creep* (the gradual, but often quite substantial, increase in strain at constant stress), in *stress relaxation* (declining stress at constant strain) and in other behavioural traits intermediate between those of elastic solids and viscous liquids. Figure 1.6 shows, again qualitatively, how temperature affects the shape of the stress–strain curve in tensile tests on amorphous thermoplastics. The polymer in the figure has a brittle–ductile transition which just happens in this case to be around ambient temperature, but it could equally be at a much higher or lower temperature.

Table 1.3 Glass transition temperatures of some resins (moisture free)

Resin	T_g (°C)
Thermoplastics	
Polyethylene[a]	−78
Polypropylene[a]	−15
Polystyrene	101
Acetal[a]	−75
Nylon 6[a]	58
Nylon 6,6[a]	50
PMMA (poly methyl methacrylate)	104
Polyethylene terephthalate[a]	68
Polybutylene terephthalate[b]	82
Polycarbonate	145
PVC (unplasticized)	87
Polyethersulfone	190
PEEK (poly ether ether ketone)	143
PEK (poly ether ketone)	162
Polyamide-imide	212
Polyetherimide	200
Polyphenylquinoxaline[b]	298
Thermosetting resins (representative values only)	
Kerimid® FE70003 bismaleimide	400
(Rhône Poulenc)	
DGEBA epoxy[c]	145
TGDDM epoxy[d]	240
Unsaturated polyester, isophthalic type	
(typical)	
Phenol formaldehyde resin	decomposes below T_g (>230)

[a] Semi-crystalline: some semi-crystalline polymers can be used above their T_g because they retain their modulus above that temperature, while amorphous polymers do not.
[b] Other quinoxaline-type polymers can have higher T_g values, e.g. 400°C.
[c] DGEBA stands for diglycidyl ether of bisphenol A. Epoxy resin properties can depend considerably on the hardener used.
[d] TGDDM refers to high temperature epoxy resins. It stands for tetraglycidyl diaminodiphenyl methane. Epoxy resin properties can depend considerably on the hardener used.

Figure 1.7 shows the converse effect, namely that the strain rate controls the brittle–ductile transition when the temperature is held constant. Very high strain rates are needed to convert ductile behaviour into brittle behaviour.

Figure 1.8 illustrates creep. Under constant stress, whether in tension or bending, the strain increases with time. The rate at which the strain in-

Young's modulus

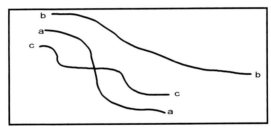

Temperature

Mechanical
damping
(tan delta)

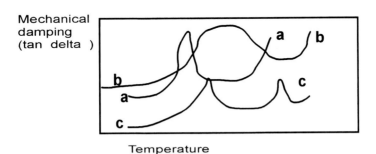

Temperature

Figure 1.5 Change in modulus and mechanical damping in the region
of the glass transition temperatures for (a) amorphous thermoplastic,
(b) crosslinked thermoset, (c) a blend of two thermoplastic polymers.
The T_g corresponds to the steepest slope in the modulus curves and
(more approximately) to the peaks in the damping curves, assuming
that the damping vibrations are of low frequency

creases depends on the temperature and on the stress. Sometimes creep
ends in sudden fracture (*creep rupture*).

Creep and stress relaxation accelerate once the temperature exceeds the
T_g of the matrix. Viscoelastic behaviour is obviously relevant to durability,
but fortunately the addition of suitably oriented fibre reinforcement can
dramatically decrease or suppress viscoelastic behaviour, to an extent that
depends on fibre direction, fibre volume fraction, and so on. This is an
important reason for using fibrous reinforcement, even when it seems unnecessary from a consideration of short term mechanical behaviour.

1.15 Predicting creep behaviour

Predicting viscoelastic behaviour over long timescales has been attempted
using various models, such as developments from the Maxwell and Kelvin/
Voigt models and from time–temperature superposition theory [4]. One
durability study on unreinforced thermoplastics [18] involved a loading

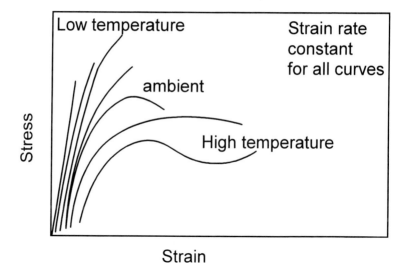

Figure 1.6 Schematic representation. Effect of temperature on the brittle–ductile behaviour in tension of an acrylic polymer, at a constant strain rate

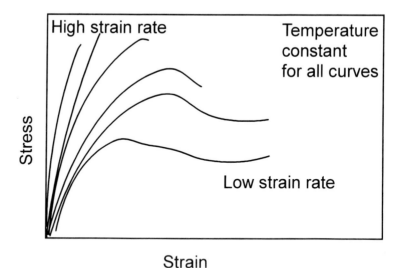

Figure 1.7 Schematic representation. Effect of strain rate on the brittle–ductile behaviour in tension of an acrylic polymer, at a constant temperature

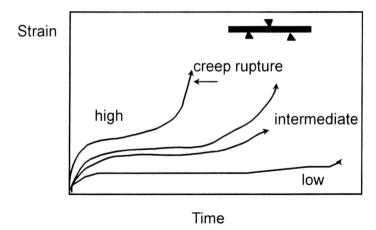

Strain

creep rupture

high

intermediate

low

Time

Figure 1.8 Schematic diagram of a thermoplastic polymer under constant bending load. The words 'high', 'intermediate' and 'low' can be taken to mean the temperature or the magnitude of the applied stress; both have qualitatively similar effects

duration of 26 years (230000 h). It was concluded that the creep strain in uniaxial tension at time t could be predicted by the power function

$$\varepsilon = \varepsilon_0 + m(t/t_0)^n$$

where ε_0 is the initial strain at time t_0, and m is a time-dependence coefficient given by

$$m = m_0' \sinh \sigma/\sigma_0$$

for low values of the stress σ. The term m_0' is a constant, independent of stress, strain and time, but dependent on the material and on the temperature. The work has recently been extended to reinforced plastics, using samples made by the pultrusion process. The conclusion after 10000 h was that the long term creep behaviour of structures such as pultruded beams made of glass fibre reinforced plastics could be predicted using a power law model. It was furthermore proposed that carbon and aramid/epoxy composites were amenable to the same treatment. Predictions of strain recovery after unloading were modelled with reasonable accuracy, using the Boltzmann Superposition Principle [4].

1.16 Cracks and notches

Many unreinforced polymers (especially amorphous ones) are susceptible to crack propagation under impact or tensile loading. They are vulnerable

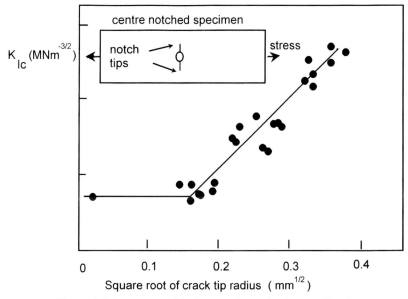

Figure 1.9 Apparent critical stress intensity factor, K_{lc}, of an unreinforced, crosslinked, unsaturated polyester resin containing notches of various degrees of sharpness [19]

to defects such as holes, porosity, edge notches and surface irregularities. The ease of crack propagation depends on crack tip geometry, stress, temperature and the presence of liquids.

One of the main roles of fibre reinforcement is to stop cracks from propagating. Considerable data is now available about the fracture mechanics both of unreinforced plastics and of reinforced materials. The sensitivity to geometry (edge notch sharpness) of an unreinforced, crosslinked polyester resin is shown by Fig. 1.9, which displays the apparent value of K_{lc}, the critical stress intensity factor, obtained by using centre notched specimens with various notch tip radii, that is, with blunt and sharp cracks [19]. When reinforcement is added, a crack propagating normal to the fibres is considerably hindered by the fibres, provided that the adhesion between fibres and resin is not too strong (otherwise the crack will propagate across the interface as if through an homogeneous material). Consequently, fracture toughness is an area where reinforcement can be extremely beneficial. Fibres cannot be arranged simply in one direction, if the cracks are expected to initiate and propagate in several directions.

1.17 **Physical ageing** (see also Chapter 4)

Amorphous, unreinforced resins are examples of glasses. They are therefore subject to the same long term ageing processes as any other glass, such as window glass. The structure of a glass is not constant but changes with time, as if striving for greater regularity and order, that is, it is not in thermodynamic equilibrium. Whereas the changes in ordinary bottle-glass take several centuries, those in organic resins can make noticeable progress in just a few years.

Glasses become very slightly denser over a period of time, as the structure progressively packs together more and more closely. The size of the effect depends on the rate of cooling from the liquid state; fast cooling results in a loose structure with scope for later consolidation. The consolidation process is known as *physical ageing*. At first it might be thought that the changes are caused by some outside agency, such as atmospheric oxygen or heat. But changes can occur even during ambient temperature storage in a dark cupboard. The impact strength and the ultimate tensile elongation at break of unreinforced plastics both decline. Raising the temperature towards the T_g increases the rate at which physical ageing occurs. The process of physical ageing could be relevant to composite spare parts which are fabricated and then warehoused for several years (although other changes, such as moisture pick-up, would also have to be considered under those circumstances).

However, the mechanical consequences of physical ageing are fortunately reduced considerably by fibre reinforcement. Moreover it is found that the alterations in at least some of the mechanical properties, such as modulus and damping, caused by physical ageing are confined to a limited temperature range around the ageing temperature and do not affect ambient temperature performance unless ageing has been carried out at ambient temperature [20] in which case the deterioration would probably take decades.

1.18   Reinforcements

The word *reinforcement* will refer in this book exclusively to strong, stiff fibres. They can be made of glass, aramid (e.g. Kevlar$^{T®}$(DuPont)) or high molecular weight polyethylene (e.g. Dyneema® (DSM)), carbon/graphite, polyamide (nylon), jute, and so on. The fibres can be long, virtually continuous or short (e.g. 1mm).

The word reinforcement will not be applied to particulate fillers, such as quartz or glass beads, because although these additives do increase the modulus, they cannot be relied on to improve the strength of the resin, nor will the word be applied to speciality rubber toughening agents that are used to improve the impact strength only.

Fibres differ in many important respects. Some fibres in the list above are inorganic and heat resistant. Carbon is heat resistant in the absence of oxidizing agents such as air. Some fibres are organic polymers, just like the matrix resin and have similar temperature limitations. Unlike glass and carbon, aramid fibres absorb moisture.

Glass is isotropic, but most other fibres have orientated structures and they possess properties (electrical conductivity, thermal expansion coefficient and degree of solvent swelling) which differ according to the direction of measurement.

Reinforcements are available in various physical forms – filament bundles, short chopped fibres, woven fabrics, three-dimensional fabrics, resin-impregnated filaments, and so on. There is often a surface treatment applied, which fulfils several purposes, such as to protect against abrasion and promote adhesion to the matrix.

1.19   Why use reinforcement?

Ordinary unreinforced plastics lack strength and stiffness. Some have very poor impact resistance. Creep is a more serious problem with plastics than with most other materials and is greatly reduced by reinforcement.

When reinforcing fibres are added, and orientated parallel to an applied uniaxial tensile stress, the modulus of the reinforced composite material, E_c, in the fibre direction is given by a simple law of mixtures:

$$E_c = E_f V_f + E_m V_m$$

where E is the modulus, V is the volume fraction, and the subscripts m, f refer to matrix and fibres, respectively. Obviously the matrix modulus contributes almost nothing to the stiffness of the composite.

The strength and modulus of the reinforced material falls rapidly from those predicted by the law of mixtures for unidirectional laminates, as the angle between the fibre direction and the stress is increased from zero towards 90°. We assumed above that the fibres are all parallel to the applied uniaxial stress, but if they are lying normal to the stress, the equation below applies and gives a much lower value for the composite modulus:

$$E_c = E_{m'} E_f / [E_f (1 - V_f) + V_f E_{m'}]$$

where

$$E_{m'} = E_m / [1 - v_m^2]$$

and v is Poisson's ratio.

The above discussion assumes first that the fibres all lie parallel to each other and second that they are continuous. In practice, reinforced

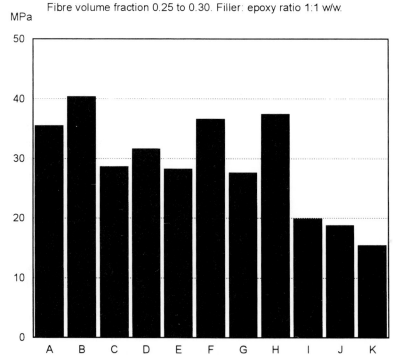

Figure 1.10 Effect of various particulate fillers on the interlaminar shear strength of glass fabric reinforced epoxy resin laminates [23]. A, no filler; B, 50 μm glass beads, untreated; C, same but silane treated; D, 7 μm glass beads, untreated; E, same but silane treated; F, 15 μm glass flakes; G, 8 μm calcium carbonate; H, 15 μm quartz; I, 15 μm alumina trihydrate, fire retardant; J, 20 μm mica; K, 60 μm thin-walled hollow-glass microspheres

plastics usually have the fibre reinforcement oriented in several different directions.

The volume fractions V_f, V_m are used in many equations and can be related to the more easily measured weight fractions, for simple two-component composites, by the equations:

$$V_m = [W_m/\rho_m][W_m/\rho_m + W_f/\rho_f]$$

and so on where ρ is the density.

Hull [21] discusses an expression by Nielsen and Chen [22] for the calculation of the modulus of laminae containing randomly oriented fibres. The issues involved in computing the mechanical properties of short, random fibre composites, and the mechanics of fibre reinforced plastics are also discussed.

1.20   Fibres combined with fillers

The best mechanical properties are obtained by using resins reinforced with the maximum amount of reinforcing fibres. Adding fillers as well tends to cause a deterioration. Nevertheless in some applications it is necessary to use both. Long fibre reinforced resins may need to be made fire resistant by incorporating, for example, alumina trihydrate, which is favoured when halogen additives are considered inappropriate. Secondly, there are many short fibre reinforced mixtures or 'moulding compounds' containing fillers as well, either for cost reduction or to impart a specific property, such as electrical conductivity.

The use of fillers tends, with few exceptions, to reduce the damage tolerance, tensile and flexural strength of laminates. The effect on the interlaminar shear strength of glass fabric/epoxy laminates depends on the filler and is shown in Fig. 1.10. Note that glass beads and quartz had no adverse effect whereas the extremely thin-walled hollow glass microspheres were damaging [23]. Thicker walled spheres would have improved the performance.

In the next chapter, we shall be concerned specifically with reinforced plastics rather than the resins themselves. We shall consider how reinforced plastics parts are fabricated and how they are inspected to check quality and likely durability. Finally we shall survey the main threats to their durability in service.

References

1 J W Braddock, 'Safety and Health', *International Encyclopedia of Composites*, ed. S M Lee, New York, VCH, 1989, Volume 5.
2 G Pritchard (ed.), *Plastics Additives: an A–Z Reference*, London, Chapman and Hall/Dordrecht, Netherlands, Kluwer Academic, 1998.
3 A W Birley, B Haworth and J Batchelor, *Physics of Plastics*, Munich, Hanser, 1991.
4 N G McCrum, C P Buckley and C B Bucknall, *Principles of Polymer Engineering*, Oxford, OUP, 1988.
5 P C Powell, *Engineering with Polymers*, London, Chapman and Hall, 1983.
6 L Holloway, *Handbook of Polymer Composites for Engineers*, Cambridge, Woodhead, 1994.
7 J M Margolis (ed.), *Engineering Thermoplastics–Properties and Applications*, New York, Marcel Dekker, 1985.
8 A Brent Strong, *High Performance and Engineering Thermoplastic Composites*, Lancaster PA, USA, Technomic, 1993.
9 R G Weatherhead, *FRP Technology*, Barking, UK, Applied Science, 1980.
10 N L Hancox, *Fibre Composite Hybrid Materials*, London, Applied Science, 1981.
11 M G Bader, 'Reinforcing fibres: the strength behind composites', *Mater World*, 1993 **1**(1) 22–26.
12 S M Lee, *International Encyclopedia of Composites*, New York, VCH, 1989.

13 A A Hodgson, *Fibrous Silicates*, Lecture Series 4, London, Royal Institute of Chemistry, 1965.

14 A J Kinloch and R J Young, *Fracture Behaviour of Polymers*, London, Applied Science, 1983.

15 C S Smith, 'Applications of fibre reinforced composites in marine technology', *Conference Proceedings, Composites–Standards, Testing and Design*, National Physical Laboratory, Guildford. IPC Science and Technology Press, April 1974, p. 54.

16 R Frissen, L Govaert and T Peijs, 'Modelling of the ballistic impact behaviour of polyethylene-fibre reinforced composites', *Proceedings ICCM-10*, Whistler, BC, Canada. Cambridge, Woodhead, 1995, Volume 5, pp 759–766.

17 R J Young, In *Developments in Reinforced Plastics, Volume 1*, ed. G Pritchard, London, Applied Science, 1980, Chapter 9.

18 A S Mosallam and R E Chambers, 'Design procedure for predicting creep and recovery of pultruded composites', *Proceedings of the 50th Annual Conference, Composites Institute*, Cincinnati, USA, Paper 6-C. New York, The Society of the Plastics Industry, January 1995.

19 H P Abeysinghe, *The Fracture Toughness of a Polyester Resin after Immersion in Aqueous Liquids*, PhD Thesis, Kingston Polytechnic, 1980.

20 S L Maddox and J K Gillham, 'Isothermal aging of a fully cured epoxy-amine thermosetting system', *J Appl Polym Sci* 1997 **64**(1) 55–67.

21 D Hull, *An Introduction to Composite Materials*, Cambridge, Cambridge University Press, 1981.

22 L E Nielsen and P E Chen, 'Young's modulus of composites filled with randomly oriented fibres', *J Mater Sci* 1968 **3** 352–358.

23 Q Yang, *Damage Tolerance of Filled Glass-epoxy Laminates*, PhD Thesis, Kingston University, 1995.

2

Fabrication, inspection and durability

GEOFFREY PRITCHARD

2.1 Fabrication of reinforced plastics products

There are obvious links between fabrication procedures, inspection methods and subsequent product durability. Some methods of manufacturing reinforced plastics produce better quality products than others, because they introduce fewer defects, or allow better control over fibre placement and orientation, or enable a higher volume fraction of fibre reinforcement to be used, or lend themselves to better quality control monitoring.

This chapter will discuss fabrication and its relevance to durability. It will also survey briefly the main threats to reinforced plastics durability, for the benefit of readers who do not need all the detailed treatments contained in the later chapters.

2.1.1 Principles of fabrication

The following description covers the more important procedures, but a quick-reference summary of fabrication procedures is provided in Table 2.1. The most attention will be given to processes which enable high fibre content load bearing articles to be made.

To make reinforced plastics objects, the resin and fibres must be combined intimately, so that they adhere together and distribute the fibres uniformly, with the desired orientation, with a minimum of porosity and with no dry, resin-free reinforcement. The correct fibre volume fraction must be ensured. In the case of thermosetting resins, there is an additional requirement to ensure that the hardening reaction, or cure, occurs at the right stage of the process and is eventually completed. The methods available depend on the size, overall shape and complexity of the article being produced, the materials involved and the number of mouldings required.

2.2   Fibre reinforcement formats

Fibre reinforcements are available in many different forms, designed for various fabrication procedures, as described below.

Continuous bundles of glass filaments are termed *roving*. They can be used directly in some of the most widely used fabrication procedures, such as filament winding, as well as in pultrusion (see later in the chapter for details of these techniques). The roving is unidirectional.

Other forms of glass fibre reinforcement include bidirectional *woven roving* fabric, in which one set of filaments is interwoven with another set at right angles to the first. Roving can be supplied already chopped into very short lengths for use in injection moulding, spraying processes or compression moulding compounds. The chopped fibres can alternatively be sprayed onto a former and glued together while still in random orientations, to form *chopped strand mat*, see Fig. 1.1 in Chapter 1.

Non-crimped fabrics have straight fibre bundles rather than the undulating ones typical of woven roving. They allow greater stiffness per unit weight of laminate than woven roving. This is because the filament bundles in non-crimped fabrics cannot respond to a stress simply by straightening out, because they are straight already.

Unidirectional tapes consisting of flat bundles of filaments already impregnated with catalysed resin can be used to increase stiffness and strength in specific places, and whole parts can be made from tape, applied under computer control.

Prepreg is virtually tape but very much wider. It is reinforcement supplied in a form already impregnated with resin and, in the case of reactive thermosetting resins, containing any necessary hardener. It is supplied as a large roll with protective backing and is usually stored at refrigerator temperatures, because otherwise the shelf life of the resin/hardener system could be too short. The operator cuts the roll into thin sheets or *plies,* which can be stacked to produce any desired thickness and fibre orientation, before heating in a press or high pressure autoclave.

Superior mechanical properties, including impact strength and resistance to delamination, are claimed for reinforced plastics made from three-dimensional fabrics [1]. Other reinforcement forms include continuous swirl mat, designed to have a minimum of fibre ends, and hybrid fabrics containing two different reinforcing fibres, for example, glass and aramid. There are many different ways in which two different fibres can be combined together.

The selection of the fabrication process is also a technical judgment as well as an economic one and it is integral to the design process, being best determined at a very early stage. Baruch *et al.* [2] point out for example that the stiffness coefficients required for the analysis of a laminated structure can be calculated only after considering the fabrication process.

Table 2.1 Fabrication methods for reinforced plastics

Method	Principle	Comments
Injection moulding	Premixed resin and short fibres are heated and forced into a split mould cavity	Medium size, intricate shapes, high throughput, mainly thermoplastics, low fibre content
Reaction injection moulding (RIM)	Premixed thermosetting resin and short fibres are injected using vacuum or pressure into cavity where cure reaction takes place	Large mouldings e.g. vehicle body panels
Resin transfer moulding (RTM)	As above, but long fibre reinforcement preform cut to shape and placed in cavity instead of using premixed short fibres. One half of mould cavity can be flexible	Large, complex shapes for automotive and aerospace use. Relatively fast cycles
Compression moulding (hot press moulding)	Moulding compounds (often complex mixtures, e.g. dough or bulk moulding compounds) prepared in Z-blade mixer and heated under pressure in closed mould	Small to medium articles, especially electrical mouldings; large capital outlay
Vacuum bag moulding (autoclave process)	Prepreg is encased in flexible bag; a vacuum is applied to draw the prepreg against a mould surface. Heat is applied. Process carried out in autoclave	Very large structures; prototypes; short production runs; high fibre content
Thermoforming	Thermoplastics sheet containing fibres is heated and forced by vacuum/pressure against a mould surface and then cooled	Low draw (shallow) articles; can be large
Stamp moulding	'Blank' reinforced sheets are preheated in oven and placed between matched metal mould halves; press then closed, heat/pressure applied	See above

Process	Description	Applications / Characteristics
Pultrusion	Resin-impregnated continuous fibres are pulled through heated zone where resin (if thermosetting) cures. Thermoplastics can be used	Constant cross-section profiles, e.g. channels for cabling, guard rails. High fibre content. Curved shapes possible
Filament winding	Continuous fibres are fed through catalysed resin bath onto rotating mandrel at predetermined angle and cured	Product must be symmetrical about long axis of mandrel. Pipe, tanks, leaf springs, air ducts. High fibre content
Sheet moulding compound (SMC)	Continuous sandwich of thermoset resin slurry/glass reinforcement between two polythene layers is first matured with thickening process, then cut to size and moulded in matched metal moulds	Large but not intricate shapes: vehicle parts, spoilers, thin body panels, aircraft interior parts, good surface quality.
Hand lay-up	Individual plies of reinforcement placed in open mould and impregnated with catalysed resin by hand tool, e.g. roller; unidirectional tape can be used	Large structures, e.g. boat hulls, chemical tanks
Spray moulding	Chopped fibres and catalysed thermosetting resin sprayed simultaneously onto open mould and cured at ambient temperature	As for hand lay-up. More rapid but short fibres provide less reinforcement

2.3   Injection processes for thermoplastics and thermosetting resins

2.3.1   Injection moulding

This method is the most familiar and widespread of the techniques used with unreinforced thermoplastics. It has spawned a number of related procedures which are increasingly used to produce reinforced plastics products.

In the most basic method, resin is mixed with short glass fibres and converted by means of an extruder into small cylindrical granules. The granules are heated and when molten, forced under pressure by means of an Archimedean screw into a split cavity mould, where cooling takes place. The method is suitable for rapid mass production of small and medium sized articles including complex shapes, such as electric drill housings. The rotating screw that drives the granules forward can sometimes damage the fibres, reducing their length and reinforcing power. Conversely, glass fibres can cause expensive screw wear.

Machinery was successfully designed many years ago for injection moulding thermosetting resins. It is essential to ensure that gelation, or solidification, takes place only in the mould cavity and not in the barrel of the injection moulding machine. Hardening takes time. Exactly how much time it takes depends on the resin, the hardening system and the temperature. In theory, the hardening reaction could slow the process down unacceptably, but modifications have been made to the technique to deal with this, giving reasonable throughput rates. The cooling stage required by thermoplastics need not be so great with thermosetting compositions which retain their rigidity at higher temperatures.

2.3.2   Reaction injection moulding (RIM)

This technique is a more recent process, designed for thermosetting resins. The method takes the principle of the ordinary injection moulding process, together with features of the traditional polyester laminating procedures mentioned later in this chapter. It makes large mouldings, for example, vehicle body panels.

Premixed resin and short fibres are injected into a large closed cavity, where cure takes place over a significant period of time. Early machinery could make only four or five large mouldings per cavity per day. The same mixing and injecting mechanism is used to fill a series of cavities, unlike conventional injection moulding where one cavity is emptied in each cycle and immediately reused. The injection process in RIM can be carried out using a vacuum rather than hydraulic pressure and the mould tools do not have to be made of particularly expensive materials.

A further development is resin transfer moulding (RTM), in which long fibre reinforcement fabrics are cut to shape as *preforms* and placed in the mould cavity, before the resin is injected. This is potentially a slow process and one alternative is for the preforms to be made rapidly by spraying short fibres onto a suitably shaped pattern coated with an adhesive or binder.

One half of the mould cavity can be flexible. Because the method is not restricted to short fibres it can ideally achieve good mechanical properties. Large, complex shapes have been made for the automotive and aerospace industries. Automated and computer controlled processes of this kind can achieve rapid cycles.

2.4   Compression moulding and moulding compounds

Compression moulding or hot press moulding is one of the oldest processes, dating from the early history of the rubber industry, before plastics were developed. In recent years, much of the innovation has been stimulated by the use of new materials.

A *moulding compound* consists essentially of a resin, mixed with short fibres, a filler and a lubricant, and a catalyst, using a Z-blade mixer. The resulting material will often be a pliable dough if the resin is liquid and the compound becomes known as a dough moulding compound (DMC) or a bulk moulding compound (BMC). The mixture is heated under pressure in a closed mould in the press for 30–200 s, or much longer with special high performance resins. Articles are limited in size by the press. Electrical components are typical examples of compression moulded products, although they can also be made by injection moulding.

Compression moulding requires considerable capital outlay. It follows that presses cannot be monopolized by a few mouldings for very long periods and the cooling stages required for thermoplastics mean that the method is not generally considered cost effective for them.

A successful development from DMC/BMC is SMC, or sheet moulding compound (see Fig. 2.1). There are many variants. In one, a continuous layer of thermosetting resin slurry and glass reinforcement is sandwiched between two polyethylene layers, and matured with a special thickening agent to remove all surface stickiness and to increase the viscosity. Then when the mix has reached the easy-handling stage, it is cut to the appropriate shape, the polyethylene backing removed and the compound is moulded in matched metal moulds. The method is convenient for large, but not particularly intricate shapes: vehicle parts, thin body panels, spoilers, aircraft interior parts such as baggage lockers and passenger transport seats. After considerable problems in the early years, good surface quality can

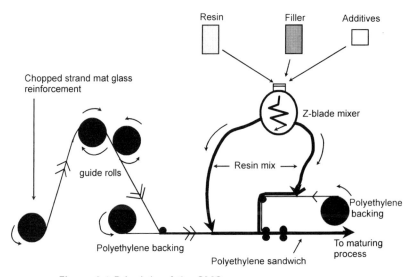

Figure 2.1 Principle of the SMC process

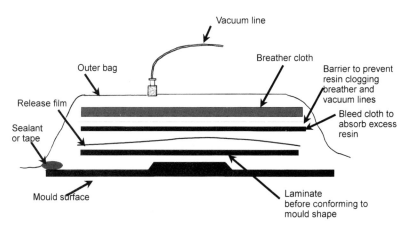

Figure 2.2 Main constituent items of vacuum bag assembly for use in autoclaves. (After Brent Strong [3])

now be achieved. The formulation or 'recipe' of the mix is important if porosity and surface roughness are to be avoided.

For very large parts that cannot be accommodated in ordinary presses, the autoclave process can be used in conjunction with a vacuum bag [3]. In one example, prepreg is encased in a flexible plastic bag, which is situated in an autoclave (Fig. 2.2). A vacuum is applied to draw the bag against the mould surface and heat is applied. The process can be used to produce very large articles, but it can also be scaled down and has advantages for short production runs or prototypes of smaller articles requiring high fibre volume fractions. The autoclave is filled with an inert gas during the moulding operation, for safety reasons.

2.5   Thermoforming and hot stamping

Thermoforming is another mass production method normally associated with unreinforced or short fibre reinforced thermoplastics, but because of recent developments in reinforced plastics technology, discussed below, there is a greater availability of thermoplastics sheet reinforced by long or continuous glass fibres, so it may become more important for these materials as well.

A thermoplastic sheet or 'blank' is horizontally clamped and heated by infrared heaters until flexible and then forced by various means (pressure, vacuum, pneumatic ram) against a mould surface and cooled. The mould can be low cost.

The process is used for large, shallow (low draw) articles. There must be no excessively thin parts in the finished moulding. Unless the sheet is heated on both sides, the usable thickness is limited to a few millimetres.

Similar results can be achieved by hot stamping, in which blanks of reinforced thermoplastics sheets are preheated in an oven and placed between matched metal moulds in a heated press.

2.6   Filament winding

In filament winding, continuous filament roving is fed from spools through a bath of resin containing catalyst and then wound onto a rotating mandrel at predetermined angles to the mandrel's long axis. The resin hardens and the resulting laminate is then removed from the mandrel (Fig. 2.3(a)). This last operation can be surprisingly difficult, and considerable ingenuity has been directed to devising mandrels from which the reinforced plastics product is easily removed, such as inflatable or acid-soluble ones.

The product must be symmetrical about the long axis of the mandrel. Pipes, tanks, leaf springs, air ducts, pressure vessels, and so on are typical filament winding products. High strength articles with high fibre contents can be obtained.

If the filaments are not wound onto a mandrel but simply gathered into relatively large bundles and led slowly into a heated consolidation zone, emerging from a die partly or completely cured, the primitive essentials of pultrusion are present (Fig. 2.3(b)). This process began as the thermosetting resin equivalent of extrusion, that is, it was a continuous operation for manufacturing profiles, such as rod and channels. The method has been extended to fibre reinforced thermoplastics.

glass roving

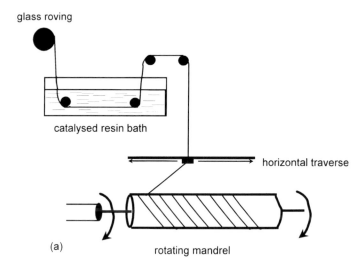

catalysed resin bath

horizontal traverse

(a)   rotating mandrel

continuous fibre strands

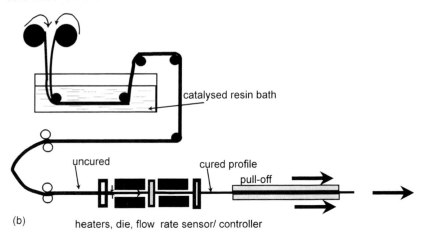

catalysed resin bath

uncured   cured profile

pull-off

(b)   heaters, die, flow rate sensor/ controller

Figure 2.3 (a) Principle of filament winding. (b) Principle of pultrusion with thermosets (schematic)

2.7   Hand lay-up and spray processes

Hand lay-up is a low technology (but in a craft sense, surprisingly skilled) process. The individual layers of reinforcement are placed manually in an open (one-sided) mould, previously coated with a thin layer of unreinforced resin known as the gelcoat, and these layers are then thoroughly impregnated with catalysed resin by means of a hand roller. The resulting fibre content depends on the reinforcement format, being typically 30% w/w with chopped strand mat and 50% with woven roving. Large structures such as boat hulls and storage tanks can be made to individualized shapes, with very little capital outlay. The quality depends to an unusual extent on the skill of the fabricator. Good quality products are certainly possible if good practice is followed. Marks of poor fabrication include, among others, patches of dry reinforcement where there is inadequate resin impregnation, unevenly distributed catalyst, inappropriate gelcoat thickness, wrong time interval between gelcoat and laminate application and serious undercure caused through inadequate temperature and humidity control, or through omission of a genuine postcure stage. Unskilled operators have been known to make gross errors either in the application of hardeners or by causing gross contamination, but competently made hand lay-up articles have demonstrated excellent durability over periods in excess of 40 years.

A mechanized development of hand lay-up is spray moulding, in which the gelcoat, resin and chopped fibres are applied by spraygun rather than by hand. The process is much quicker than hand lay-up, but the capital costs are no longer rock bottom. The flow characteristics and hence the styrene content of the resins are different. The fibres cannot be very long and the mechanical properties of the products will reflect this.

2.8   New fabrication methods for long fibre thermoplastics

Chapter 1 mentioned that the use of reinforced thermoplastics in load-bearing applications is comparatively new because good mechanical properties require (a) high fibre contents and (b) continuous fibres. Appropriate methods of fabrication have only become available in the last few years and process development is still continuing. The fundamental problem has been the very high viscosity of molten thermoplastics compared with liquid, uncured thermosetting resins. A low viscosity liquid is preferable so that the reinforcement can be completely wetted out, or impregnated, and good interfaces obtained. Otherwise the fibres cannot reinforce the matrix properly.

Miller and Gibson [4] divide the available techniques into two types:

1   Pre-impregnation processes, in which impregnation happens before consolidation under heat and pressure;
2   Those involving intimate mingling of resin and reinforcement, during the final shaping stage.

One example of pre-impregnation is melt impregnation, in which roving is pultruded through a resin melt and over cylindrical pins. The pins are designed to spread the fibre bundle or tow, and to entrain a film of liquid resin which becomes drawn into the tow. Another approach is solvent impregnation, which can only be used with soluble thermoplastics. The fibres, whatever form they are in, are easily wetted out by the low viscosity solution, but the difficulty is the subsequent removal of all the solvent, including the last traces. Failure to remove all the solvent leads to porosity in the product. This in turn means inferior mechanical and electrical properties. The measurement of liquid absorption has been recommended as a rough way of comparing nominally similar samples for void content.

Miller and Gibson also mention several procedures for intimate mingling of resin and reinforcement at the late stage. In film stacking, alternating layers of polymer film and reinforcement are stacked together and heat is applied, with pressure (Fig. 2.4(a)). The process is slow and cycle times are not particularly economical. High pressures can result in fibre breakage at crossover points in woven fabrics. One development is the glass mat thermoplastics process for producing prepreg to be used in hot stamping, as mentioned above.

Commingled fibre technology involves supplying the matrix in filament form and mingling it with the reinforcing fibres (Fig. 2.4(b)). The two materials can even be cowoven in the same strands. Thus the material is available in the form of continuous roving, chopped fibres, fabrics, etc. Excellent drape qualities can be obtained with fabrics. The technique can be used for pultrusion, but not for filament winding.

Powder impregnation is a technique in which dry resin powder is applied to the fibres and normally then melt fused. The fibres are spread out mechanically or pneumatically. In various developments, the particles may be electrically charged; the fibre tow can be passed through a fluidized bed and aqueous powder dispersions can be used.

2.9   Quality control and non-destructive evaluation of reinforced plastics during manufacture or in service

Quality control is an important element in the optimization of material durability. Achieving consistent output involves sound operator training, regular routine raw materials screening, competent processing machinery maintenance, good mould tool design and vigilant oversight of the actual processing operation, including atmospheric conditions such as humidity, temperature and dust content. It also requires the use of up-to-date NDT/ inspection methods on the final products. The following discussion deals only with this last topic.

2.9.1 Checking for gross errors

Only a few reinforced plastics materials are sufficiently translucent for visual inspection. Glass fibre reinforced resins can in some instances be

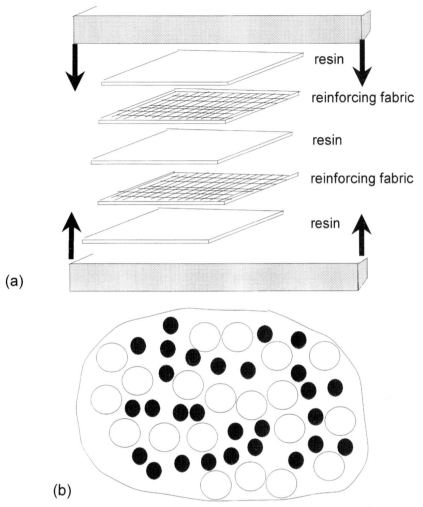

Figure 2.4 (a) Commingling by film stacking: alternate layers of resin film and reinforcing fabric. (b) Cross-section of commingled reinforcement and resin, both in fibre form, before consolidation. See also Miller and Gibson [4]. ○, Matrix in fibre form; ● reinforcing fibres

examined by transmitted light, whereas carbon fibres are not amenable to this approach. It is possible for major errors to escape notice unless instrumental monitoring methods are employed.

The first consideration is whether the operator has omitted any plies, laid one of them in the wrong orientation, or inadvertently introduced an extra layer of foreign material such as release film or packaging film.

Instruments are available for determining the thickness of a component accessible from only one side. A laser-based technique has been developed

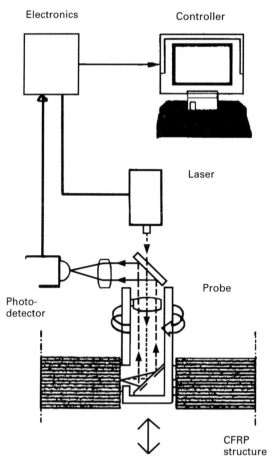

Electronics

Controller

Laser

Probe

Photo-
detector

CFRP
structure

Figure 2.5 Laser-based technique to check ply orientation in reinforced plastics laminates (By permission: S Hill, 'Rapid non-destructive testing of carbon fibre reinforced plastics'. *Materials World*, August 1996, Institute of Materials, London)

to check ply orientation (see Fig. 2.5). The probe is inserted through a hole in the composite, such as a rivet hole, and measures the amount of light reflected back from the wall. The hole surface has first to be made smooth. The fibres oriented at right angles to the laser beam produce the most reflection, while the ones parallel to the beam reflect the least. The whole process is completed within a few minutes [5].

Small deviations from the intended fibre orientation within a nominally unidirectional ply can reduce the mechanical strength and stiffness of continuous fibre laminates considerably, especially with aramid reinforcement. Wisnom [6] has investigated the reduction in strength caused by misalignment of unidirectional carbon fibres in XAS/914 carbon/epoxy laminates. The compression strength was reduced by roughly half for less than one degree misalignment.

Control of fibre orientation with short fibres, such as those sometimes used in injection moulding or reaction injection moulding (RIM) is also important. The flow in the mould may be constrained in two directions such as width and thickness, giving linear flow only in one direction. Or, the flow can instead be radial, that is, constrained in just one direction, for example the thickness, in which case the usual shear forces due to the mould walls will operate, but there will be an additional extensional force, caused by the expanding flow front. This results in fibre orientation being normal to the flow direction. The orientation angle is usually very small at the surface, but it can be almost 90° at the mid-plane [7].

If the orientation can be examined by optical microscopy, it can be quantified using computerized image analysis techniques. Anisotropy can be studied by applying an ultrasonic, electrical or radiation signal, which penetrates into the test material in different directions. Gross fibre waviness can be checked by visual inspection.

2.9.2 Fibre volume fraction

It is useful to check the fibre volume fraction. Many of the procedures are destructive, and cannot be used to inspect a finished product before use, but they can be employed on dummy parts or, in the case of low cost mass produced articles, on a small percentage of randomly selected production output. When the laminate consists simply of glass and organic resin, the latter can be removed from a sample by heating it at about 600°C, leaving only the fibres, which are calculated by weight difference. This procedure is not suitable for carbon fibre reinforced plastics (CFRP) because the fibres and resin would both be burnt off together. Instead a chemical oxidizing reagent is used which oxidizes the resin but not the carbon. Preferential oxidation of an epoxy matrix can be carried out by heating with concentrated sulfuric acid and hydrogen peroxide, leaving the fibres. Different procedures again are required for reinforced plastics materials that contain mineral fillers. Some fillers can be dissolved away by acids, leaving the way clear for the burn-off method.

Indirect methods for fibre content, such as thermal conductivity, can be used, but they require calibration against direct methods.

2.9.3 Local internal defects

Local ply delaminations, poorly impregnated regions and areas of porosity need to be identified. Non-destructive methods are available which can also be adapted to the assessment of internal damage such as delaminations and

Table 2.2 Non-destructive testing techniques for reinforced plastics. Adapted from Matiss [8]. Courtesy of RAPRA Technology Ltd., UK.

Method	Direct application	Indirect application
Ultrasonic	Velocity and attenuation factor	Modulus of elasticity, geometry, degree of cure, stress–strain state, delamination, density, reinforcement ratio, porosity, flaw size
Vibrational	Resonance frequency and damping factor	Modulus of elasticity, geometry, degree of cure, density, reinforcement ratio, porosity, damage accumulation, delamination
Mechanical impedance	Complex mechanical impedance	Density, modulus of elasticity, geometry, delamination, reinforcement ratio
Acoustic emission	Number and intensity of acoustic pulses, amplitude and energy distribution, spectral envelope	Accumulation of damage, fracture, adhesion, load bearing capacity
Electromagnetic (eddy current)	Complex electrical conductivity, complex magnetic permeability	Density, reinforcement ratio, porosity, damage accumulation, geometry, structure, flaw size
Electrical	Complex electrical conductivity, dielectric permittivity, loss factor	Chemical and structure characterization, degree of cure, reinforcement ratio, ageing, moisture absorption, geometry, density, electromagnetic wave, transparency
Electromagnetic waves	Electromagnetic wave reflection, absorption and transmission coefficients	Dielectric properties (permittivity, loss factor), moisture absorption, delamination, geometry, flaw size
X-rays	X-ray absorption, reflection and transmission coefficients	Internal structure, internal stresses, delamination, reinforcement pattern, density, imperfection sizing
Heat	Thermal conductivity, capacity	Delamination, flaw size, porosity, reinforcement ratio
Optical	Light absorption, reflection and transmission coefficients	Reinforcement or matrix damage, reinforcement ratio, fatigue damage, geometry
Holography	Linear displacement	Stress–strain state

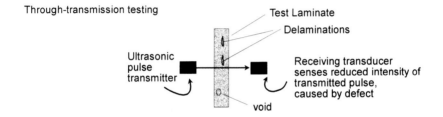

Through-transmission testing

Test Laminate

Delaminations

Ultrasonic pulse transmitter

Receiving transducer senses reduced intensity of transmitted pulse, caused by defect

void

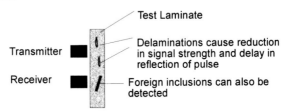

One-sided pulse-echo test

Test Laminate

Transmitter

Delaminations cause reduction in signal strength and delay in reflection of pulse

Receiver

Foreign inclusions can also be detected

Figure 2.6 Principle of ultrasonic scanning of reinforced plastics laminates for defects. Access to both sides is not always possible. Output gives a 'map' of the laminate

matrix cracking. This is useful for predicting the future durability of damaged articles already in service.

The chosen method depends on the geometry and properties of the article. Table 2.2 lists many of the available non-destructive test methods for composites for initial quality control and for assessing damage and repair [8]. The following discussion highlights just a few of the more important techniques.

Figure 2.6 illustrates the principle of inspection for defects by ultrasonic pulses. The intensity and attenuation of ultrasonic signals, transmitted by transducers, can be monitored on their emergence from the sample. The intensity of the exit signal is reduced by such defects as delaminations or foreign material. The output is usually displayed as a two-dimensional map of the sample, called a C-scan, showing the location of the defects, such as regions of high void content.

Delaminations reduce the time required for the reflected signal to arrive at the surface in pulse-echo setups (also shown in Fig. 2.6) which have both the transmitting and detecting transducers on the same side of the test sample. The same method can distinguish between flaws at different depths. A slightly different ultrasonic resonance method utilizes the fact that delaminations reduce the sample stiffness and hence produce a resonant frequency shift.

X-radiography is used to detect problems in the honeycomb cores of bonded sandwich structures. Thermography can locate flaws in continuous

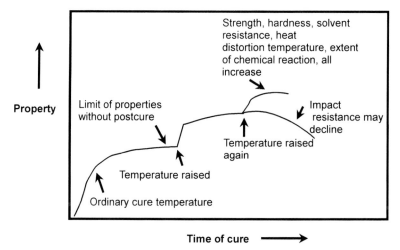

Figure 2.7 Effect of postcure on thermosetting laminate quality

tow thermoplastics components by rapid inspections after the positioning of each individual tow, rather than at the completion of the fabrication process when it is much more difficult to eliminate any defects observed (one of the advantages of thermoplastics is that small defects can be corrected immediately by locally applied heat).

Acoustic emission utilizes the fact that a material's response to stress is accompanied by the release of energy as sound or ultrasound. Different kinds of events – notably matrix cracking, debonding and fibre fracture – can be distinguished. Thus the mechanical condition of a composite can be examined.

2.9.4 Cure of thermosetting resins

It is common practice to check the degree of cure of thermosetting resin laminates by measuring mechanical or physical properties. When the key properties stop changing, it is assumed that cure is complete. (This is often the only practical approach, but it is no more logical than assuming that a train has arrived at its destination because it has stopped!). Complete cure requires the utilization of all potentially reactive chemical groups involved in the process. This is very unlikely in thermosetting resins without stepwise elevated temperature postcure (Fig. 2.7). The degree of cure of a thermosetting resin matrix should ideally be measured by an on-line technique that has previously been calibrated against a chemical procedure. In low technology systems, cure can be monitored in low technology products by very crude tests such as surface hardness measurement, but in more demanding areas ultrasonic, spectroscopic and dielectric methods are preferred. Some mass production processes have built-in instrumentation for cure monitoring. Kranbuehl *et al.* [9] have described the use of dielectric properties and viscosity in intelligent control of the resin transfer moulding (RTM) process. The control of the injection moulding of thermosetting BMCs can be achieved by dielectric methods. One of the difficulties is that dielectric properties and viscosity do not depend simply on the degree of cure. They are extremely sensitive to temperature as well.

It is instructive to compare the heating processes used in the aerospace composites industry with those used in the polyester boat-building industry. Granted that the resins used and the process economics are different, the fact remains that boats are rarely given sufficient heat treatment for cure to approach completion because of the lack of investment in postcure facilities. This inevitably means that the water resistance of the hull, although adequate for 15 years, is insufficient to achieve the long term durability with minimal maintenance that could otherwise be delivered. Well-made glass fibre polyester laminates with elevated temperature cure in heated sheds should be able to resist 'softening' by water for much longer than commonly observed.

It has been suggested that the advent of long fibre reinforced thermoplastics materials and advanced fabrication techniques will result in reinforced thermoplastics boat hulls replacing the now traditional unsaturated polyesters and offering cleaner fabrication processes, with better durability and easy recycling. There is not much firm evidence, whether based on case histories or laboratory data, about the long term behaviour of glass reinforced polypropylene in seawater, but it should be durable, given a good interface. The problems associated with such materials may turn out to be quite different.

2.9.5 Cooling rates after fabrication of crystalline thermoplastics

The rate at which thermoplastics articles are cooled in the mould can vary enormously from one process to another, with a significant effect on the crystallinity and properties of the final product. Cooling rates ranging from less than $1°C\ \text{min}^{-1}$ to more than $1000°C\ \text{min}^{-1}$ are discussed by Cattenach and Cogswell [10]. The changes in crystallinity caused by changes in cooling rate are sufficient to affect the mechanical properties (such as impact strength) and the solvent resistance. These considerations do not apply to amorphous resins.

Cooling does not occur at a uniform rate across the moulding. Consequently, the extent of crystallization can also vary through the thickness. Internal stresses can be generated as a result of temperature differentials and orientation effects.

2.10   Property retention as a guide to durability

The best initial mechanical properties are achieved with continuous fibre composites. Table 2.3 gives some representative values. The numbers refer to the properties of the materials in their original (as-manufactured) condition. After several years in service, the properties will not be the same, even in the absence of obvious mechanical damage. It is customary to cite the change in properties with time as a measure of the extent of deterioration, and '% retention' has become by implication a measure of durability. Not all properties change equally rapidly and the selection of significant properties requires careful consideration.

There are several potential causes of deterioration. Among the most important are moisture, chemicals, high temperature, ultraviolet light and mechanical fatigue. All these factors will be discussed in depth in other chapters, so we shall simply identify some of the main issues here and leave the detail for later.

The durability of a reinforced plastics article is sometimes a function of its appearance, including colour changes, loss of surface quality, and so on. Sometimes the changes are more obvious; there is gross cracking or swelling for instance. But at the other extreme, there may be no visible signs whatsoever. Deterioration might only be apparent when using advanced diagnostic equipment, whether this is thermography, acoustic emission equipment, ultrasonic instrumentation, an electron microscope or spectrometers for chemical analysis. If this is the case, there may be no cause for concern and materials suppliers will be able to furnish data tables showing that there is little or no change with time in key properties such as flexural strength and modulus, dielectric properties, dimensions, and so on.

2.11   Moisture

All organic resins and organic reinforcing fibres (but not glass or carbon) absorb water to varying extents, usually at a very low level, and are water permeable. Water absorption into glass or carbon reinforced plastics is slow. Changes occur in the physical properties, for example, electrical properties dimensions. Moisture migrates through the resin and eventually reaches the fibre–resin interfaces. It is often said that moisture migrates by 'wicking' along the interface by capillary action, starting from exposed fibre ends, but hard evidence is usually absent, and a well-bonded fibre is not so

Table 2.3 Representative properties of unidirectional, continuous-fibre composites (stress parallel to fibre direction) at room temperature

Fibres	Matrix	V_f	Property	Value	Units	Ref
Boron	Epoxy	0.50	Tensile strength	1600	MPa	11
Boron	Epoxy	0.50	Tensile modulus	210	GPa	11
Boron	Epoxy	0.50	Compressive strength	2900	MPa	11
Boron	Epoxy	0.50	Poisson's ratio	0.21	—	11
Boron	Epoxy	0.50	Thermal exp. coeft.[a]	4.5	$\mathrm{mm\,mm^{-1}\,K^{-1}} \times 10^{-6}$	11
Carbon	Polyimide	0.61	Tensile strength	1730	MPa	11
Carbon	Polyimide	0.61	Tensile modulus	142	GPa	11
E-glass	PPS[c]	0.57	Tensile strength	835	MPa	12
S-glass	PPS[c]	0.57	Tensile strength	1117	MPa	12
Carbon	PEEK[f]	0.63	Tensile modulus	140	MPa	12
Carbon	PEEK	0.63	Tensile strength	2040	MPa	12
Carbon	Epoxy	0.60	Compressive strength	1231	MPa	13
Kevlar 49	Epoxy		SBSS (Note 2)	62	MPa	14
Kevlar 49	Epoxy		Compressive strength	255	MPa	14
Glass	Epoxy	0.60	Flexural strength 3-pt	1300	MPa	15
Glass	Epoxy	0.60	Water absorption	1.2	%	15
Kevlar 49	Epoxy	0.60–0.65	Thermal exp. coeft.[a]	−2.3 to −4[d]	$10^{-6}\mathrm{\circ C^{-1}}$	16
Kevlar 49	Epoxy	0.60–0.65	Thermal exp. coeft.[b]	35[e]	$10^{-6}\mathrm{\circ C^{-1}}$	16
Carbon	Epoxy	63	In-plane shear strength	95	MPa	16

[a] Longitudinal direction.
[b] Transverse direction.
[c] Polyphenylene sulfide.
[d] In range −79°C to +100°C.
[e] In range −195°C to +120°C.
[f] Polyether ether ketone.

easily separated from the resin that it allows migration along a long and continuous section of its length. Weak interfaces are a different matter.

The effects of moisture, once absorbed, are complex, and have been the subject of much research. Changes in the appearance and properties of the reinforced plastics product may be slight or severe, chemical or physical, permanent or reversible. The more moisture absorbed, the more deterioration in properties is likely to be found and the less reversible are the changes on drying. Reductions in strength and modulus are observed, although an initial increase in strength is not unknown, because of the relief of internal stresses. The increase is followed by a decline with further absorption.

To protect the interface against moisture, glass fibres are given 'finishes', such as silane surface coatings, chosen to match the resin matrix being used. Nevertheless partial and localized separation of the resin matrix from the fibres after prolonged immersion in aqueous liquids at high temperatures is fairly common. Debonding can sometimes be seen in translucent polyester–glass laminates even with the naked eye, because of the reflection of light by the debonded interface. It is more easily detected with a low powered optical microscope.

The individual constituents of reinforced plastics, immersed separately, can be susceptible to water damage. Glass fibres can be severely attacked by prolonged contact with hot water, and if they are under stress at the time, surface flaws can develop in the fibres, with a consequent loss of strength. This does not apply to carbon fibres. Glass fibres unless specially formulated are also vulnerable to acid. The main corrosion process is thought to be the removal of calcium and aluminium ions from the glass. The rate of ion removal is dependent both on the hydrogen ion concentration in the acidic solution and on the nature of the acid. Acids which form insoluble salts or complex ions, such as oxalates or sulfates, greatly accelerate the corrosion process compared with acids that form soluble salts, such as nitrates from nitric acid or chlorides from hydrochloric acid. So at pH 1, oxalic acid is far more damaging than nitric acid. The difference reduces as the pH rises and is negligible above 3.5 [17].

The more susceptible resins (polyesters, polyester urethanes, some epoxies) are attacked by boiling water fairly quickly, but could still resist cold water for very long periods. Other resins, with different chemical structures, are unaffected at all temperatures within their normal use range.

The combination of absorbed moisture and temperature fluctuations can cause cracking and other damage. Sometimes, the matrix itself absorbs only a small amount of moisture and at first, the laminate behaves similarly, but at a late stage, when cracking in the interface region occurs, or there is delamination or translaminar cracking between plies, an unexpected increase in the laminate's overall absorption is noticed.

2.11.1 Predicting deterioration rates

The question is often asked, can we deduce the rate of deterioration over long timescales at ambient temperatures, from laboratory data obtained quickly at higher temperatures? It must be admitted that the consensus view at present is against relying on any such predictions, although the temptation is difficult to resist. The reasons for caution are: (i) that the resins undergo subtle changes on heating and can become much more susceptible to deterioration than they ever would at ambient temperatures and (ii) the degradation processes have several constituent elements which accelerate to different and unknown extents on raising the temperature. Nevertheless we should probably be safe to assume that the predictions of high temperature experiments, when applied to lower temperatures, tend towards the pessimistic rather than the optimistic side.

Tests on short-fibre marine laminates made from chopped strand mat and polyester resin in the 1950s suggest that ocean service in various different climates causes about 20% fall in strength within 12 months or less, followed by surprisingly little further change over the next 20 years [18]. Later studies are not inconsistent with this early study. For example Chiou and Bradley [19] reported that the burst strength of glass/epoxy filament wound tubes was reduced by 20% by ageing in synthetic seawater for six months at ambient temperature.

It is widely believed that distilled water is more harmful than seawater which contains dissolved salts. Some workers have used artificial seawater made from 'Instant Ocean', a mix sold to pet shops for the benefit of tropical fish keepers and containing trace elements [20]. Marine laminate durability is discussed in detail in Chapter 7.

In an investigation of damage propagation in glass–polyester tubes exposed to hot water at 60°C for 3000 h, Ghorbel et al. [21] concluded that the final failure of the tubes was governed by the critical strain to failure of the tubes' lining, a resin-rich chopped strand mat layer, and therefore it was difficult to predict the lifetime of pipes from short term tests and consideration of damage accumulation.

Any water-soluble additive is liable to be dissolved and leached out from a laminate over a period of time when exposed to wet environments, and if the additive is a particulate solid, dispersed in the resin, the result can be widespread porosity. Before the additive is washed out altogether, the trapped pockets of additive solution constitute osmotic pressure sites, which induce microcracks in the matrix. Figure 2.8 shows a cross-section through a unidirectional glass epoxy laminate, in which the dissolving of residual unused dicyandiamide hardener by a hot, wet atmosphere caused porosity and osmotic cracks. Prolonged exposure at 100°C caused fragmentation of the laminate [15].

Figure 2.8 Scanning electron micrograph of unidirectional glass epoxy laminate containing residual dicyandiamide curing agent. After prolonged exposure to hot water, cracks propagated from where the particles had been before they dissolved in the water

Thick laminates are obviously much less affected than thin ones in a given period and this explains the durability of many early reinforced plastics structures. It has been calculated that an epoxy-based reinforced plastics material with a typical diffusivity towards moisture of $10^{-13}\,\mathrm{m^2\,s^{-1}}$ would require 13 months to reach saturation if left in a tropical climate at 35°C and 95% relative humidity (RH) if the thickness was 2 mm, but a 90 mm thick section would need 1342 years. During the approach to saturation, there is a through-the-thickness variation in moisture content and therefore in properties.

The use of carbon fibres rather than glass improves water resistance considerably, although it does not guarantee immunity from water damage, because both the resin matrix and the interfacial bond can still be affected. The fatigue life of carbon/epoxy cross-ply laminates is actually shorter when the fatigue process is carried out on dry specimens than on those previously immersed in seawater until they were saturated, but longer than when the fatigue process took place while the samples were actually immersed in water [22].

2.12   Chemicals (see also Chapter 3 and Chapter 9)

A surprising number of reinforced plastics applications involve occasional or prolonged contact with chemicals. Water, already discussed, is of course a chemical itself, although because it is so very widespread and has been so extensively investigated, it has been allocated a separate section. Many reinforced plastics articles are routinely placed in contact with detergents, cleaning solvents, acids, alkalis, strong oxidizing agents, bleach, cleaning and degreasing agents, fuels, hydraulic and brake fluids, de-icers, paint strippers (methylene chloride based ones are known to be damaging), lubricants, etching chemicals, flue gases, or food and drink, including wine.

It must be stressed that the resistance of reinforced plastics to highly reactive chemicals is generally very good. This explains their widespread use in the chemical process equipment industry, where it is often difficult to find any other affordable, processable materials capable of withstanding the very harsh conditions. It is very rare indeed for reinforced plastics articles to be attacked as rapidly as some common metals are when placed in contact with acids. A few chemicals that are handled in chemical factories, such as powerful oxidizing agents, strong caustic alkalis, bromine and wet chlorine still pose severe problems for general purpose organic matrix resins. Otherwise, the well-informed selection of materials, in consultation with the suppliers and after reference to the relevant data banks, means that complete disaster is a very rare occurrence.

2.12.1 Effects caused by a chemical reaction

A *reaction* between the reinforced plastics material and the surrounding medium is defined here as an event which breaks and makes new covalent chemical bonds in the participating substances. One illustration is the hydrolysis reaction which breaks ester and amide groups in polyesters, polyamides and certain polyurethanes, when they are exposed to dilute acids or alkalis. Another example is the oxidation reaction that attacks carbon fibres whenever they are in contact with dilute nitric acid, acid/dichromate, hypochlorite solutions, and so on. Such reactions can easily be foreseen and avoided by the use of more appropriate resins, fibres, and so on. The rate of chemical reactions involving resins is much more difficult to predict and they can sometimes be so slow that the effects are not necessarily unacceptable at ambient temperatures. One obvious illustration is the successful use of polyester resins in marine applications despite their theoretical vulnerability to hydrolysis in reactions with water.

2.12.2 Galvanic corrosion

Galvanic corrosion does not occur at all with ordinary organic plastics, because they are insulators. In reinforced plastics the phenomenon is confined to situations where carbon fibre reinforcement is situated in very close proximity to certain metals. This problem can easily be overcome by inserting some insulating material to break the electrical circuit.

2.12.3 Absorption (uptake) and subsequent drying out of absorbed liquid

The process is slow except at high temperatures. Considering unreinforced resins first, absorption causes dimensional changes and softening, that is, a reduction in surface hardness and in modulus. The glass transition temperature and associated heat distortion temperature of the resin fall substantially, lowering the maximum permissible use temperature of the resin [23].

2.12.4 Reversibility

The changes in mechanical properties caused by absorption are more or less reversible on drying out after short periods of contact with the environment, but more prolonged immersion tends to make full recovery of initial properties much more difficult. The mechanical property changes are accompanied by corresponding alterations to electrical conductivity, dielectric loss and other physical properties.

The same trends can still be observed in fibre reinforced plastics, although the reinforcement will moderate the changes in mechanical properties. However, there are other possibilities, such as glass fibre–resin debonding, caused by water absorption from aqueous liquids. This can mean a reduction in translucency. Load transfer between fibres is also less effective. Carbon fibres, in contrast, are unaffected by water below 1000°C.

Aramid fibres absorb about 5% water, which can sometimes affect their performance in aqueous liquids [24]. Some reinforced plastics applications involve repeated soak–dry cycles and it can be the drying stage, rather than absorption that causes damage. If the second absorption cycle produces more absorption in a given time than the first cycle did, it usually indicates permanent damage.

Changes in the microstructure of reinforced plastics can be observed under scanning electron microscope examination. When debonding is ex-

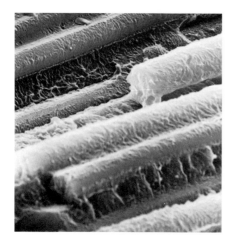

Figure 2.9 Left: Good adhesion between fibres and resin is demonstrated by the residual resin on the fibre surfaces after fracture. The specimen was a carbon fibre reinforced PEEK (polyester ether ketone) material. Right: Exposure to hot water has caused some loss of adhesion between the glass fibres and the unsaturated polyester resin, as shown by the smooth regions of the fibre where little resin adheres after fracture

tensive, a fracture surface of the specimen will show a tendency towards smoother fibre surfaces, with less resin adhering (Fig. 2.9).

2.12.5 Environmental stress cracking (see also Chapter 3)

Until now, all the effects we have mentioned have been gradual. Environmental stress cracking (ESC) can occur in reinforced and unreinforced plastics and can have sudden consequences.

ESC is the term used to describe the failure mechanism when a material such as a resin or unidirectional glass fibre reinforced laminate fractures by rapid brittle crack propagation, as a result of being subjected simultaneously to a mechanical stress and a chemical environment, both being necessary for failure to occur. Only small quantities of the chemical or solvent are required and the time of contact with the solvent can be of any duration, from near instantaneous to very long timescales. In unreinforced plastics, a well-known example is the tendency of polycarbonate crash helmets to fail as a result of the combined effects of moulded-in stresses and the solvents in the paint used by owners to decorate the helmets. The significance of ESC for reinforced plastics structures (mainly those in which all of the fibres are aligned parallel to each other) is that ESC failures are sudden and catastrophic, whereas the other kinds of damage induced by liquids take considerable time. Usually the ESC mechanism is evident from the characteristic planar fracture surfaces. In Chapter 3, a closely related term ESCC has been used to denote cases where chemical corrosion is known to occur, which is not always the situation; sometimes the liquid simply has a physical role.

2.13   Temperature (see also Chapter 4)

Organic matrix resins and their reinforced equivalents have only a limited ability to withstand high temperatures. As a crude guide, resins able to withstand more than 250°C for long periods in air are still regarded as exotic, still expensive and are not much used in every day civil applications. High temperature resins are certainly available with superior heat resistance, but they often have the disadvantage of long processing or cure schedules.

Enormous research effort has been directed to improving the availability of high temperature resistant organic matrix resins. Polyimides and bismaleimides [25] are just two examples of resins with glass transition temperature, T_g values of up to 400°C, and in addition, a wide range of high temperature thermoplastics [12] are now gradually achieving acceptance after a long gestation period.

Adverse effects of high temperatures include:

1   Chemical decomposition of the resin by heat alone,
2   Reduction in modulus, strength and other properties at high temperatures,
3   Acceleration of other degradation processes such as those involving chemicals (including oxygen in the air) or radiation,
4   Damaging effects of thermal expansion and thermal mismatch between two materials with different expansion coefficients. Debonding is one possible consequence. 'Thermal shock' can cause cracking after a series of rapid temperature changes (see Chapter 4).
5   Repeated rapid heating of reinforced plastics articles which have previously absorbed moisture or solvents causes various forms of damage to the composite, related to the absorbed liquids.

2.14   Outdoor use

Outdoor use means exposure to several influences simultaneously: ultraviolet light, fluctuating temperatures, wind and moisture. If the article is buried in the soil, there may be microbiological activity, but this will not usually be a problem with the materials used in conventional reinforced plastics.

Resins vary a great deal in their ability to withstand outdoor use for long periods. Those with poor outdoor performance can sometimes be completely transformed by trace additives, so the problem becomes one of using the right grade of resin. For example, polypropylene is commonly used as a durable outdoor plastics material, but when stored in bulk without any stabilizer, it has been known to decompose so vigorously that the heat produced from the chemical decomposition reaction was sufficient to melt it! [26].

The consequences of sunlight will be discussed in more detail in the chapters concerned with weathering (Chapter 6) and with marine applications (Chapter 7). We shall simply mention here a few general points.

First, the effects of outdoor use on structural reinforced plastics such as glass/polyester or carbon/epoxy laminates are confined to the surface and do not often involve a serious threat to their structural integrity, unless perhaps there is a reduction in impact strength as a result of surface cracking. Fortunately carbon is a well-known UV absorber and therefore the fibres act as a good stabilizer. The problems are mainly cosmetic.

The main cause of property losses during weathering in sunlight and in dry atmospheres is a combination of the effects of sunlight and atmospheric oxygen.

Many resins are prone to photooxidative attack. Moisture will be a further complication in all except desert climates. Outdoor exposure in Australia over a period of several years caused unpainted carbon/epoxy sandwich beam specimens, representative of an aircraft structure, to lose weight, because the degraded resin surface was readily removed by wind and rain [27]. In some locations, the weight loss eventually stabilized, but in others, erosion removed loose fibres from the surface as well and the weight loss continued. Painted samples were not affected in the same way but it was recommended that aircraft should be kept in hangars as much as possible to extend paint life when used in similar climatic conditions.

Urban situations have the added complication of pollution. It has been shown that sulfur dioxide has a role in the crosslinking of polypropylene by ultraviolet light in presence of oxygen [28]. This is unlikely to be the only example.

Predicting the weathering performance of reinforced plastics articles is carried out by considering first the theoretical vulnerability of the known material constituents and then confirming the position with (a) artificially accelerated laboratory weathering experiments and (b) field trials. The latter take several years and relatively few organizations have been able to generate large data banks, but there are now sufficient case histories of reinforced plastics products to give us performance data extending over three decades or more.

Changes in appearance occur because of (i) changes in the matrix (ii) debonding or (iii) light-induced changes in additives such as pigments or fire retardants. Yellowing is usually the first change to appear in the matrix itself, caused by the radiation of shortest wavelength in sunlight. Yellowing is not reflected in corresponding decreases in strength.

One reason for debonding can be the difference between the thermal expansion coefficients of glass and resin, which results in high stresses at the interface, or stresses can derive from the swelling and subsequent shrinkage of the resin through moisture absorption. Signs of surface deterioration in unpigmented polyester/glass laminates include a loss of translucency of the laminate, caused by erosion of the resin surface, the exposure of fibres, increased roughness of the surface, an increase in dirt pickup and continued microcracking of the resin. Pigment discoloration can be induced by ultraviolet light.

2.15 Radiation (other than UV) [29]

Electron beam radiation damage is usually reflected in a deterioration in the strength and modulus of reinforced plastics. The amount of the loss depends on the radiation dose. The interlaminar shear strength can increase during radiation.

Neutron irradiation of glass epoxy and glass polyimide composites causes more damage than gamma radiation, but no such difference is found when carbon is the reinforcement. The effect of radiation on ultra high molecular weight polyethylene has been studied in a medical context, with changes in structure and morphology being observed for five years after the irradiation dose [30].

Glass and carbon fibres and other inorganic materials resist electron beam radiation better than organic fibres or matrix resins. Organic matrix composites undergo microcracking when subjected to high doses. Rubbers are sensitive to radiation and rubber-toughened epoxy resin composites in particular are more affected. The radiation damage is mainly in the matrix and (in the case of glass epoxy), at the interface. The best organic matrix composites, using inorganic reinforcement, can withstand 2×10^8 Gy and some have found use in applications such as space vehicles and magnets for particle accelerators.

Ionizing radiation tends to break the macromolecules in the matrix ('chain scission') or else to crosslink them. These processes can in turn lead to gas voids or to dimensional changes.

Aromatic resins such as polyimides survive better than aliphatic ones. Aromatic epoxy resin systems based on the high temperature TGDDM (tetraglycidyl diaminodiphenylmethane: this category of epoxy resin is designed to withstand continuous use at higher temperatures than ordinary epoxy resins, typically 175°C) formula and crosslinked with aromatic hardeners will be superior to DGEBA/DDS (diglycidyl ether of bisphenol A/ diphenyldiaminosulfone) systems. (DGEBA is the same as BADGE (bisphenol A, diglycidyl ether of) and refers to an ordinary standard epoxy resin. DDS is a common curing agent) For the same reason, polyethersulfone (PES) and polyether ether ketone (PEEK) composites would be expected to perform relatively well. PEEK insulated cables have been specified for use in nuclear reactors because of their high resistance to irradiation. While this is not a composite application, the resin is also recommended for use in high performance reinforced plastics structures.

2.16   Mechanical stress (see Chapters 5, 11 and 12)

2.16.1 Fibre, matrix and interface roles

Reinforced plastics containing continuous fibres rely on the load being carried almost entirely by the fibres. The direct contributions of the matrix to the tensile or flexural strength and modulus of the material are trivial in comparison. Most reinforcing fibres have excellent durability towards stress in a wide range of conditions, so it might be assumed from the above statements that mechanical durability is easy to achieve in reinforced plastics. But deterioration of the matrix can have important indirect effects on mechanical properties because the matrix must continue to facilitate load transfer between fibres and must protect individual fibres from mechanical abrasion, as well as from penetrating fluids. Surface weathering of the matrix could expose fibres to mechanical damage. A soft matrix is easily eroded or scratched (Fig. 2.10).

If the interfacial bond between fibres and resin is broken, whether by mechanical stress, hot water immersion or thermal cycling, the fibres will not be able to support the load adequately. As a consequence, there can be localized matrix cracking and the cracks in the matrix eventually propagate by a sequence of events which ultimately leads to large scale fibre fracture or to delamination, or both.

Certain types of laminate are susceptible to creep, as discussed in Chapter 1. The use of 0° plies greatly reduces the creep rate and the problem is not usually severe with long fibre composites, although laminates consisting entirely of ±45° plies can show a scissor-closing motion, that is, a reduction in the angle to ±43° or thereabouts if held under a sustained load with a sufficiently ductile matrix.

2.16.2 Minor impact damage

A common hazard for reinforced plastics is damage from stones on vehicles, from aircraft runways on airplanes, from collisions between boats and driftwood or between boats and the quayside, or from dropped tools on chemical plant and aircraft wings, and so on. The resulting damage is often difficult to see with the naked eye, but it can include delamination, matrix cracking, fibre debonding and in severe cases, fibre fracture. Particulate

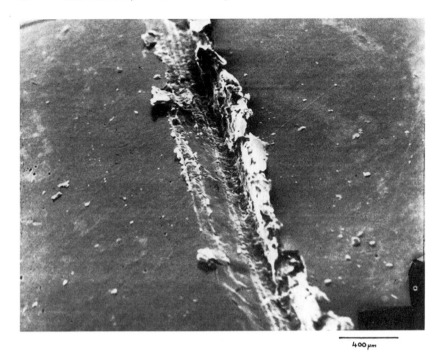

400 µm

Figure 2.10 The scratch resistance of reinforced plastics is dependent
on the resin matrix. Here, a paper reinforced kitchen worktop has
been scratched using a diamond, to show a scratch under electron
microscope examination

filled materials are often more easily damaged than unfilled ones [31]. Most
impacts occur in practice at an oblique angle which tends to reduce the
severity of normal incidence, no matter whether damage is measured by the
damage area, indentation depth or residual strength [32].

The fact that there is scope for on-site repair of impact damage in reinforced plastics, even in remote areas, is an important favourable consideration in their durability. This subject is considered in Chapter 10.

2.16.3 Matrix toughness

It follows from the above and from other considerations, including fatigue
life, that matrix toughness is important. The toughness of the matrix influences mode II (sliding) interlaminar fracture toughness. According to
Alstädt *et al.* [33] the quasistatic fracture energy of the matrix controls
interlaminar crack growth in composites, under mode I (crack opening)
conditions.

Thermosetting resins are not known for their toughness. They typically have a fracture toughness value (critical stress intensity factor, K_{lc}) of only about $0.6\,MN\,m^{-3/2}$ when measured at ambient temperature and low strain rates. Two main methods have been used to improve the toughness. One is to add small quantities (say 10% by weight) of impact modifiers, such as rubbers or thermoplastics, to the resin before fabrication. This is not easy, as the increase in toughness is only achieved under carefully controlled conditions where the dispersed phase has the right particle size distribution. The best known toughening additives (or 'impact modifiers') for thermosets have traditionally been nitrile rubbers containing specific modifying chemical groups. Another approach is to insert energy-absorbing (elastomeric) layers between the reinforcing plies.

A more radical course of action is to abandon thermosets altogether in favour of intrinsically tougher thermoplastics. This solves the toughness problem, but it still leaves questions about possible inadequate modulus retention at high temperatures, and about notch sensitivity, creep and ESC.

The matrix cannot be neglected when considering compression strength, as the fibres have to be prevented by the matrix from buckling whenever they are oriented parallel to a compressive stress. Other areas of importance for the matrix are abrasion resistance, and erosion, for example by rain impinging on the nose-cones of low-flying high speed aircraft. Tough thermoplastics perform better here than brittle thermosets.

2.16.4 Mismatches between layers in laminates made from stacked plies

The properties of composite materials cannot be predicted adequately by considering the fibre and resin constituents one by one. An important mechanism of composite failure under stress is delamination caused by differences between the engineering properties of successive plies or layers. These differences arise from the fact that successive layers may have different fibre orientations [34] or, occasionally, different fibres. It is a feature of laminates made by stacking pre-impregnated layers of reinforcement and is not an issue with, for example, unidirectional pultrusions. The process of delamination has been reviewed by Davies [35]. The fabrication of three-dimensional composites is an important step towards reducing or eliminating unwanted delaminations. Such materials are at an advanced stage of development.

2.17 Joints

Joints are necessary in large composite structures such as ships because of production and design considerations. These articles are too large to be fabricated in one piece, so several parts have to be joined and stiffeners are necessary.

Joints are potential failure sites. This applies whether they are adhesively bonded or mechanically fastened, and whether they join two reinforced plastics sections, or one reinforced plastics component and one constructed from another material.

In the construction industry, joint failure in reinforced plastics is likely to mean leakage of water into the building, rather than structural collapse.

Leggatt [36] identified joints as the most likely failure points in reinforced plastics buildings. He advised that the total number of joints should be minimized and suggested spending 50% of the total design effort on them. Water penetration follows repeated thermal movements or load deflections, or could be traced to errors at the manufacturing or erection stages. He recommended that joints should be located well away from supplies of water. On roofs, they should be on ridges and not in gutters. Adequate seal pressure is necessary. Joints should be readily accessible for inspection and replacement.

2.17.1 Adhesively bonded joints

If two sections of reinforced plastics material are simply butt-bonded together, adhesively, using the same matrix resin as the adhesive, mechanical failure is almost inevitable because of the absence of any load-bearing fibres bridging the bonded interfaces. Any geometrical irregularities at the interface will constitute stress concentrations. According to early US naval studies, carried out when reinforced plastics were still in their infancy [37], the effective tensile and shear strengths of the polyester resins used then in marine applications could be as low as 1.5 MPa and 3 MPa, respectively. It is therefore necessary to design joints so as to provide bridging reinforcement. The design of adhesive joints for composites has been reviewed by Lee and McCarthy [38].

Adhesive joints in reinforced plastics structures are capable of achieving higher strength than mechanical ones and may be preferred for that reason, but once made, adhesive joints cannot be disconnected and they can be vulnerable to prolonged high humidity. Their durability depends more on the flexibility and toughness of the resin used in the adhesive than on its strength [39].

A larger number of failure modes is thought to be possible with composite materials bonded by adhesive joints than with equivalent metal joints. Failure in the adherends can be tensile, interlaminar or transverse. The last two can be either in the resin or in the interface. Finally, cohesive failure can occur in the adhesive.

Prediction of joint strength can be carried out by performing a stress–

strain analysis and applying an appropriate failure criterion. Stresses in the adhesive bonds can be predicted using finite element analysis and closed form or continuum mechanics.

2.17.2 Mechanically fastened joints

Mechanically fastened joints have the advantage that they can be disconnected if desired. Godwin and Matthews [40] studied the (mostly experimental) data on mechanically fastened joints. These joints fail in the same way as metals, that is, in tensile, shear or bearing modes, but with the difference that the failure load is not so obvious with composites. Bolts offer the greatest mechanical strength obtainable without adhesives, especially when the bolt is a good fit to the hole. Metal bolts must be protected against corrosion and the use of special materials such as stainless steel can be cost effective. The edges of drilled holes need to be coated if the joint is exposed to liquids that attack the fibres.

A subsequent review of stress analysis and strength prediction of fibre reinforced plastic (FRP) mechanically fastened joints [41] identified five common failure modes: tension, shear, bearing, cleavage and pull-through (See Fig. 2.11). The tensile strength of a single hole joint is said to depend

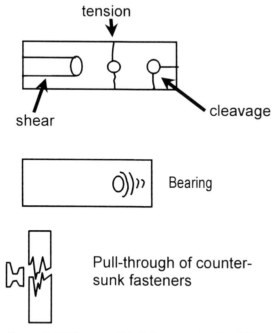

Figure 2.11 Five possible failure modes for FRP mechanical joints (after Camanho and Matthews [41])

very much on the ply orientation and the addition of $\pm 45°$ plies to $0°$ ones considerably increases the strength.

There have been several recent studies of composite joints. The following examples are illustrative.

Chamis and Shiai [42] have recently described a methodology for probabilistic composite design which evaluates *inter alia* the stresses in a composite bolted joint. A study of key-lock joints in reinforced plastics pipes has shown that failure can occur at high pressures on the male component of the joint, if the axial load is high [43]. Hong-Sheng Wang *et al.* [44] investigated bearing failure in double-lap metal-composite–metal joints. The composite layers were cross-ply and quasi-isotropic carbon fibre reinforced epoxy. It was found that bearing damage could be catastrophic without lateral support (shear cracks were induced). A second, related paper [45] interpreted the findings using an analytical model.

According to Hart-Smith, a multihole joint is not necessarily much better than a single hole joint and several bolts would be needed to get a 25% improvement in strength [46].

Out-of-plane joints have been reviewed by Junhou and Shenoi, mainly in the context of marine structures [47]. The most common marine topologies are the monocoque or top-hat stiffened structures, using single skin glass/polyester or vinyl ester laminates, and the sandwich or double wall structures with a foam core. The authors illustrate six different kinds of out-of-plane joints used in this context with FRP.

References

1 She Chou and Hong-En Chen, 'Effects of structure and microcracks on the flexural, interlaminar shear and impact properties of three-dimensional composites', *Polym Polym Compos* 1996 **4**(6) 387–395.

2 M Baruch, J Arbocz and G Q Zhang, 'Laminated conical shells – considerations for the variation of the stiffness coefficients,' Technical Report LR-671, Faculty of Aerospace Engineering, Delft University of Technology, April 1992; also, *Proceedings of the 35th AAIA/ASME/ASCE/ASC Structures, Structural Dynamics and Materials Conference,* Hilton Head, SC, USA, pp 2505–2516.

3 A Brent Strong, 'Manufacturing,' in *International Encyclopedia of Composites,* ed. S M Lee, New York, VCH, 1989, Volume 3.

4 A Miller and A G Gibson, 'Impregnation techniques for thermoplastic matrix composites', *Polym Polym Compos* 1996 **4**(7) 459–481.

5 S Hill, 'Rapid non-destructive testing of carbon fibre reinforced plastics', *Mater World* 1996 **4**(8) 450.

6 M R Wisnom, 'The effect of fibre misalignment on the compression strength of unidirectional carbon fibre/epoxy', *Composites* 1990 **21**(5) 403–407.

7 V P Serwinski, 'The effect of flow patterns and gate parameters on fiber orientation', *Proceedings ANTEC '92,* Detroit, MI. Society of Plastics Engineers, Brookfield, CT, USA. May 1992, Volume 1, pp 1277–1278.

8 I Matiss, 'NDT of composite materials – problems and solutions. 1. NDT as a source of information', *Polym Polym Compos* 1996 **4**(3) 181–187.

9 D Kranbuehl, P Kingsley, H Rhodenizer, G Hasko, B Dexter and A Loos, 'Sensor-model in-situ monitoring and intelligent control of the resin transfer moulding composite cure process', *Proceedings ANTEC '92*, Detroit, MI. Society of Plastics Engineers, Brookfield, CT, USA, May 1992, Volume 2, 2049–2051.

10 J B Cattenach and F N Cogswell, 'Processing with Aromatic Composites', *Developments in Reinforced Plastics*, ed. G Pritchard, London, Elsevier Applied Science, 1986, Volume 5, Chapter 1.

11 R P Caruso, 'Repair of metallic aircraft structures utilizing boron/epoxy composites', *International Encyclopedia of Composites*, ed. S M Lee, New York, VCH, 1989, Volume 5.

12 D C Leach, 'Continuous fibre reinforced thermoplastic matrix composites', *Advanced Composites*, ed. I K Partridge, Barking, UK, Elsevier Applied Science, 1989, Chapter 2.

13 R J Lee and A S Trevett, 'Compression strength of aligned carbon fibre reinforced thermoplastic laminates', *Proceedings ICCM-VI*, Imperial College, London. Elsevier Applied Science, London, 1987, Volume l, p 278.

14 M W Wardle and D A Steenmaker, 'Special considerations in testing advanced composite materials reinforced with Kevlar aramid fibers', *Proceedings ICCM-VI*, Imperial College, London. Elsevier Applied Science, London, 1987, Volume 1, p 199.

15 K A Kasturiarachchi *Hygrothermal Degradation of Fibre Reinforced Epoxide Resins under Stress,* PhD Thesis, Kingston Polytechnic, UK, 1983.

16 B D Agarwal and L J Broutman, *Analysis and Performance of Fibre Composites*, New York, Wiley, 1980.

17 Q Qiu and M Kumosa, 'Corrosion of E-glass fibers in acidic environments', *Compos Sci Technol* 1997 **57** 497–507.

18 C S Smith, 'Applications of fibre reinforced composites in marine technology', *Proceedings, Composites – Standards, Testing and Design,* National Physical Laboratory,Teddington, UK. IPC Science and Technology Press, 8/9 April 1974, p 54.

19 P-L Chiou and W L Bradley, 'Moisture-induced degradation of glass/epoxy filament-wound composite tubes', *J Thermoplastic Compos Mater* 1996 **9** 118–128.

20 C A Wood and W L Bradley, 'Determination of the effect of sea-water on the interfacial strength of an interlayer E-glass/graphite/epoxy composite by in-situ observation of transverse cracking in an environmental SEM', *Compos Sci Technol* 1997 **57** 1033–1043.

21 I Ghorbel, D Valentin, M C Yrieus and J Grattier, 'Damage propagation law in glass fibre reinforced polyester tubes under hygrothermal loading conditions', *Proceedings Conference on Durability of Polymer-based Composite Systems for Structural Applications, Brussels*, eds A H Cardon and G Verchery, London/New York, Elsevier Applied Science, 1990.

22 R Kosuri and Y Weitsman, 'Sorption process and immersed-fatigue response of graphite/epoxy composites in sea-water', *Proc. ICCM-10*, Whistler, BC, Canada. Cambridge, UK, Woodhead, 1995, Volume 6, pp 177–184.

23 E L McKague, J D Reynolds and J E Halkias, 'Swelling and glass transition relations for epoxy materials in humid environments', *J Appl Polym Sci* 1978 **22** 1643–1654.

24 J R M D'Almeida, 'Effects of distilled water and saline solution on the interlaminar shear strength of an aramid/epoxy composite', *Composites* 1991 **22** 448–449.

25 J A Parker, D A Kourtides and G M Fohlen, 'Bismaleimides and related maleimido polymers as matrix resins for high temperature environments', *High Temperature Polymer Matrix Composites*, ed. T T Serafini, Park Ridge, NJ, USA, Noyes Data Corporation, 1987, pp 54–75.

26 R F Becker, L P J Burton and S E Amos, 'Additives', *Polypropylene Handbook*, ed. E P Moore, Munich, Hanser, 1996, Chapter 4.

27 R J Chester and A A Baker, 'Environmental durability of F/A-18 graphite/ epoxy composite', *Polym Polym Compos* 1996 **4**(5) 315–323.

28 A Davis and D Sims, *Weathering of Polymers*, Barking, UK, Applied Science 1983.

29 R L Clough, K T Gillen and M Dole, 'Radiation resistance of polymers and composites', in *The Effects of Ionizing Radiations on Polymeric Materials*, eds A A Collyer and D W Clegg, Amsterdam, Elsevier, 1990.

30 M Goldman, R Gronsky, R Ranganathan and L Pruitt, 'Effects of gamma radiation sterilization and ageing on the structure and morphology of medical grade ultra high molecular weight polyethylene', *Polymer* 1996 **37** 2909–2913.

31 G Pritchard and Q Yang, 'Microscopy of impact damage in particulate-filled glass-epoxy laminates', *J Mater Sci* 1994 **29** 5047–5053.

32 S. Madjidi, W S Arnold and I M Marshall, 'Damage tolerance of CSM laminates subject to low velocity oblique impacts', *Compos Struct* 1996 **34** 101–116.

33 V Alstädt, R W Lang and A Neu, 'The influence of resin, fibre and interface on delamination fatigue crack growth of composites', *Durability of Polymer-based Composite Systems for Structural Applications*, ed. A H Cardon and G Verchery, London, Elsevier Applied Science, 1991, p 180.

34 J C Thesken and F Brandt, 'Experimental investigations of mode I and mixed-mode delamination growth', *Proceedings of ICCM-10,* eds A Poursartip and K Street, Whistler, BC, Canada. Cambridge, UK Woodhead, 1995, Volume 1, pp 149–156.

35 P Davies, 'Delamination', *Advanced Composites*, ed. I K Partridge, Barking, UK, Elsevier Applied Science, 1989, Chapter 9.

36 A. Leggatt, *'GRP and Buildings, A Design Guide for Architects and Engineers',* Butterworth, 1984.

37 A. Rufolo, *A Design Manual for Joining of Glass Reinforced Structural Plastics*, August 1961, US Naval Material Laboratory Report Navship 250-624-1.

38 R J Lee and J C McCarthy, 'Design of bonded structures', *Advanced Composites*, ed. I K Partridge, Barking, UK, Elsevier Applied Science, 1989, Chapter 8.

39 H Hamada, Z Maekawa, T Morii, A Gotoh and T Tanimoto, 'Durability of adhesive bonded FRP joints immersed in hot water', *Durability of Polymer-based Composite Systems for Structural Applications*, eds A H Cardon and G Verchery, London, Elsevier Applied Science, 1991, p 418.

40 E W Godwin and F L Matthews, 'A review of the strength of joints in fibre reinforced plastics. Part 1. Mechanically fastened joints.' *Composites* 1980 **11**(7) 155–160.

41 P P Camanho and F L Matthews, 'Stress analysis and strength prediction of mechanically fastened joints in FRP: a review', *Composites* 1997 **28A** 529–547.

42 C C Chamis and M C Shiao, 'Probabilistic composite analysis design', Paper

11-D, *Proceedings of the 50th Annual Conference, Composites Institute, The Society of the Plastics Industry*, Cincinnati, Ohio, USA. Society of the Plastics Industry, New York, January 30–February 1, 1995.

43 A Fahrer, A G Gibson and P Tolhoek, 'A study of the failure behaviour of key-lock joints in glass fibre reinforced pipework', *Composites* 1996 **27A** 429–435.

44 Hong-Sheng Wang, Chang-Li Hung and Fu-Kuo Chang, 'Bearing failure of bolted composite joints. Part 1: Experimental characterization', *J Compos Mater* 1996 **30**(12) 1284–30.

45 Chang-Li Hung and Fu-Kuo Chang, 'Bearing failure of bolted composite joints. Part 2: Model and verification', *J Compos Mater* 1996 **30**(12) 1359–1342.

46 L J Hart-Smith, 'Mechanically fastened joints for advanced composites – phenomenological considerations and simple analysis', *Douglas Paper* 1978 **6748** 1–32.

47 P Junhou and R A Shenoi, 'Examination of key aspects defining the performance characteristics of out-of-plane joints in FRP marine structures', *Composites* 1996 **27A** 89–103.

3

Durability of reinforced plastics in liquid environments

FRANK R JONES

3.1 Introduction

Reinforced plastics, sometimes called polymer composites, consist of reinforcing fibres or particles embedded in a polymeric matrix. It is now recognized that a third component, called an interphase region, can exist at the interface between the fibre and resin. The properties of the interphase are probably not constant but vary to give a graded region. Thus the environmental durability of a composite material is a complex interplay between the various microstructural aspects of the material, which are:

1 The polymeric matrix or resin. It can be thermosetting or thermoplastic. It can sometimes consist of a homogeneous or heterogeneous blend of two or more resins.
2 The fibres such as E-glass, ECR (see Section 2.8) or other kinds of glass; carbon, aramid, or polyethylene; and/or the particulate reinforcement, which is often a mineral filler.
3 The interface between the fibre and interphase or matrix.
4 The structure of the interphasal material.
5 The 'interface' between the interphase region and the matrix.

Many industrial composite materials contain both fibres (often of different type or form) and particulates, so that their interaction with an environment can be complex. In order to discuss these issues, a simplistic approach will be used to indicate how a durable material can be achieved. One of the most important aspects is the chemical inertness of the polymer matrix. For most purposes, aqueous environments are of greatest significance and they will form the main part of the discussion in this chapter. In the case of non-aqueous environments, thermodynamic considerations can be used to assess the resistance of the matrix to solvent attack.

3.2   Hygrothermal conditions

3.2.1 Introduction

Moisture diffuses into all organic matrix composites, to varying extents and at various rates, leading to changes in their mechanical and thermophysical properties [1]. Usually, the matrix has a greater sensitivity to moisture than the fibre. Therefore matrix dominated properties will be most affected. For a unidirectionally reinforced composite subjected to moisture absorption, the strength and stiffness in the fibre (0°) direction are largely unaffected, whereas the same properties in the transverse direction (90° to the fibres) will change significantly. The failure of the material in shear is matrix dominated and also highly affected.

3.2.2 Predicting moisture diffusion

Of particular importance is the timescale over which diffusion occurs under various conditions of relative humidity (RH) and temperature. The RH determines the equilibrium moisture concentration, whereas higher temperatures will accelerate the moisture sorption process. In order to predict the moisture profile in a particular structure, it is assumed that Fickian diffusion kinetics operate. It will be seen later that many matrix resins exhibit non-Fickian effects, and other diffusion models have been examined. However, most resin systems in current use in the aerospace industry appear to exhibit Fickian behaviour over much of their service temperatures and times. Since the rate of moisture diffusion is low, it is usually necessary to use elevated temperatures to accelerate test programmes and studies intended to characterize the phenomenon. Elevated temperatures must be used with care though, because many resins only exhibit Fickian diffusion within certain temperature limits. If these temperatures are exceeded, the steady state equilibrium position may not be achieved and the Fickian predictions can then be inaccurate. This can lead to an overestimate of the moisture absorbed under real service conditions.

3.2.3 Fickian kinetics

Compared with heat transfer, the process of moisture transport is slower by a factor of approximately 10^6. For example, moisture equilibration of a 12 mm thick composite, at 350 K, can take 13 years whereas thermal equilibration only takes 15 s. Fick adapted the heat conduction equation of Fourier, and his (Fick's) second law is generally considered to be applicable to the moisture diffusion problem. The one-dimensional Fickian diffusion law, which describes transport through the thickness, and assumes that the moisture flux is proportional to the concentration gradient, is:

$$\frac{\partial c}{\partial t} = \frac{\partial(D\partial c)}{\partial x(\partial x)}$$

[3.1]

where c is the moisture concentration at time t in the x-direction.

For most composite materials, the diffusion coefficient appears to be independent of moisture concentration and of its through thickness location. Therefore equation [3.1] becomes

$$\frac{\partial c}{\partial t} = D_x \frac{\partial^2 c}{\partial x^2}$$

[3.2]

where D_x is the diffusion constant in the direction normal to the surface and x is the distance from the edge of the material.

In practical experiments, c is replaced by M, where M is the moisture content at time t, obtained from the weight change occurring during immersion in the environment, whether gaseous or liquid.

$$M = \left(\frac{W_w - W_d}{W_d}\right)100$$

[3.3]

W_w and W_d are the wet and dry weights of the material. Equation [3.2] can be solved by finite difference, giving

$$D = \frac{\pi d^2}{16 M_\infty^2}\left(\frac{M_2 - M_1}{t_2^{1/2} - t_1^{1/2}}\right)^2$$

[3.4]

where d^2 is the material thickness. M_1 and M_2 are moisture concentrations, by weight, after times t_1 and t_2. The term $\left(\frac{M_2 - M_1}{t_2^{1/2} - t_1^{1/2}}\right)^2$ is effectively the slope of the linear part of the M versus $\sqrt{t}$ curve. M_∞ is the equilibrium or maximum moisture concentration observed at a specific relative humidity at time $t = \infty$.

It is difficult to carry out experiments in such a manner that diffusion occurs in one direction, through one face of the material or laminate. Shen and Springer [2] obtained the following expression which enables a corrected diffusion constant, D_x, to be calculated from D, which accounts for diffusion through all six faces or edges of the coupon.

$$D = D_x\left[1 + \left(\frac{d}{b}\right) + \left(\frac{d}{l}\right)\right]^2$$

[3.5]

where d, b and l are the coupon thickness, breadth and length, respectively.

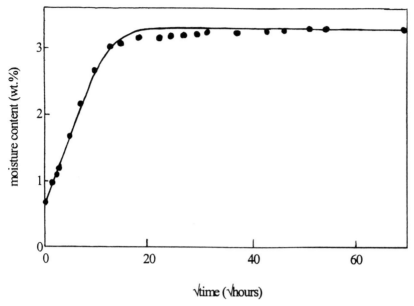

Figure 3.1 Typical Fickian moisture diffusion curve for an equilibrated resin casting of a bismaleimide modified epoxy resin which had previously been subject to thermally enhanced moisture absorption, followed by desorption prior to reabsorption. The sorption shown does not begin at zero because of a small residual quantity of absorbed moisture from the first conditioning cycle [4]. ●, experimental; ——, Fickian model

Equation [3.4] requires a knowledge of M_∞ for the calculation of D, and this can require several months to obtain experimentally. Ellis and Found [3] suggested that an estimate could be obtained by assessing the half-life of the process and thus making an accelerated assessment of M_∞. The half-life time for sorption is given by Equation [3.6]

$$M_{t_{\frac{1}{2}}} = M_\infty \frac{4}{\pi^{1/2}} \left(\frac{D t_{1/2}}{d^2} \right)^{1/2} \qquad [3.6]$$

where $M_{t_{\frac{1}{2}}}$ is the moisture concentration at $t = t_{1/2}$.

Figure 3.1 shows an example of a Fickian diffusion curve [4]. In this case, the resin had been previously conditioned and subjected to moisture absorption at 50°C and 96% RH, followed by desorption in vacuum at 50°C, prior to resorption [4]. One-dimensional diffusion can therefore be described by the following characteristics of the M_t versus $\sqrt{t}$ curve:

1   The diffusion curve is linear over the first 60% of the process.
2   An equilibrium moisture concentration is achieved.

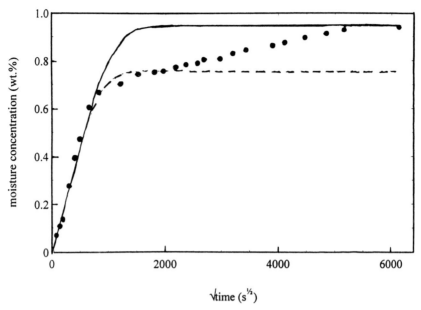

Figure 3.2 Comparison of Fickian prediction with experimental sorption data for Fibredux 927 resin [4]. ——, Fickian prediction, M_{max} = 0.95 wt%, ---, M_{max} = 0.75 wt%. ●, experimental points

Figure 3.2 gives an example [4] of a moisture absorption curve for an advanced epoxy resin and the theoretical curve obtained from equation [3.4]. This illustrates the difficulties of assessing the diffusion coefficient when deviations from Fickian behaviour occur and an equilibrium moisture concentration has not been achieved.

3.2.4 Moisture distribution into a laminate

Figure 3.3(a) shows the predicted moisture absorption through the thickness of a carbon fibre laminate [5] after 3 days in 96% RH at 50°C. This illustrates the timescale of the diffusion process. Thus it can take 13 years in a composite laminate of thickness 12 mm before saturation is reached. In service, materials are subjected to variations in temperature and kinetic processes such as diffusion follow the Arrhenius law:

$$D = A\exp(-E_a/RT) \qquad [3.7]$$

where E_a is the activation energy and A is the pre-exponential factor.

Thus at higher temperatures the rate of moisture ingress will increase and the moisture profile through the material will change. Figure 3.3(b) gives

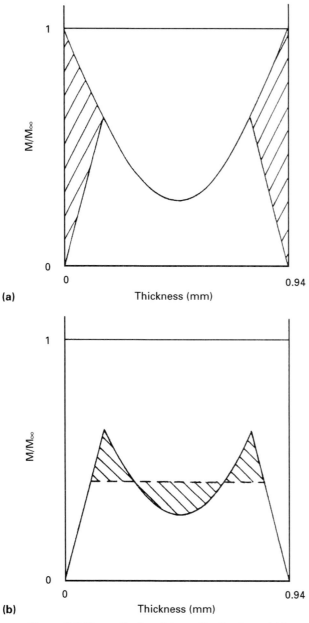

(a)

(b)

Figure 3.3 Theoretical moisture distribution within a unidirectional bismaleimide epoxy (Narmco Rigidite 5245C) laminate, (a) subjected to 96% RH at 50°C for 3 days; (b) as (a) but following a rapid thermal excursion (thermal spike) to 150°C. The shaded areas indicate the movement of water in and out of the specimen [5]

the calculated moisture profile through the laminate in Fig. 3.3(a) after it had been subjected to a thermal excursion to 150°C.

3.2.5 Contribution of the fibre–matrix interface to diffusion

There is much concern that interfacial transport may occur along imperfect fibre–matrix interfaces in composite materials. These effects can be studied by preparing coupons in which the fibres traverse the thickness. In this way diffusion in the fibre direction is made to dominate the transport process. By comparison with the data for conventional materials in which the fibres lie parallel to the major surface, the role of interfaces in the diffusion process can be identified. The drawback of this approach is the potential for interfacial damage during cutting and polishing of appropriate sections containing fibres at 90° to the major surface. These difficulties can be reduced by preparing materials of differing length and breadth and using an extrapolation technique. Equation [3.5] is a contracted form of equation [3.8], which assumes that the material is homogeneous. If we consider that a unidirectional laminate is anisotropic with three differing diffusion constants, then

$$D = D_x \left[1 + \frac{d}{l} \sqrt{\frac{D_y}{D_x}} + \frac{d}{b} \sqrt{\frac{D_z}{D_x}} \right]^2 \qquad [3.8]$$

where D_x, D_y, and D_z are the diffusion coefficients through-thickness, along-the-length and across-the-breadth of the material. For a unidirectional laminate, the y-direction is equivalent to transport parallel to fibres, whereas the D_x and D_z represent transport at 90° to the fibres. Therefore equation [3.8] becomes

$$D = D_x \left[1 + \frac{d}{b} + \frac{d}{l} \sqrt{\frac{D_y}{D_x}} \right]^2 \qquad [3.9]$$

and

$$\sqrt{D} = \left(1 + \frac{d}{b} \right) \sqrt{D_x} + \frac{d}{l} \sqrt{D_y} \qquad [3.10]$$

From a plot of $\sqrt{D}$ against d/l, values of D_x and D_y can be obtained from the slope ($\sqrt{D_y}$) and intercept $\left[\left(1 + \frac{d}{l} \right) \sqrt{D_x} \right]$. A series of experiments with laminates of differing geometry allows differences between the diffusion coefficients parallel (D_y) and transverse (D_x) to the fibres to be measured. For example, the ratio of D_y/D_x for Narmco Rigidite 5245C was found to be 0.21, indicating that preferred interfacial transport was not

Table 3.1 Typical values of moisture diffusion coefficients for resins and composites [3–7]

Resin or composite	M_∞ (wt %)	D $(10^{-7}\,mm^2\,s^{-1})$	RH (%)	T (°C)
Resins: Unsaturated Polyester	1.6	32	100	50
−isophthalic [6]	1.4		90	50
	0.9		75	50
−orthophthalic [3]	2.0	30	100	50
Epoxy [4]	6.9	2.8	96	50
Advanced epoxy resin −	1.96	11.0	96	45
Narmco 5245 [7]	1.38	14.7	75	45
	0.80	19.4	46	45
	0.54	17.2	31	45
Composites: Carbon Fibre	0.84	1.94	96	45
($V_f = 0.6$) Narmco 5245C [7]:	0.6	2.37	75	45
Unidirectional (O_{16})	0.35	2.9	46	45
	0.24	3.3	31	45
Cross-ply ($O_2/9O_4/O_2$)	0.92	2.91	96	45
Composites: random glass fibre ($W_f = 0.33$) (CSM)−orthophthalic [3]	2.7	0.7	100	50

occurring to any significant extent. Typical values of diffusion coefficients and saturation moisture contents for resins and composites are given in Table 3.1 [3–7].

3.2.6　Moisture sensitivity of resins

Moisture diffuses into polymers to differing extents, depending upon a number of molecular and microstructural aspects:

1　Polarity of the molecular structure
2　Degree of crosslinking
3　Degree of crystallinity (in the case of a thermoplastic)
4　Presence of residual 'monomers' and/or hardeners and/or other water attracting species, e.g. glass surfaces.

For composite materials in contact with aqueous liquids, the resins most frequently considered include epoxy, unsaturated polyester and vinyl ester resins, together with advanced thermosets such as polyimides and bismaleimides and a number of thermoplastics. The latter can be engineering thermoplastics of the polyamide type, or one of a group of advanced thermoplastics with superior properties (PEEK (polyetheretherketone), polyetherimide, etc). These materials can have a range of sensitivities to moisture absorption, depending on the individual contributions of the above factors. For example, cured epoxy resins without fibres can absorb from 2–10% w/w depending upon the base resin and the curing system [8–10]. Translated into a composite with $V_m = 0.4$, the equilibrium moisture M_∞ content of the composite can still exceed 3%.

3.3   Epoxy resins

3.3.1 Structural effects

The differing degrees of equilibrium moisture or water absorption arise from variations in the molecular structure of the network of the cured epoxy resin. Thus the nature of the base epoxy resin and the hardener or catalytic curing agent employed plays an important role. The family of diglycidyl ethers of bisphenol A before cure can be represented by Structure (I). Anhydride hardeners form copolymers with the epoxide endgroups on the resin molecules, giving rise to something which can strictly be regarded as polyester networks, see Structure (II). The polarity is for the most part determined by the degree of polymerization (n) within the base epoxide resin.

(I) DGEBA epoxy resin

(II) Ester group from anhydride cure

Catalytic curing (e.g. using boron trifluoride as BF_3.amine complex) occurs via a ring-opening polymerization of the terminal epoxide groups, to form a network chain of polyether, see Structure (III). As with the anhydride cured systems, the resin polarity is principally a function of the hydroxyl group concentration in the base resin. Without the anomalies associated with the conversion of unreacted epoxy groups to hydroxyl groups and of unreacted anhydride to acid groups, these resin systems tend to have 'low' moisture concentrations at equilibrium.

$$OH$$
$$-CH_2-CH-O-CH_2-CH-O-CH_2-CH-O-CH_2-CH-O-CH_2-CH\wedge\wedge\wedge$$

(III) Ether groups in cured epoxy

Other hardener systems rely on active hydrogen atoms from groups such as amine groups, which react with the epoxy rings and generate hydroxyl groups to act as polar centres for hydrogen bonding with molecules of absorbed moisture, see Structure IV. The reaction of primary amine groups results in a secondary amine and a β-hydroxyl amine network chain. The reaction of a secondary amine with an epoxide group occurs at a lower rate, so residual secondary amine and epoxide groups can contribute to variations in moisture concentrations and deviations from Fickian behaviour.

$$OH \qquad\qquad OH$$
$$\wedge\wedge CH-CH_2 \diagdown \qquad\qquad \diagup \wedge\wedge$$
$$N-R-N$$
$$\wedge\wedge \diagup \qquad\qquad \diagdown \wedge\wedge$$
$$OH \qquad\qquad OH$$

(IV) Amine cured epoxy

3.3.2 Modified epoxy resins

As discussed in Section 3.5, moisture plasticizes epoxy resins, lowering their glass transition temperature, T_g and reducing their maximum service temperature significantly. As a result, resin blends are employed, containing modifiers to reduce the sensitivity of the matrix to moisture. Both thermo-

Table 3.2 Effect of thermosetting and thermoplastic modifiers on the moisture content of carbon fibre composites based on epoxy resins in comparison with the advanced thermoplastic (PEEK) and thermoset (BMI) resin matrices [11]–[13]

	Modified epoxide resins [11]				PEEK [12]	PMR-15 [13]
Resin system	924E	924C	927C	5245C	APC-2	BMI
Modifier	None	PES	Cyanate/PI	BMI	—	—
Moisture content (%)	2.44[a,d]	1.72	0.98	0.82	0.02[b]–0.23[c]	1.32[d]

PES = Polyethersulfone, PI = Polyimide, BMI = Bismaleimide.
[a] Estimated from data on cast resins (6.95%).
[b] 23°C/50% RH/350 h.
[c] Immersion/100°C/360 h.
[d] 50°C/96% RH.

setting and thermoplastic modifiers are used. Table 3.2 demonstrates potential improvements in moisture sensitivity [11–13]. For a more detailed discussion of the structure of cured epoxy resins, see Ellis [14].

3.3.3 Advanced resin systems

Table 3.2 also includes data for the advanced thermoplastic resins (PEEK) and for a thermosetting resin, an end-capped bismaleimide (BMI) called PMR-15. Moisture contents tend to be lower for these advanced materials [12,13]. One way to overcome the environmental sensitivity of epoxide resins is to employ these advanced resins, as demonstrated in Table 3.2. However, changing to other resin systems brings with it other concerns. For example, PEEK relies on crystallinity for its higher temperature performance. Its glass transition temperature is only 143°C and a change in modulus can be observed at that temperature. In addition, higher process temperatures are required both for high performance thermosets and thermoplastics. The consequent higher residual thermal stresses can off-set some of the advantages of a higher service temperature, in comparison with advanced epoxy resins.

3.3.4 Styrenated resins

Unsaturated polyester resins and vinyl ester resins are styrene-based matrices cured by free radical copolymerization of the reactive diluent (solvent) styrene and the unsaturated groups in the dissolved polymeric ester. In the case of the polyester resins the unsaturated groups are within the molecular backbone of the polyester (Structure (V)).

In the case of the vinyl esters, Structure, (VI) and incidentally also vinyl urethanes, Structure (VII), the polymer has terminal unsaturated groups, that is, they are only at the chain ends.

The chemical reactivities of the unsaturated groups at the ends of vinyl ester resin and vinyl urethane resin chains are different in several respects from those of the same groups when situated in mid-chain positions, as they are in polyester resins. As a consequence of the different reactivity ratios, the two kinds of cured resin behave differently from moisture and chemical resistance points of view. The structural differences are also reflected in the mechanical properties, such as fracture toughness.

Generally, saturation moisture concentrations for these resins are lower than those of the epoxide systems by a significant amount, because of the lower polarity of the ester groups and of the aromatic hydrocarbon network. The moisture content will be similar to that of the anhydride cured epoxy resins, which have an essentially polyester-like network. Typically the moisture concentration can reach 2.5%, depending on the chemical

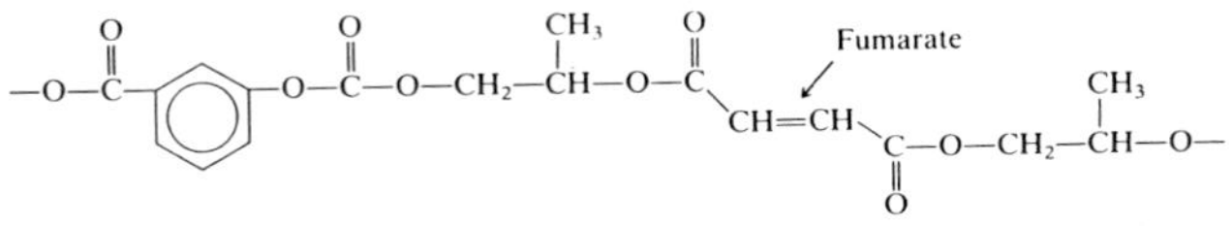

(V) Polyester

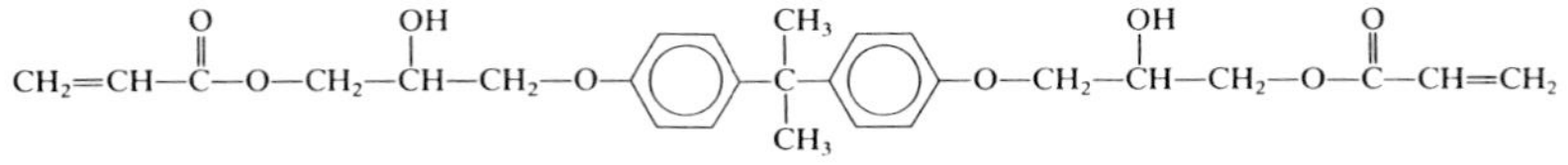

(VI) Vinyl ester

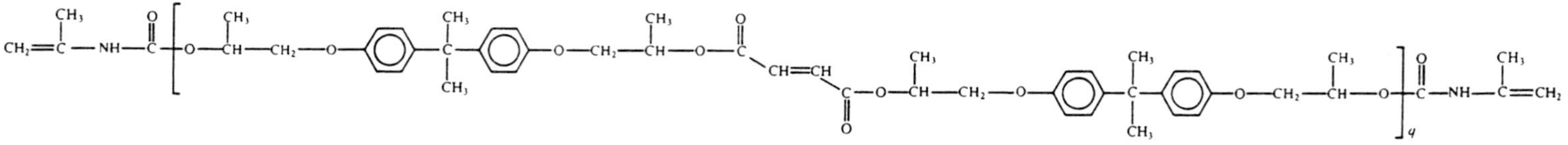

(VII) Vinyl urethane

structure of the resin. Excessive moisture absorption is generally a consequence of osmotic effects associated with trace residues in the resin.

The actual moisture absorption by a specific polyester resin is a complex function of the components used to synthesise the unsaturated polyester. For example, all the syntheses rely upon the isomerization of maleic anhydride to unsaturated fumarate links. The degree of isomerization is a function of the polyesterification temperature and the nature of the glycol and the saturated acids employed. Generally, the higher the degree of isomerization, the greater the reactivity of the polyester unsaturation towards copolymerization with styrene. As a result of the greater reactivity, the crosslink density is increased. Other important aspects are the chemical nature of the glycol and saturated acid. Polyesters from aromatic acids have lower moisture absorption. The orthophthalic resins typically absorb 0.4% moisture, in comparison to 0.25% for isophthalic resins. The nature of the glycol also influences moisture absorption. Thus a diethylene glycol-orthophthalate resin will absorb 2.5% whereas the equivalent ethylene glycol resin absorbs 0.15%. Thus ether groups in the polyester backbone increase the polarity of the resin. It is quite difficult to rationalise these effects because of the complications which arise from subtle changes in polyester resin structure as a result of side reactions. Etheration of the glycol can occur to differing extents, depending on the synthesis conditions and the recipe.

The vinyl esters have similar levels of moisture absorption. Differences do exist and the base resin tends to be more polar, with hydroxyl groups on the network, but polar contaminants are much less likely. The premium chemically resistant resins are also sometimes based on vinyl urethane terminated polyesters.

3.4   Anomalous effects

3.4.1 Osmotic impurities

One of the most important aspects of the durability of resin and composites in aqueous environments is the fact that crosslinked resins are effective semi-permeable membranes. Water can permeate through the resin, but the permeation of larger molecules or inorganic ions is hindered. As a consequence, large polar ions and molecules can act as osmotic centres, providing a thermodynamic driving force for the transport of moisture into the composite. The combination of moisture plasticization, or resin softening, and the osmotic pressure generated can lead to blister formation. The osmotic pressure π generated within the resin at an inclusion is a function of the concentration of the solubilized inclusion, as shown by the Van't Hoff equation

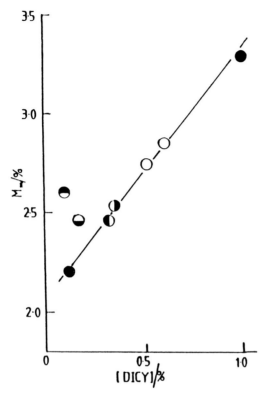

Figure 3.4 Effect of residual dicyandiamide (DICY) in GRP (glass reinforced plastic) laminates from Fibredux 913, on the maximum moisture absorption (M_m) in water at 50°C. The concentration of residual DICY was reduced by increasing the cure and postcure temperature and employing lower concentrations in the prepreg . The laminates from prepreg containing 6% DICY (○), 4% and 8% DICY (●) were cured at 120°C. The laminates from prepreg containing 6% DICY were also postcured at 140°C (◐); cured at 140°C (◑); and postcured at 160°C (◓); cured at 160°C (◒) [10]

$$\pi = RTc \qquad\qquad [3.11]$$

where c is the molal concentration of solute, R is the gas constant and T is the absolute temperature. Clearly at higher temperatures there is a greater propensity towards blistering because of (i) the higher osmotic pressure, and (ii) the higher probability of the glass transition temperature (T_g) of the wet resin being close to ambient. This is why the so-called 'boat pox' is more prevalent in hotter climates, especially where salinity of the sea or river water is low.

The consequence of soluble inclusions for moisture absorption is shown in Fig. 3.4, where the effect of residual dicyandiamide curing agent (DICY) on the apparent moisture concentration of a DICY cured resin [10] is given. In this case, the concentration of residual DICY in the cured resin was a function of the original hardener concentration and the curing schedule employed.

The osmotic effect is not specific to a particular resin system, but it can be more prevalent in general purpose unsaturated polyester resins because of the higher probability of an occluded osmotic centre. Typical polar molecules which can act as osmotic centres in polyester resins are residual glycols and phthalic anhydride. Furthermore, with orthophthalic resins, the monoester orthophthalate end-groups thermally dissociate into the anhydride. This could be one reason for the benefits of isophthalic resins. For applications where contact with aqueous environments is expected, care must be taken to select resins which have been formulated accordingly. Similarly, general purpose emulsion bound E-glass chopped strand mat (CSM) can provide the conditions where interfacial transport can occur, which in turn can lead to aqueous solutions of glass corrosion products, especially in the case of resins and fibre sizes with significant acidity. As a general rule, isophthalic polyester resins in combination with powder bound mats or well formulated sized rovings can provide excellent aqueous resistance. The selection of resin and reinforcement is important [15]. Another source of hydrophilic sites can be contaminants within poorly refined pigments.

3.4.2   Other mechanisms of blister formation

During the manufacture of a GRP laminate, a backing resin and glass fibre mat are applied sequentially to a gelcoat which has been partially cured. The styrene in the backing resin will diffuse into the partially cured gelcoat, causing it to swell at the boundary between the two. The styrene then polymerizes, leaving the boundary layer in a swollen and stressed state because the swollen gelcoat and the backing resin cure and shrink at different rates. Additional swelling caused by moisture enhances the stress at the boundary layer, increasing the formation of microcracks, which develop into blisters on further ingress of water. Chen and Birley [16–18] have classified blisters by their mechanistic origins as shown in Table 3.3.

3.4.3   Designing for blister resistance

The chemical compositions of the gelcoat and backing resins are the most important factors for the provision of blister resistance. Isophthalic gelcoats are generally superior to orthophthalic ones. Isophthalic acid/neopentyl glycol based gelcoats have even better blister resistance than those based on

Table 3.3 Mechanisms of blister formation in GRP [16–18]

	Agent	Mechanism
1	Contaminants	Contaminants in the gelcoat can cause blistering by osmosis
2	Bubbles	Bubbles in the gelcoat may develop into blisters
3	Precracks	Cracks, which form during or immediately after manufacture, under a residual boundary-layer stress, can develop into blisters on water immersion
4	Fibre bundles	Blisters form along the glass fibre bundles immediately behind the gelcoat, from a combination of internal stress and subsequent osmosis

isophthalic acid/polypropylene glycol. Matching the chemical structure of the gelcoat and backing resins reduces the potential for internal stress development and blister formation.

Careful control over the thickness of the gelcoat is very important. Very thin gelcoats are much less effective. Choice of fibre finish is crucial and laminates containing powder bound CSM display lower water absorption and higher blister resistance than those made from emulsion bound CSM. The use of surface tissue immediately behind the gelcoat can significantly reduce the chance of blistering.

In addition to the above considerations, osmotic contaminants can arise from poor control over initiator, accelerator concentrations and from the choice of additives such as pigments and fillers. Good manufacturing practice can mitigate some of these aspects.

3.4.4 Enhanced moisture absorption through thermal spiking

The subject of spiking falls between this chapter and the one on temperature effects (Chapter 4) and the reader will find some overlap in the discussion of this topic but the present approach is focussed on the effects of moisture absorption.

In comparison with isothermal moisture absorption, when reinforced plastics articles are periodically subjected to a rapid thermal excursion or spike and returned to the humid environment, they absorb more moisture than they did before [9,11,19–22]. This is illustrated in Fig. 3.5 for a polyethersulphone modified epoxy resin (Fibredux 924). This is by no means unusual and the effect has been observed for a number of epoxy resin systems (see also Chapter 4). The enhancement appears to be associated with the resin rather than an artefact of the composite material. Fur-

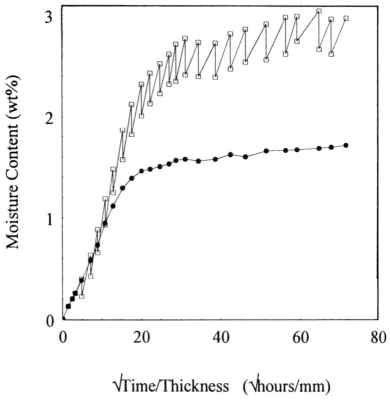

Figure 3.5 Effect of thermal spiking to 140°C on the moisture absorption of Fibredux 924C-O$_8$ laminate, at 50°C and 96% RH. The moisture content immediately before and after 140°C spiking (□); isothermal control data (50°C) (●), [11]

thermore, there appears to be a temperature at which moisture enhancement is maximized. This is illustrated in Fig. 3.6(a), where the enhanced moisture absorption is plotted against thermal spike temperature, for a series of carbon fibre composites employing modified epoxy resins as matrices. Fibredux 927 and Narmco 5245C represent the latest generation of advanced composite matrices. Figure 3.6(b) shows that the matrix is responsible for the effect and that the enhancement cannot be attributed to the modifier since the base epoxy resin (924E) also exhibits the same phenomenon. It is also apparent that normalization of composite data (924C) to account for the fibre volume fraction is in good agreement with independent studies on the commercial resin. The difference between the data can be explained by an uncertainty in the value of the fibre volume fraction and the additional presence of sizing resins on the carbon fibres.

It has been suggested that the mechanism of this enhancement involves

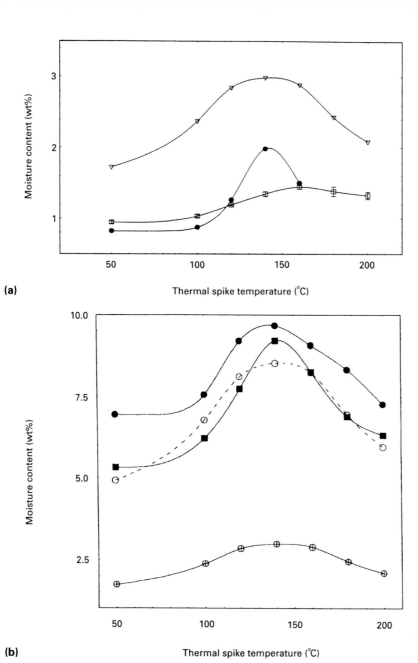

(a)

Thermal spike temperature (°C)

(b)

Thermal spike temperature (°C)

Figure 3.6 (a) Moisture enhancement effect on thermal spiking for a series of carbon fibre unidirectional laminates; ●, Narmco Rigidite 5245C; □, Fibredux 927C and ▽, Fibredux 924C [22]. (b) Moisture enhancement on thermal spiking for a range of matrix resins; ●, base epoxy resin-924E; ■, thermoplastic modified epoxy matrix resin-924T; ○, calculated matrix and composite-924C (normalized); and ⊕, 924C composite. There are 22 thermal spikes with conditioning for 5000h at 50°C, 96% RH [22]

microcrack formation in which the excess water is located [19–21]. In detailed studies Hough and Jones [11] and Xiang and Jones [9], could find no
evidence for resin microcrack formation. However, microcrack formation
in a fibre reinforced plastic composite is a complex function of the fracture
toughness of the 'wet' matrix or interphase and of the hygrothermally
induced stresses (see Sections 3.6 and 3.7.4.4). Whilst the mechanism responsible for the enhancement is still uncertain, the most important observation is that the moisture effect is accompanied by an amplification of the
low temperature relaxation peaks in the thermomechanical spectra. It is
also observed that the maximum enhancement temperature is close to the
onset of the relaxation phenomena for the wet resin. It has been argued that
a redistribution of the unoccupied volume between free volume and
microvoids is responsible for an increase in the amount of space available
for the water molecules to be accommodated and can contribute to the
plasticization phenomenon.

3.5   Thermomechanical response of composite to moisture absorption

3.5.1 Isothermal absorption

By far the most important response of a resin to moisture is the reduction
in its glass transition temperature. This is known as plasticization and is
observed for all polymers. The effect is illustrated in Fig. 3.7. It is particularly important in the area of fibre reinforced plastics. For example, the
glass transition temperature of a cured epoxy resin is reduced, on average,
by 20K for each 1% moisture absorption. Thus a resin which absorbs 7%
moisture at saturation could have its T_g reduced by 140K. A typical aerospace epoxy resin matrix utilizes a high crosslink density and blends of non-
polar thermoplastic or thermosetting resins to limit the moisture absorption
and hence limit the maximum reduction in T_g. ΔT_g might more typically be
50K, so that the maximum service temperature allowable must be less than
the T_g by this amount plus a further margin of safety of say 15K. As with
many thermosets, the dry T_g will be approximately equal to the cure or
postcure temperature. An epoxy resin cured at 150°C may have a maximum
useful temperature significantly below 100°C.

As well as a reduction in T_g, water absorption results in a reduction in the
modulus of the matrix. Since the matrix contributes significantly to the
transverse modulus of an individual unidirectional ply, this is a significant
limitation to the use of a composite in a structural capacity at elevated
temperatures.

The absorption of moisture predominantly modifies the matrix-
dependent properties of a composite and this tends to reduce the tempera-

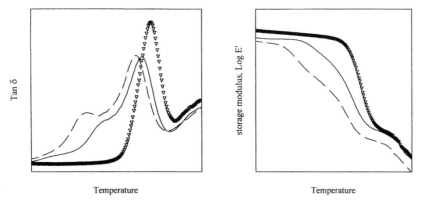

Figure 3.7 Schematic showing the effect of moisture absorption on the thermomechanical response (DMTA, dynamic mechanical thermal analysis) of a resin, (Left) tan δ (right) storage modulus; dry/as-cured resin (∇); wet from isothermal conditioning at 50°C (——), comparative wet sample from thermal spiking experiment (– –)

ture at which the structure remains stable to shear and compressive loading. Fibre-dominated properties such as the ultimate tensile strength are less affected. Time-dependent properties such as creep are generally determined by the viscoelasticity of the matrix, which in turn is influenced by the plasticizing presence of moisture.

3.5.2 Moisture absorption under non-isothermal conditions
(see also Chapter 4)

The plasticizing effect of moisture absorption on the dynamic thermomechanical response of the composite is illustrated schematically in Fig. 3.7. Other effects have been noticed, for example the thermomechanical response as a result of thermal spiking has been studied in order to assess the role of moisture induced by thermal spiking. Table 3.4 presents data for the two relaxation events, defined as T_{g1} and T_{g2}, the two maxima in the DMTA spectrum [22]. All three resins use a base epoxide modified with a thermoplastic (924) and a thermosetting (927, 5245) modifier. In all three, T_{g1} is reduced by isothermal moisture absorption at 96% RH at 50°C to a constant value. T_{g2} (the lower temperature peak which develops during conditioning), reaches a plateau in the case of the thermoplastic modified resin (924), but above a spike temperature of 120°C, the value of T_{g2} for the thermoset modified systems continues to decrease. Comparison of the moisture concentrations in all three materials strongly suggests that the

Table 3.4 Effect of thermal spiking and moisture conditioning at 96% RH and 50°C on the primary (T_{g1}) and secondary (T_{g2}) tan δ peaks for advanced epoxy resin systems [22]

Spike temp (°C)	5245C O$_{16}$ laminates			927 O$_8$ laminates			924 O$_8$ laminates		
	H$_2$O content (wt%)	T_{g1} (°C)	T_{g2} (°C)	H$_2$O content (wt%)	T_{g1} (°C)	T_{g2} (°C)	H$_2$O content (wt%)	T_{g1} (°C)	T_{g2} (°C)
dry	0	222	–	0	223	–	0	234	—
control	0.82	200	174	0.98	198	183	1.72	222	179
100	0.87	205	172	1.03	197	172	2.37	216	155
120	1.26	198	157	1.19	196	162	2.84	216	151
140	1.98	200	146	1.34	197	152	2.98	216	152
160	1.50	200	124	1.44	190	140	2.88	215	150
180				1.38	195	126	2.43	216	152
200				1.32	196	125	2.08	216	153

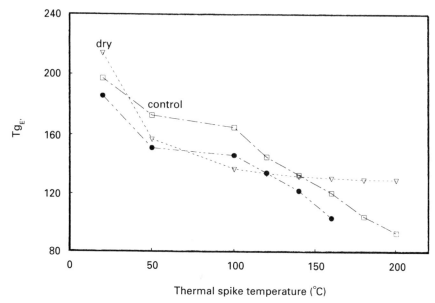

Figure 3.8 Effect of moisture absorption at 50°C, 96% RH and thermal spiking on the temperature for the onset of the relaxation region within the glassy polymer, as defined by reduction in storage modulus; 924C (▽); 927C (□); 5245C (●) [22]

spike temperature is more important than the presence of water. This is illustrated in Fig. 3.8, where the heat distortion point of the laminates (defined as the onset of the relaxation region from the temperature dependence of the storage modulus) is plotted against thermal spike temperature. The reduced stability of the resins at higher thermal spike temperatures appears to reflect the chemistry by which the modifiers interact with the base epoxy resin.

3.6   Residual strains in laminates

3.6.1 Concepts

Thermal residual strains are generated in composite materials on the microscale and macroscale. The former arise from a mismatch in the coefficients of thermal expansion of the fibre and the matrix. This effect is magnified by the presence of fibre bundles, where the radial stresses can change from compressive to tensile and influence interfacial failure under load. On the macroscale, in laminated structures, the residual stresses arise from a mismatch in the expansion coefficients of the individual plies. The simplest system is a crossply or 0°/90°/0° laminate where the expansion coefficients of the 0° and 90° plies differ significantly. The thermal residual strains induced on cooling from temperature T_1 to T_2 can be predicted from equations [3.12] and [3.13].

$$\varepsilon_{t/l}^{th} = \frac{E_l b (\alpha_t - \alpha_1)(T_1 - T_2)}{E_t b + E_t d} \tag{3.12}$$

$$\varepsilon_{t/l}^{th} = \frac{E_l d (\alpha_t - \alpha_1)(T_1 - T_2)}{E_t b + E_t d} \tag{3.13}$$

where $\varepsilon_{t/l}^{th}$ and $\varepsilon_{l/t}^{th}$ are the thermal strains in the transverse ply in the longitudinal direction and in the longitudinal ply in the transverse direction, respectively. E_l and E_t are the longitudinal and transverse moduli of the individual plies in the fibre direction and at 90° to the fibres, b and d are the outer ply and semi-inner ply thicknesses, respectively, α_l and α_t are the linear expansion coefficients of the longitudinal (or 0°) and transverse (or 90°) plies, respectively. The residual strains arise from a constrained contraction of the 90° ply, from the strain free temperature, T_1 to the service temperature, T_2. Thus tensile strains are induced at 90° to the fibres. The values of $\varepsilon_{t/l}^{th}$ are typically higher for carbon fibre reinforced plastics (CFRP) (~0.3%) than glass reinforced plastics (GRP) (~0.1%) because the expansion coefficient of carbon fibre, in the fibre direction is negative (−0.4 to −1.1 × $10^6 K^{-1}$), whereas for isotropic glass fibres it is 5 × $10^{-6} K^{-1}$. The radial expansion coefficient of a carbon fibre is 10–20 × $10^{-6} K^{-1}$. Typically, the linear expansion coefficient of a cured resin (α_m) is ~50 × $10^6 K^{-1}$. These residual strains increase the propensity for microcrack formation parallel to the fibres (i.e. transverse cracking) because they are additive to the external load

$$\varepsilon_{t/u} = \varepsilon_{tu} - \varepsilon_{t/l}^{th} \tag{3.14}$$

where $\varepsilon_{t/u}$ is the transverse cracking strain and ε_{tu} is the transverse ply failure strain. The values of α_l and α_t can be obtained from the Schapery equations [23]. Clearly α_l is a fibre-dominated property because $V_f > V_m$, whereas α_t is resin matrix dominated.

Thus α_m has a strong influence on the magnitude of the thermal strain induced. Since water absorption is mainly a matrix phenomenon, the expansion coefficient of the matrix will be affected, whereas that of the fibre will remain constant. Thus moisture absorption will affect α_m and T_1 and therefore influence the thermal strain induced on cooling.

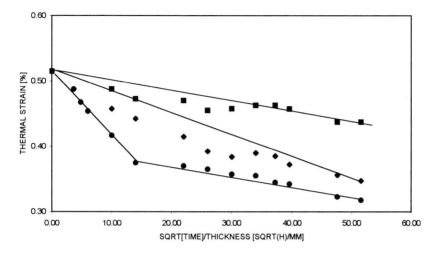

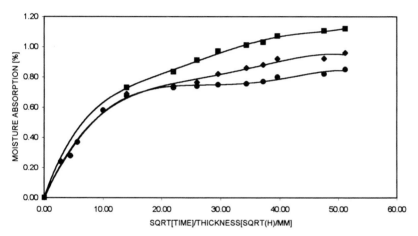

Figure 3.9 Effect of moisture absorption at 50°C and 96% RH on the thermal strain in a balanced 0°/90°/0° 927C laminate estimated from the curvature of an unbalanced 0°/90° beam[24]. ●, control; ◆, 120°C spike; ■, 140°C spike

3.6.2 Effect of moisture absorption on the residual thermal strain in cross-ply laminates

As shown in Fig. 3.9, the thermal strain in a cross-ply laminate is reduced by moisture absorption because the resin swells in the presence of water [24,25]. Thus the constrained swelling of the plies counteracts the constrained shrinkage which occurred on cooling, after manufacture. The moisture swelling coefficient for the individual plies can be assessed by studying the reduction in tensile residual strains

$$\varepsilon_{tl}^{s} = \frac{E_l b (\beta_t - \beta_l) \Delta M}{E_l b + E_t d}$$

[3.15]

where ε_{tl}^{s} is the swelling strain in the longitudinal direction of the transverse ply; β_t and β_l are the swelling coefficients in the longitudinal and transverse directions, respectively, and ΔM is the change in moisture content. For randomly orientated or woven fibre composites those effects can still operate on the microscale.

3.6.3 Effect of moisture absorption and a thermal excursion on the residual thermal strains in a cross-ply laminate

3.6.3.1 Moisture in resins prior to curing

Adventitious moisture can be present in the resins (or prepreg) used for lamination. Thus if the value of α_m and hence α_t is enhanced, then from equations [3.12] and [3.13] this will lead to an increase in ε_{tl}^{th} and ε_{lt}^{th}, and from equation [3.14] a reduction in the cracking strain of a transverse ply within a laminate. This analysis assumes that $(T_1 - T_2)$ is largely unaffected by very small levels of water in resins prior to fabrication. From equations [3.12] and [3.13], there is a competition between the efficiency of the plasticization process, which reduces T_1, and the increase in the average value of α_t, and hence the magnitude of $(\alpha_t - \alpha_l)$ [25,26]. Jones and co-workers [25] studied in detail the effect of adventitious moisture in an unsaturated polyester resin on the development of thermal strains in cross-ply laminates after curing. Their studies are summarized in Fig. 3.10.

3.6.3.2 Moisture absorption combined with a thermal excursion

Moisture absorption reduces the residual thermal strain through swelling of the matrix. On raising the temperature, the enlarged matrix expansion coefficient means that the constrained shrinkage can be significantly enhanced. This leads to the induction of a higher thermal strain, as shown in Fig. 3.9. In the context of equations [3.12] and [3.13], $(\alpha_t - \alpha_l)$ is enlarged. In this case the moisture absorption will be sufficient to reduce T_1 to a lower value. However, it is the integral of the α_t/T curve which determines the change in volume which is constrained by the stiffer 0° plies. Figure 3.11 gives the linear transverse thermal expansion curves for dry and wet (saturated) unidirectional carbon fibre composites based on advanced bismaleimide modified epoxy resins. Superimposed are the glass transition temperatures for the wet and dry matrices. It can be assumed that T_1 can be

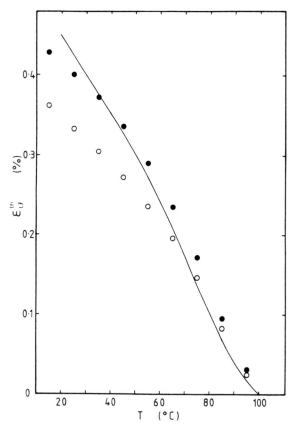

Figure 3.10 Effect of adventitious moisture on the generation of thermal strain in a polyester resin based cross-ply GRP, postcured at 130°C. The continuous line represents experimental data calculated from the continuous monitoring of the curvature of an unbalanced, as-manufactured beam. The predicted values are calculated from dry (○) and wet (●) (0.15%) laminate properties using equation [3.12] [25]

equated with T_g (T_1 is related to T_g but is not identical). It is preferable to measure T_1 directly from the curvature of an unbalanced 0°/90° beam [25,26]. We can demonstrate that the thermal strain is higher for a wet balanced laminate, despite the reduction in T_1 [25,27]. In this particular study, it was found that after a number of thermal excursions, ε_{tl}^{th} exceeded the measured ply failure strain and from equation [3.14] this led to an increase in the density of transverse cracks with each thermal cycle [27]. Based on this analysis, a critical temperature T_{CRIT} (wet) could be calculated, below which hygrothermally induced cracks would not form. These calculations were in agreement with experimentally observed transverse

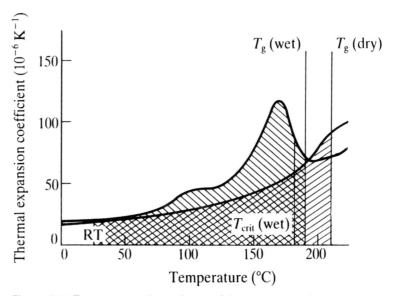

Figure 3.11 Temperature dependence of the transverse thermal expansion coefficients of the bismaleimide modified epoxy resin based carbon fibre composite (Narmco Rigidite 5245C): dry (lower curve) and wet (upper curve). The cross hatching indicates the relative thermal strain induced on cooling from T_g (dry) and T_g (wet). T_{crit} (wet) represents the temperature above which transverse cracking occurs in a wet cross-ply laminate [27], [33]

cracking under thermal loading. The effect is not limited to this matrix system and has been observed for other higher T_g systems such as PMR-bismaleimides [28].

3.7   Chemical durability of composite materials

The durability of a fibre reinforced plastic in chemical environments is largely determined by the resistance of the individual components.

3.7.1 Chemical resistance of matrix resins

3.7.1.1 Organic solvents

As with moisture absorption, solvent ingress will occur through diffusion processes which are driven by thermodynamic considerations. Thus in agreement with Flory–Rehner considerations, the degree of swelling will be a function of the Flory interaction parameter and the crosslink density [29]. For this discussion, the most important consideration is the relative polarity of the resin and the solvent. The simplest approach to this problem is the use of solubility parameters. Thus for maximum compatibility with a solvent, that is, solubility in all proportions for a linear polymer, maximum swelling of a partially crystalline polymer or crosslinked thermoset, equal thermodynamic properties are required (i.e. the simple rule-of-thumb, 'like-dissolves-like'). Thus

$$\delta_s = \delta_p \hspace{6cm} [3.16]$$

where δ is the solubility parameter of solvent (s) and polymer (p).

This arises because mixing is purely entropic and occurs without heat change. For other thermodynamic pairs, solubility requires a strong interaction between the solvent and the polymer, in which case solubility is exothermic (or sometimes endothermic). Generally the choice of a solvent is determined by matching the solubility parameter of each component, in which case the heat effect is minimized.

The solubility parameter of a polymer is generally obtained experimentally by identifying a solvent which causes maximum swelling of a lightly crosslinked polymer. However, it can be estimated from the molar attraction constants of the individual molecular groups which make up the polymer. For a linear polymer, the chemical structure is generally known, whereas for the cured thermosets such as the epoxy and phenolic resins, the detailed molecular structure is uncertain. In this case, knowledge of the hardener system can give an average molecular structure, which may be adequate for identifying groups of potentially damaging solvents. A detailed discussion of the application of solubility parameters to selection in plastics technology is given by Brydson [30].

In terms of polymer matrices for composite materials, there will be a compromise between solvent and water resistance. Thus non-polar resins are likely to be less resistant to hydrocarbon solvents, which have low polarity, but more resistant to moisture absorption. Polar resins behave in the opposite way. Strongly polar solvents, such as dimethyl sulphoxide or similar, can interact with polar structures in the resin and are difficult to resist. Crystalline thermoplastic polymers are often better for such applications. For example, polyethene will only dissolve in hydrocarbon solvents (of similar solubility parameter) at temperatures above the crystalline melting point. Polar semi-crystalline polymers such as the polyamides or nylons can be dissolved in highly polar solvents, such as cresol, because of a stronger interaction than that between molecules within the crystallites. High performance thermoplastic polymers such as polyether ether ketone (PEEK) have been promoted for their resistance to organic solvents (see Table 3.5) [12]. The chemical resistance of unsaturated polyester and vinyl ester and urethane resins is indicated in Table 3.6 [15].

Table 3.5 Resistance of carbon fibre/PEEK (APC-2) composites to immersion in common solvents for 15 days [12]

Environment	Temperature (°C)	Absorption (%)
Jet fuel	82	0.03–0.1
Skydrol LD4	70	−0.09
Ethylene glycol	70	0.02
Methyl ethyl ketone	23	0.03
Methylene dichloride	23	2.41
Water	100	0.23
50% RH	23	0.02

3.7.1.2 *Aqueous environments*

The resistance of thermosetting resins to aqueous 'reactive' solvents is determined by the chemical nature; acid, alkaline, neutral, etc. Resins containing hydrolysable groups are less resistant to aqueous solutions in the longer term and at higher temperatures. As explained already, cured epoxide resins can have significantly different chemical structures, and it is impossible to generalize about the chemical resistance of epoxy resins. Catalysed anhydride cured resins contain ester linkages and are therefore not very resistant to aqueous environments, especially acidic and alkaline ones which catalyse the hydrolysis process. But these resins are much less polar than other cured epoxy resins and absorb significantly less moisture.

The rate of a chemical reaction is determined by the concentration of reagent, in this case H_2O, H_3O^+ or OH^-, and temperature, so these resins can exhibit good short term resistance. Hydrolysis is also a function of the mobility of the network chains which control the diffusion of the reactant and the concentration of the relevant functional group. Therefore, higher T_g resins will also exhibit resistance in the short term and at ambient temperature. For these reasons, it is clearly necessary to assess each recipe individually.

These same rules apply to other resins such as the family of unsaturated polyesters. In these, the reactive diluents dilute the ester groups. A reduction in ester group concentration can be achieved by using more rigid skeletal structures and employing only terminal ester groups (vinyl ester resins) or urethane links (vinyl urethanes, urethane methacrylates). The chemical resistance of these resins is summarized in Table 3.6.

3.7.2 Reinforcing fibres

Carbon fibres are generally considered to be resistant to most chemical environments other than those which are strong oxidants such as concen-

Table 3.6 Chemical resistance of typical unsaturated polyester and vinyl ester resins as defined by a maximum use temperature (°C) [15]

Environment	Resin type						
	Orthophathalic	Isophathalic	Isophathalic/NPG	Bisphenol	HET acid	Vinyl ester	Vinyl urethane
Solvents							
Acetone	NR	NR	NR	NR	NR	NR	NR
Acetone (10% aq)	25	NR	NR	25–82	—	NR	82
Amyl acetate	NR	NR	NR	25	25	NR	NR
Aviation fuel	30	25	25	25	40	—	25
Benzyl alcohol	30	25	NR	25	25	38	25
Butyl alcohol	35	30	30	20	35	NR	25
Chemical and aqueous solutions							
Acetic acid, 70%	NR	25	55	65–70	30	70	71
Acrylic acid	—	NR	—	25–35	NR	—	25
Ammonia, 5%	NR	25	35	60	45	70	—
$Ba(OH)_2$, 10%	NR	NR	25	30–65	NR	—	66
NaCl (sat)	55	50	75	95–105	80–100	90	93
$Ca(OCl)_2$, 17%	NR	NR	45	50	25	50	—
$Al_2(SO_4)_3$(aq)	55	50	75	95–105	70–120	90	93

NR = not resistant.

trated nitric acid. There are also reports of the intercalation of halogens and certain metallic salts between the graphite basal planes. For composite materials, the resistance of the matrix and/or interphase regions may be more important. Aramid fibres, on the other hand, can absorb up to 4.5% w/w moisture, depending on the presence of coagulant residues. Since these fibres have an aromatic amide structure, there is always the potential for hydrolysis of the amorphous regions, especially under alkaline conditions. There is very limited information available on their long-term durability in extreme environments. The most degrading environment for aramid fibres is ultraviolet light, which causes a discoloration and a loss in strength, with time. There is also some evidence for a static fatigue phenomenon which could be related. Fortunately, the degradation products are self-screening, protecting the underlying polymer in the fibre, from degradation. It is therefore advisable to use matrix resins, which absorb ultraviolet light at the appropriate wavelength harmlessly, or use protective coatings. Glass fibres on the other hand are much more susceptible to corrosion in aqueous environments. The most commonly used fibres in reinforced plastics by far are made from E-glass, which is susceptible to both acidic and alkaline environments where a reduction in performance by corrosion or extraction mechanisms is observed. Consequently a range of glass fibre reinforcements exists for specific applications. ECR glass is recommended for general corrosion resistance (especially acid resistance), and for alkali resistance, AR glass is preferred. The latter is also being developed for compatibility with polymer resins to provide both acid and alkali resistance. In many situations, it is satisfactory to use E-glass as the structural reinforcement in combination with a C-glass tissue reinforced surface resin. Optimum performance can also be achieved by employing C-glass surface tissues in combination with woven rovings rather than chopped strand mat. S and R glasses are employed where high performance is required, but they also provide enhanced durability over E-glass. As with moisture absorption, for chemical resistance the fibres should be carefully selected with a recommended surface finish. (Chemical resistance is also discussed in Chapter 9, on chemical process equipment).

3.7.3 Environmental stress corrosion cracking of E-glass fibres

This is a delayed brittle fracture effect, caused by a synergism between the stress and the environment. E-glass fibres suffer from environmental stress corrosion cracking (ESCC), and in water this is referred to as 'static fatigue'. It is generally responsible for the weakening of the fibres, during their lifetime. However, ESCC can occur rapidly under acidic conditions. Typically at 0.5% strain in dilute aqueous sulphuric acid, this occurs in less than ten hours at ambient temperatures. Other glass fibres such as S, R and ECR are much less susceptible. AR glass fibres are also highly resistant.

In alkaline environments, there is no evidence for the synergism between stress and the environment which leads to ESCC, only the general weakening of the fibres. The mechanism of static fatigue of most glasses and the rapid ESCC of E-glass in acidic environments is considered to be a stress-assisted hydrolysis of the silicate network

$$\equiv Si-O-Na^+ + H_2O \rightarrow \; \equiv Si-OH + Na^+ + OH^-$$
$$\equiv Si-O-Si \equiv + OH^- \rightarrow \; \equiv Si-OH + \; \equiv SiO^-$$
$$\equiv Si-O^- + H_2O \rightarrow \; \equiv Si-OH + OH^-$$

For water, this can be attributed to the presence of a high alkaline concentration at the tip of a surface or flaw or crack. Charles [31] explains that ESCC occurs when the rates of hydrolysis of the silica network and crack propagation are similar (i.e. stress-assisted corrosion). When hydrolysis is more rapid than crack propagation then corrosion leads to a rounding of the crack tip and a reduction in its capacity to propagate. If the rate of crack growth is much larger than the rate of corrosion then the environment will have little influence on fibre strength and time dependence. It is believed that since Na^+ is directly involved, then its diffusion to the crack tip is rate determining.

Since the ESCC is clearly pH dependent, the network modifiers and alkali metal ions must play a significant role. However, it is well established that leaching of Ca^{2+}, Al^{3+}, Fe^{3+}, Na^+, K^+ and other residuals leaves a weakened sheath around a strong core. This leads to spiral cracking of the sheath when unstressed fibres are stored in aqueous acid for a short period. It is not clear whether the crack forms prior to optical or scanning electron microscope (SEM) examination or after removal from the environment but the latter is favoured. There are two confusing observations:

1 that the spiral cracks appear to form when the hydrated sheath dries out
2 that not all of the fibres in a bundle suffer from either spiral cracking or multiple fracture.

It would appear therefore that the leaching of the network modifier ions is also stress dependent and that variable residual stresses are built into the fibres as a result of the water cooling regimes employed in manufacture. The leaching can be further complicated by complexation of the leachable ions such as Fe(III) within the aqueous solutions, which can promote the process. This may be the mechanism by which ultraviolet illumination accelerates ESCC of E-glass filaments [32]. The maximum rate of ESCC of E-glass fibres occurs at pH $\approx$ 0. Thus $0.5\,M$ H_2SO_4, or $1\,M$ HCl or any other acids where the proton concentration is $1\,M$, are the most corrosive.

3.7.4 Environmental stress corrosion cracking of glass fibre composites [33–42]

3.7.4.1 Brittle fracture caused by stress and chemicals

E-glass fibre composites, in particular, suffer from a delayed brittle fracture in acidic environments, under stress. Fibre composites exhibit tough behaviour because of the large surface area which results mainly from fibre/matrix debonding (and matrix cracking). As a consequence, failure generally results in a fibrous fracture surface with extensive fibre pull-out. Under ESCC conditions, failure of GRP (principally from E-glass) results in a planar fracture surface, typical of brittle behaviour. This can be readily recognized in the SEM microscope by the presence of mirror-fractured fibres devoid of hackle and river lines. But ESCC of the fibres leads to growth in the failure crack through the adjacent resin and fibres. As the stress increases under a constant load with a decrease in cross-sectional area, the fibres exhibit hackle and the fracture surface becomes less planar (and can exhibit steps) away from the crack initiation zone. The resin matrix has to protect the fibres from corrosion, leaching and ESCC in the components under stress, in aqueous acidic environments. Resin-rich surface layers or gelcoats adequately protect the fibres from diffusion of the environment to the fibres during the lifetime of the component, and the prime concern should be the chemical resistance of the matrix resin and the interface region between fibre and matrix under an applied stress. However, different principles operate because a sharp crack can promote brittle fracture at strains as low as 0.05%, which is a result of the susceptibility of the E-glass fibres to ESCC and the subsequent crack propagation into the adjacent matrix. In Fig. 3.12, time-dependent failure exhibits three main mechanisms [33]. In region I, at high stress, failure occurs in relatively short times because the rate of crack growth is larger than the rate of diffusion of the environment to the crack tip. Therefore, the failure times are similar to those in the absence of an active environment.

In region II, stress corrosion cracking results from the synergism between the two agencies. At lower stresses in region III, the failure is less sensitive to stress and the lifetime of the material is determined by the rates of environmental diffusion and fibre corrosion.

3.7.4.2 Types of ESCC failure

Three distinct failure modes have been observed [33–35] which involve a synergism between the applied (and/or internal) stress and environment. They have been characterized as Type A, B or C.

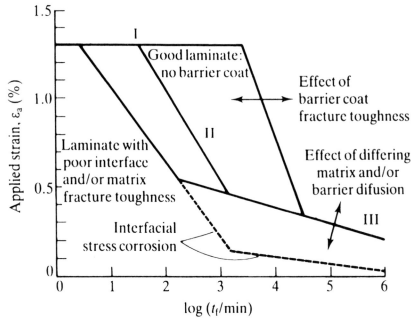

Figure 3.12 Schematic failure map for a unidirectional E-glass composite under acidic environmental stress corrosion cracking conditions, illustrating the material variables. t_f is the failure time in minutes under a constant tensile load [33]

- Type A – Brittle fracture within the environment.
- Type B – Brittle fracture of the unexposed part of the laminate in the atmosphere about 20 mm above the environment.
- Type C – Damage accumulation within the unexposed part of a cross-ply laminate, under a small or zero external stress, which is initiated near the surface of the environment.

3.7.4.3 Type A failure

This is the type of failure generally observed in service, since it occurs in the submerged composite at applied strains in excess of 0.15%. It is initiated by the stress corrosion failure of a glass fibre near the surface of the laminate, as a consequence of environmental diffusion. Attempts to confirm diffusion of the acidic environment through the resin to the fibre have proved difficult but it is likely that the applied stress increases the rate of diffusion. The statistical nature of fibre strength means that a premature fibre break can lead to an enhancement of the local stress in the adjacent fibres by stress transfer. Thus a series of stress corrosion nuclei will form in the surface of a 0° composite under tension. Brittle failure will occur once the flaw reaches critical dimensions. In flexure, the stress in the tensile face of the beam may not be uniform and this can lead to the initiation of *one* stress corrosion crack.

In angle-ply laminates, such as 0°/90°/0° composites, a transverse crack provides sufficient enhancement of the stress in 0° plies adjacent to the crack for a single crack to propagate through the laminate in tension. In flexure, cracks have also been observed to propagate from the inside to the outside of the material.

Similar stress situations can arise if surface cracking of a barrier or gelcoat layer occurs. It has been shown that a cracked gelcoat is more damaging than its absence, under ESCC conditions [36,38].

3.7.4.4   Resin matrix effects

Hogg and Hull [38] have provided a basic understanding of the role of the matrix in ESCC. Thus with a brittle matrix, the resin fillet between fibres will also immediately fracture as a result of fibre ESCC fracture, leading to further crack growth as the environment is transported to the crack tip. As a consequence, the fracture toughness of the resin will determine the rate of crack growth under stress corrosion conditions. For ductile resins, diffusion of the corroding substance through the matrix fillet needs to occur before the next fibre can fail and throw an additional load onto it. Therefore resin properties which compromise fracture toughness and environmental diffusion are required for optimal durability. Since chemical resistance is generally obtained with resins of high crosslink density and aromaticity, those resins which provide the best chemical resistance do not necessarily provide optimum stress corrosion resistance because of their low fracture toughness. It has been argued that the poorer stress corrosion resistance of a ductile polyester resin arises because of a larger moisture diffusion coefficient but it should be remembered that, with lower crosslink density, the chemical resistance is also generally lower, and even partial hydrolysis can reduce the fracture toughness of the resin.

A further complication is the effect of moisture absorption and plasticization on the fracture toughness of the matrix [33,39]. As shown in Table 3.7, the moisture content of the resin can influence its fracture toughness as determined by K_{Ic}, the critical stress intensity factor under mode I loading conditions. It is clear that the effect of moisture on the fracture toughness of the more brittle chemically resistant resins will be less, so that the chemically resistant resins will provide less protection to ESCC of the fibres, even in the presence of absorbed moisture [37,39].

As a consequence, a limiting strain for stress corrosion failure may exist for a particular glass fibre composite, which is governed by the fracture toughness of the resin but the effects of fibre/matrix interfacial strength

Table 3.7 Fracture toughness (K_{IC}) of chemically resistant polyester and vinyl ester resins [37], [39]

Resin	Type	Environment	K_{lc} (MN m$^{-3/2}$)
Crystic 272	Isophthalic/ propylene glycol	Dry	0.79
		Moist ($\sim$0.2% H_2O)	0.62
		Wet (1.5% H_2O) – tested dry	0.55
		Wet (1.5% H_2O) – tested wet	0.89
		Dry – tested wet	0.85
Crystic 272/30% Crystic 586	Flexibilized isophthalic/ propylene glycol	Laboratory environment – moisture present	0.77
Derakane 411–45	Vinyl ester	Laboratory environment – moisture present	0.75
Crystic 600 PA	Bisphenol polyester	Laboratory environment – moisture present	0.49
Beetle 870	Chemical resistant	Laboratory environment – moisture present	0.46
Atlac 382-05A	Urethane, bisphenol, vinyl ester	Laboratory environment – moisture present	0.45

and/or matrix failure strain especially for angle-ply laminates should not be ignored. For example, a limiting applied strain for an isophthalic unsaturated polyester resin-based composite of $\approx$0.4% was identified. In general therefore the maximum recommended external applied strain for good quality GRP to provide resistance to ESCC is 0.3% [34].

Provided that the interface between fibre and matrix remains intact, the fracture toughness of the wet matrix appears to be the most important factor which determines resistance to Type A ESCC. However, the interface or interphase region is also subject to synergistic effects and stress-assisted corrosion of the interface can also occur. This effect can lead to a time-dependent reduction in transverse strength under a constant load in the presence of the environment [40]. For cross-ply laminates, transverse cracks, which will form in the 90° ply at lower strains than normal in the absence of the environment, will promote a Type A ESCC adjacent to the first 90° ply crack. Similar arguments can be applied to angle-ply laminates, more typical of the configuration of filament wound pipes used in service.

3.7.4.5 Type B failure

At very low applied strains ($<$0.15%), for a particular epoxy resin/glass system, Type B failure of the unexposed part of the cross-ply laminate occurs before the environmentally induced transverse cracks can completely propagate across the 90° ply. This type of failure is also observed in 0° epoxy composites at similar strains (<0.1%) and in 0° polyester laminates at much higher strains (0.6%). For the former, catastrophic fracture occurred after 18 days at 0.09% strain and 243 days at 0.07% strain, whereas the latter required 359 days at 0.6% strain.

A time-dependent stress-assisted degradation of the interface causes wicking of the environment along the debonded fibres, drawing with it an acidic solution of corrosion products. On reaching the 'dry' part of the material, salts such as calcium sulphate, potassium sulphate, sodium aluminium sulphate (from sulphuric acid) or phosphates (from phosphoric acid) precipitate at the fibre/matrix interface, providing an additional 'crystallization' pressure which, together with a small applied stress, initiates an ESC crack which is environmentally fed by wicking of the environment. The formation of a single crack is a consequence of the higher rate of stress corrosion of the interface compared to that for diffusion of the environment through the resin. Thus Type B failure occurs under conditions that encourage crystallization of the corrosion products, and is inhibited when the non-submerged composite is exposed to a high humidity (e.g. a part-filled closed container) or in hydrochloric acid where the calcium or aluminium salts are very soluble. Similarly with highly insoluble salts, such as the phosphates, or in the presence of a more concentrated acid, for example sulphuric, where the glass corrosion products are less soluble, crystallization can occur within the environment, causing a premature quasi Type A crack.

3.7.4.6 Type C failure

Type C failure is a damage accumulation process in the unexposed laminate. It has identical origins to the Type B mechanism, except that in the absence of an external load, a failure crack does not form. This is because in a cross-ply laminate, the 0° fibres are put into compression by the thermal strains which develop during fabrication. A consequence of corrosion' product precipitation in the 'dry' half of the laminate, therefore, is the transverse cracking of the 90° and 0° plies near the surface of the environment, which progresses up the unexposed part of the composite. Whereas the Type A/B mechanisms operate only in acidic environments, Type C stress-assisted environmental damage accumulation can also occur in alkaline environments. For example, it can occur more rapidly with aqueous sodium hydroxide, since the rate of corrosion of the glass is greater than in the presence of an aqueous acid.

Table 3.8 Generalized trend in corrosion and stress resistance of E-glass laminates with various matrix resins [33][a]

Unsaturated polyester resin		Vinyl ester resin	Chemical resistance	ESC resistance
Orthophthalic	Diethylene glycol ethylene glycol propylene glycol			
Isophthalic	Diethylene glycol ethylene glycol propylene glycol neopentyl glycol	Vinyl esters		
Fumarate	Bisphenol A			
		Urethane modified bisphenol A		

[a] Chemical resistance increases downwards and ESCC resistance upwards, through the list of resins shown.

3.7.4.7 *Laminate construction for ESCC resistance*

The highest probability of ESCC failure of GRP is in dilute aqueous acids of pH $\approx$ 0. ESCC of E-glass fibre composites can be effectively inhibited by careful choice of resins, fibres and laminate construction (see Fig. 3.12). The resin can be selected according to its fracture toughness and chemical resistance as illustrated in Table 3.8 [32]. Careful laminate construction should prevent the formation of sharp resin cracks. Thus flexible resin gelcoats, with optional random mat reinforcement, preferably from chemically resistant glass or other (e.g. polyester) fibres, form a barrier coat. The structural wall can still be provided by the cheaper E-glass fibres. For example, in hand lay composites, the barrier layer consists of an appropriate gelcoat and C-glass veil. The structural composite must employ powder bound CSM. It is not recommended to use different resins for the gelcoat and structural layers, since interlaminar failure can occur.

Woven rovings and CSM in optimum combination can provide excellent chemical resistance to barrier layers of laminates used for chemical plant. For filament-wound vessels the barrier layer is a resin-rich layer of laminating resin reinforced with chopped E-glass, C-glass or polyester fibre veil. This layer is overwound with layers of hoop windings and 90° woven tape or with helical windings of continuous fibres at ±53°. Some commercial systems employ alternate layers of graded sand fillers in the structural wall. Long-term studies on filament wound pipes have confirmed that with a maximum flexural stress of 20–13 MPa, a 50 year life would easily be achieved [41,42].

Centrifugal casting has the advantage that a highly ductile surface resin (failure strain ≈25%) can be graded into the less ductile structural resin. In this case, there is no need to employ chemical-resistant fibre veils since the volume fraction of chopped fibre can be graded from zero at the surface through to 50% in the structure and simultaneously from random to circumferential by varying the speed at which the mould is spun. Furthermore, alternate layers of compacted sand-filled resin can be incorporated in the structural laminate.

ECR glass flake can be used to provide reinforcement to the barrier layers. Resistance to highly corrosive environments can be further improved by employing ECR, Chemglass, S or R, glass fibres instead of E-glass for the structural wall.

3.8   Conclusions

The susceptibility of resins, reinforcing fibres and composites to aqueous and non-aqueous environments has been reviewed. It is clear that with careful material selection and composite construction, reinforced plastics are very durable in aggressive conditions. One of the most severe environments, aqueous acids, can be designed to use the ideas presented here. This practice is implemented in industry, as demonstrated by the durability of GRP in environments encountered in chemical plant and conduit for corrosive effluents.

References

1 G S Springer, *Environmental Effects on Composite Materials*, ed. G S Springer, Westport, USA, Technomic, 1981, Volume 1.
2 C H Shen and C S Springer, *J Compos Mater* 1976 **10** 2–20.
3 B Ellis and M S Found, *Composites* 1983 **14** 237.
4 J A Hough, *Enhanced Moisture Absorption in Advanced Composites*, PhD Thesis, University of Sheffield, 1997.
5 F R Jones, in *Epoxy Resins*, ed. B Ellis, Glasgow, UK, Blackie, 1991, pp 258–302.
6 P M Jacobs and F R Jones, *J Mater Sci* 1989 **24** 2343.
7 Z D Xiang and F R Jones, *Final Report to RAE Farnborough Agreement* D/ER1/9/4/2031/139 XR MAT, 1992.
8 W W Wright, *Composites*, 1981 **12** 201.
9 Z D Xiang and F R Jones, *Compos Sci Technol* 1997 **57** 451.
10 F R Jones, M A Shah, M G Bader and L Boniface, in *Sixth International Conference on Composite Materials (ICCM VI/ECCM 2)*, eds F L Matthews *et al*, London, Elsevier Applied Science, 1987, Volume 4, pp 443–456.
11 J A Hough and F R Jones, 'Effect of thermoplastic additives and carbon fibres on the thermally enhanced moisture absorption by epoxy resins', in *Eleventh*

International Conference on Composite Materials (ICCMIII) vol V, Textile Composites and Characterisation, ed. M L Scott, Cambridge, UK, Woodhead/Australian Composites Structures Society, 1997, pp 421–431.

12 J Barnes, in *Handbook of Polymer Fibre Composites*, ed. F R Jones, Harlow, UK, Longman, 1994, Chapter 2.1, pp 69–74.

13 P M Jacobs, M Simpson and F R Jones, *Composites* 1991 **22** 99.

14 B Ellis, *Epoxy Resins*, Glasgow, UK, Blackie, 1991.

15 R G Weatherhead, *FRP Technology, Fibre Reinforced Resin Systems*, London, Applied Science, 1980.

16 F Chen and A W Birley, *Plastics Rubber Compos Proc Appl* 1991 **15** 161.

17 F Chen and A W Birley, *Plastics Rubber Compos Proc Appl* 1991 **15** 169.

18 F Chen and A W Birley in *Handbook of Polymer-Fibre Composites*, ed. F R Jones, Harlow, UK, Longman, 1994, pp 376–379.

19 C E Browning, *Polym Eng Sci* 1978 **18** 16.

20 C E Browning, G E Husman and J M Whitney, in *Composites Materials Testing and Design, ASTM STP 617*, Philadelphia, American Society for Testing and Materials, 1977, p 481.

21 J M Whitney and C E Browning, in *Advanced Composite Materials – Environmental Effects, ASTM STP 658*, Philadelphia, American Society for Testing and Materials, 1978.

22 J A Hough, F R Jones and Z D Xiang, 'Thermally enhanced moisture absorption and related degradation mechanisms in composite materials', *Proceedings 4th International Conference Deformation and Fracture of Composites*, London, Institute of Materials, 1997 pp 181–190.

23 R A Schapery, *J Compos Mater* 1968 **2** 380.

24 E Calota and F R Jones, 1997 unpublished data.

25 M Mulheron, F R Jones and J E Bailey, *Compos Sci Technol* 1986 **25** 119.

26 F R Jones, in *Handbook of Polymer – Fibre Composites*, ed. F R Jones, Harlow, UK, Longman, 1994, Chapter 4.12, pp 254–260.

27 P M Jacobs and F R Jones, 'The mechanism of enhanced moisture absorption and damage accumulation in composites exposed to thermal spikes', in *Composites, Design, Manufacture and Application (ICCM VIII)*, eds S W Tsai and G S Springer, Vol 2, Corina, CA, USA, SAMPE, 1991, Volume 2, Chapter 16, paper G.

28 M Simpson, P M Jacobs and F R Jones, *Composites* 1991 **22** 105.

29 P J Flory, *Principles of Polymer Chemistry*, New York, Cornell University Press, 1953.

30 J A Brydson, *Plastics Materials*, 5th Edn, London, Butterworth, 1988.

31 R J Charles, *J Appl Phys* 1958 **29** 1549.

32 F R Jones, P Marson and P A Sheard, 'UV sensitive enviroment stress cracking of E-glass fibres', in *Proceedings International Conference, Fibre Reinforced Composites '88 (FRC 1988)*, London, Plastics and Rubber Institute, 1988, paper 34.

33 F R Jones, in *Handbook of Polymer-Fibre Composites*, ed. F R Jones, Harlow, UK, Longman, 1994, Chapter 6.5, 6.6, pp 379–391.

34 F R Jones, J W Rock and A R Wheatley, *Composites* 1983 **14** 262.

35 F R Jones, J W Rock and J E Bailey, *J Mater Sci* 1983 **18** 1059.

36 P J Hogg, *Progr Rubber Plastics Technol* 1989 **5** 112.

37 P J Hogg, *Composites* 1983 **14** 254.

38 P J Hogg and D Hull, *Metal Sci* 1980 **14** 120.

39 F R Jones, *J Strain Anal* 1989 **24** 223.

40 J W Rock and F R Jones, in *Interfacial Phenomena in Composite Materials 1989*, ed. F R Jones, London, Butterworth, 1989, pp 163–168.

41 S-W Tsui and F R Jones, *Plastics Rubber Processing Applic* 1989 **11** 141.

42 S-W Tsui and F R Jones, *Compos Sci Technol* 1992 **44** 137.

4

Temperature – its effects on the durability of reinforced plastics

JOHN J LIGGAT, GEOFFREY PRITCHARD AND
RICHARD A PETHRICK

4.1 Introduction

Plastics and reinforced plastics have many virtues, but in comparison with other structural materials, outstanding high temperature resistance is not one of them. Users must consider their likely response to high temperatures. The same goes for rapid changes in temperature, even when the maximum use temperature expected comfortably exceeds the service temperature.

It would be wrong to give the impression that everything can and will go wrong whenever reinforced plastics are heated. It is nevertheless a sensible precaution to get access to good, meaningful data and good case histories, and to note what is already achievable in practice with well-established, commercially available materials. This account is designed to provide some background for making judgments. The authors will mention many ways by which heat can damage reinforced plastics. However the problems can be minimized by a knowledge of the pitfalls.

The word 'high' as used in 'high temperatures' will be interpreted flexibly; suffice it to say that not many commercial plastics can withstand 325°C, even for a few hours and several cannot be used continuously even at 120°C. Thermal shock problems are not confined to exotic applications, they even affect moulded bath tubs.

The maximum use temperature is a term coined by the US organization, The Underwriters' Laboratory, and it is the maximum temperature at which a polymer can be used continuously, under low stress conditions, with the loss of no more than 50% of the original useful properties (tensile and impact strength, for example, or dielectric strength in the case of cable insulation). For the engineer unfamiliar with plastics, the low values of typical maximum use temperatures can come as a shock. For example, glass-filled nylon 6,6 has a heat distortion temperature of 252°C, but the resin can embrittle as a result of thermal oxidation within 2 h at 250°C. Even at 70°C, embrittlement will still occur within two years [1].

4.1.1   Fibre reinforcements

Glass fibres are thermally stable materials that give no anxiety to users of reinforced plastics. They withstand a higher temperature than any of the current generation of commercialized matrix resins. This can be confirmed by the fact that the standard procedure for measuring the amount of glass fibres in glass/organic resin composites is simply to heat the samples at 600°C until all the resin has been burnt off, leaving the glass unaffected, at least as far as chemical change or weight loss is concerned. However, the surface treatment applied to glass fibres for use with standard epoxy and polyester resins will not necessarily withstand the processing temperatures required for the latest high temperature matrix resins. The result is that unless special surface treatments are used, there will be inadequate transverse tensile and flexural strength, low shear strength and probably a deterioration in all interface-related mechanical properties after water immersion.

Asbestos fibres withstand much higher temperatures still, although strictly they lose a small percentage of their weight in the form of water at 200°C and above [2]. Because of health hazards, they are not often used nowadays in composites and will not be considered further.

Polyethylene fibres are made by several different processes and their thermal characteristics vary, but lack of high temperature resistance is very noticeable compared with other reinforcing fibres. Most polyethylene fibres begin to lose their mechanical properties at about 120°C and they melt soon afterwards.

Aramid (e.g. Kevlar 49®) fibres do not melt, but they degrade on heating in vacuum above 375°C and will oxidize at 400°C. The tensile strength of commercial Kevlar fibres declines with temperature, to about 75% of the ambient value by the time that 177°C is reached. Unprotected yarns lose 25% of their tensile strength and 7% of their modulus, measured at room temperature, after heating for 96 h in air at 205°C. Data on bare fibres can be misleading; impregnation with a resin protects the fibres, and there is no observed adverse effect [3] when a normal cure cycle is carried out for Kevlar/epoxy at 180°C.

Carbon fibres have excellent heat resistance in inert, that is, oxygen-free atmospheres, but they oxidize slowly when heated in air, although in practice a heat-resistant matrix (such as a ceramic or a metal) will protect them for a considerable time. The rate at which the fibres in reinforced plastics are attacked in air depends on the fibre accessibility, that is, on the permeability of the matrix to oxygen, and on the presence of exposed fibre ends. Experiments with unidirectional carbon fibres in an aluminium matrix have demonstrated that the fibres are progressively oxidized away at high temperatures, starting at their exposed ends and working along the fibre length.

Table 4.1 Changes in carbon fibre reinforced Xylok resins on heating in air at 227°C (50 days) [4]

Fibre type		Resin type	% Volume increase	% Weight increase	% Retention flex modulus	% Retention strength
Type 1 HM	Grafil	Xylok 210	+1	−3	90	37
Type 2 HT	Grafil	Xylok 210	−3	−4	82	40
Type 3 HT-S HM-S	Grafil A	Xylok 210 Xylok 210 Xylok 210	−4 — —	−2 — —	88 — —	68 70 62

Note: Xylok was a high temperature phenol–aralkyl resin. The fibre suffix S denotes surface treated.

Carbon fibres come in several varieties that differ in the extent to which they resemble graphite, that is, in their degree of structural order. The rate of oxidation in oxidizing conditions can be reduced by specifying a highly graphitic grade of carbon fibres, although the usual motivation for selecting highly graphitic fibres is the higher modulus they impart. Hart [4] measured the reduction in Young's modulus and tensile strength of three different kinds of single carbon fibres (all early, untreated grades manufactured around 1969) after heating in air for 100 days at 250°C and 300°C. Two of the grades showed modest reductions in tensile properties, but the extent of the reduction after such a short time is misleadingly high, because there was no protective matrix. It is more significant to consider Hart's composite data, shown in Table 4.1. The table gives dimensional, weight and mechanical property changes for Xylok 210® (phenol–aralkyl) composites after heating in air. Neither the fibres nor the matrix mentioned are currently available and improvements have since been made in carbon fibres, but the results show the significant effect that grade differences can have.

4.1.2 Service conditions

When choosing reinforced plastics for use in extreme temperatures we have to consider:

1 Whether the heating is continuous or intermittent,
2 The maximum and minimum temperatures in a working cycle,
3 The heating and cooling rates,

4   Other factors such as mechanical stress, or fluids (which may or may not
    be chemically reactive, but will probably permeate into and dissolve in
    the matrix, changing its characteristics) and
5   Its required lifetime, which may be a matter of minutes, or decades.

4.1.3 The concept of failure

It is not always disastrous if a material fails to withstand high temperatures.
Failure of a material is taken here to mean that it does not behave as
intended; so a resin that is actually intended to decompose sacrificially, for
example in heat shields, is fully expected to be destroyed by heat the first
time it is used. Changes in physical properties, induced by temperature
changes, are essential elements in the design of certain devices.

Maximum use temperatures are governed by two main factors: the resin's
glass transition temperature, T_g, and the temperature at which chemical
decomposition starts to become significant. The latter cannot easily be
sharply defined as the decomposition processes usually begin almost imper-
ceptibly and accelerate very gradually as the temperature rises.

Decomposition temperatures, whether sharp or vague, are seldom actually reached in service life. Reinforced composites are preeminently load-
bearing materials, and it is their temperature-dependent mechanical
properties, such as T_g, or the closely related heat distortion temperature,
that usually determine the maximum use temperature, at least for short or
intermediate term use. Strength, yield stress and modulus all decline with
increasing temperature, reflecting the increasing mobility of the molecular
structure, and unacceptable levels of physical property loss will often occur
well before the onset of thermal or thermo-oxidative degradation.

4.1.4 Aerospace applications drive the demand for improved materials

Subsonic civil aircraft, such as the Boeing 747, experience temperatures on
the ground of between $-10°C$ and $+40°C$ but at 39 000 feet (11 887 m) and
550 miles h^{-1} (885 km h^{-1}) the external fuselage temperature is usually
between $-50°C$ and $-55°C$. The outer skin temperature of supersonic
aircraft is much higher.

The next generation of high speed civil transports (HSCTs) is designed to
carry 250–300 passengers at supersonic speeds [5]. The European version of
this aircraft (a second generation Concorde) has a proposed speed of Mach
2.05, consistent with transatlantic operation, whereas the US plane has a
proposed speed of Mach 2.4 for transpacific operation. Estimated surface
skin temperatures will increase from $110°C$ to $177°C$ simply as a result of
the speed difference. The operational life at these elevated temperatures

for these types of aircraft will need to be about 60000 hours, that is, 6.8 years of actual flight. To achieve the required weight to power ratio, 52% of the gross take-off weight of 340000 kg (750000 pounds) will need to be fuel, which explains the need to use light materials for about 50% of the structural weight. Polymeric materials are likely to be used for structural components, decorative fixtures and elastomers. It is essential that these structural materials maintain their high modulus, their dimensional stability and (in the case of elastomers) their flexibility over the entire service life of the aircraft.The temperature of the edges of the wings, tail fins and nose during climbing and descent can exceed 100°C even in subsonic flight, whereas supersonic aircraft can briefly exceed 300°C. Concorde increases in length by several tens of centimetres during flight, because of thermal expansion.

In other aerospace applications, the demands made on the materials can be even more extreme. Walls of missile engines using solid propellant [6] are exposed to gas flow under pressures that can reach 150 atmospheres, and to temperatures as high as 4000°C, with velocities of tens to hundreds of metres per second.

Rocket motors last only for periods of a few minutes and their requirements are very different from those for supersonic or subsonic aircraft. Space re-entry vehicles can reach temperatures as high as 450–500°C, or even higher, for a period of 5–10 min. 'Black box' flight recorders used in civil aircraft have to be able to withstand fires, with temperatures exceeding 1000°C for several minutes. Each of these materials has a different design specification and composition.

These high temperature materials may appear esoteric, but understanding how they fail is relevant to being able to produce any products with a very long life at lower temperatures. In order to understand the problems of durability in composites it is important to consider the physical and chemical processes causing changes in material properties.

4.2   Physical mechanisms

4.2.1 Thermal expansion coefficients

The thermal expansion coefficients of dry reinforced plastics are a function of the matrix, the fibres, the fibre volume fraction and the fibre orientation. Matrix expansion coefficients are far higher than those of inorganic fibres.

4.2.1.1 Fibre anisotropy

Individual glass fibres are isotropic, but carbon and the organic fibres (aramid and polyethylene) are anisotropic and therefore have different

Table 4.2 Thermal expansion coefficients of various materials (measured in the region of ambient temperature)

Material	Linear expansion coefficient $(K^{-1} \times 10^{-6})$
E glass fibres	6
C glass fibres	7.2
Carbon fibres, high strength (parallel to fibre axis)	−0.99
Carbon fibres, high strength (normal to fibre axis)	16.7
Kevlar 49 (aramid) fibres (parallel to fibre axis)	−2
Kevlar 49 (aramid) fibres (normal to fibre axis)	+59
Polycarbonate	68
Polyetherimide	49
Crosslinked epoxy resin, DGEBA type	60
50:50 v/v epoxy/solid glass beads	30
Glass/epoxy, unidirectional, (depends on glass content)	5–15
Typical polyester/chopped strand mat, 30% glass w/w	30
Typical polyester/woven roving laminate, 50% glass w/w	23
Carbon/epoxy, unidirectional, parallel to fibres[a]	0.4
Carbon/epoxy, unidirectional, transverse to fibres[a]	30
Carbon/PEEK, unidirectional, parallel to fibre	~0
Carbon/PEEK, unidirectional, transverse to fibres	29
Carbon steel	11
Stainless steel	16
Aluminium	24

[a] Laminates made from prepreg tape.
PEEK, polyetheretherketone.
DGBA, diglycidyl ether of bisphenol A.

linear expansion coefficients in the longitudinal and transverse directions. Carbon fibres possess sizeable positive transverse coefficients, α_t, but small and negative thermal expansion coefficients in the longitudinal (fibre) direction, α_l. This enables carbon fibres to be used in making zero expansion mouldings. It is also a potential disadvantage and according to Weatherhead [7] some brittle resins such as epoxy-novolacs can have such low strains to failure that the negative coefficient of carbon fibres causes interlaminar shear failure in their cross-plied laminates.

4.2.1.2 Mismatch between material constituents

The thermal expansion coefficients of organic polymer matrix resins are much larger than those of mineral fillers and of inorganic reinforcing fibres such as glass. Some figures are given in Table 4.2. and refer to data near to ambient temperatures.

The measurement temperature has a substantial effect, as has been

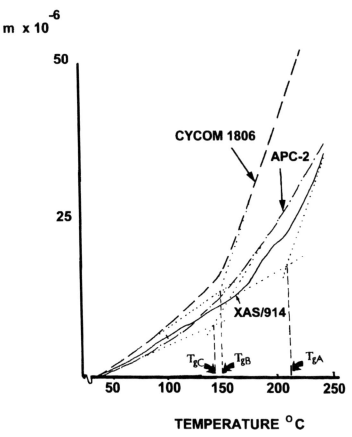

m x 10^{-6}

Figure 4.1 Thermal expansion of three different composite materials, measured through the thickness, using 2 mm thick specimens at 25°C, through the temperature range occupied by the T_g (from Stansfield [8])

demonstrated in Fig. 4.1 for three composite materials: APC-2 (carbon-PEEK (polyetheretherketone), ex ICI), Cycom 1806 and XAS/914 (both carbon/epoxy, supplied by Cyanamid and Ciba Geigy, respectively).

The figure shows the through-thickness expansion coefficients of ±45° laminates as a function of temperature [8]. The first effect of the large difference between the thermal expansion coefficients of the matrix and the fibres is that large stresses can develop within the composite during its manufacture. After gelation, the matrix contracts only slightly as a result of further crosslinking, but when the cure is finished, subsequent cooling of the moulding from the press or autoclave produces a much smaller reduction in the fibre dimensions than it does in the surrounding resin dimensions. The result is (i) a compressive stress on the fibres, as a result of which isolated filaments can become wavy; (ii) a tensile stress on the matrix, which according to Short and Summerscales [9] is given quantitatively by the expression:

$$\sigma_m = E_m[\Delta T(\alpha_m - \alpha_f) + \varepsilon_f]$$ [4.1]

where ΔT is the difference between the temperature of cure and the ambient temperature, α is the expansion coefficient, E is the Young's modulus in tension, and m and f refer to matrix and fibre, respectively. Sometimes σ_m exceeds the tensile strength of the resin and causes microcrack formation. These microcracks may be readily visible or they may be quite small and yet provide failure initiation sites in subsequent service life.

Hybrid composites have two different fibre reinforcements, with dissimilar thermal expansion coefficients. The reinforcements could be arranged in separate layers, or the two different fibres could be chopped and intimately mixed.

The axial and longitudinal expansion coefficients of unidirectional carbon fibre composites, α_l and α_t, respectively, can be computed [10,11] using equations [4.2] and [4.3], where the subscripts f, m, l, t refer to fibre, matrix, longitudinal and transverse, respectively:

$$\alpha_l = \frac{E_m \alpha_m V_m + E_{lf} \alpha_{lf} V_f}{E_m V_m + E_{lf} V_f}$$ [4.2]

$$\alpha_t = \alpha_m +$$

$$\frac{2(\alpha_{tf} - \alpha_m)V_f}{\upsilon_m(F - 1 - V_m) + (F - V_f) + \left(\frac{E_m}{E_{lf}}\right)(1 - \upsilon_{lf})(F - 1 - V_m)}$$ [4.3]

E, υ and V refer to Young's modulus, Poisson's ratio and volume fraction, respectively. The term F is a fibre packing factor, equal to the theoretical maximum volume fraction for a given fibre configuration, for example, 0.906 for a hexagonal array. The coefficient of linear thermal expansion for unidirectional laminates in the direction making an angle with the axis of fibre orientation is given in terms of the above mentioned coefficients by:

$$\alpha_\Theta = \alpha_t \sin^2 \Theta + \alpha_l \cos^2 \Theta$$ [4.4]

4.2.1.3 Angle-ply laminates

It can be shown theoretically, as well as by experiment, that multilayer angle-ply laminates have negative or zero thermal expansion coefficients at certain angles. The reader is referred to Zweben [12] for expressions for the linear thermal coefficients of expansion in two directions of a transversely isotropic composite (having two phases that are elastically and thermally isotropic).

4.2.1.4 Trapped moisture

Liou *et al.* [13] have measured the effect of absorbed moisture on the thermal expansion coefficients of carbon fibre epoxy laminates. Moisture lowers the transverse coefficient, but has little effect on α_l at room temperature. Failure in some epoxy laminates used to make multilayer PCBs has been traced to high z-axis thermal expansion stresses on marginal plated-through holes. Moisture is a factor, and complete drying can sometimes eliminate the problem [14].

4.2.1.5 Fillers

The addition of mineral fillers such as silica to a resin usually reduces the thermal expansion coefficient considerably. One electrical consequence of thermal expansion in particulate filled resins has been demonstrated by Strümpler *et al.* [15]. Epoxy resin filled with the hard filler, titanium diboride, TiB_2, show enormous but reversible changes in electrical resistivity (by eight orders of magnitude) on heating from ambient temperature to the cure temperature. This is a consequence of thermal expansion affecting interparticle contacts.

4.2.2 Thermal shock

A sudden large change in temperature, even if confined within the recommended use range, can be damaging to brittle materials. Repeated temperature cycling can act like a fatigue process.

When a material obeys linear elastic fracture mechanics, its tendency to undergo crack initiation or propagation as a result of mechanical stress can be assessed in terms of fracture toughness parameters, such as K_{Ic} (critical stress intensity factor) or G_{Ic} (strain energy release rate). Analogous parameters can be used with thermally induced cracking.

The majority of reinforced plastics have decidedly low thermal conductivities, in the range 0.15–$0.35\,\mathrm{W\,m^{-1}\,K^{-1}}$. Rapid changes in temperature can induce damage in particulate filled resins and fibre reinforced plastics.

The maximum use temperature does not have to be reached in order for cracking to occur and the size of the temperature change need not be very large. Sinks made of reinforced plastics containing fillers have frequently

Table 4.3 Thermal shock resistance indices

Property values	Carbon–carbon	Graphite	Steel
K (W m^{-1} K^{-1})	80	100	40
σ (MPa)	400	35	900
α ($\times 10^{-6}$ K^{-1})	1	2	12
E (GPa)	90	10	200
TSRI (kW m^{-1})	355	175	15
TSRI relative to steel	24	12	1

Note: for meaning of symbols, see text below.

suffered from thermal shock. Montes [16] reports cracking after only 300 cycles between ambient temperature and 79°C. This is hardly a severe test. In this case, the cause is repeated exposure to hot and cold water, but temperature is probably the main factor. Some research-grade 'heat-resistant' resins can withstand continuous use at high temperatures, but they are so brittle that they cannot withstand the volumetric shrinkage caused by removal from a hot press, without cracking.

Kubouchi *et al* [17] have devised a test method to assess thermal shock resistance, in which a sharply notched disk 60 mm in diameter and 10 mm thick is transferred from a hot oven to a cooling bath, while thermally insulated on both flat sides. A thermal shock fracture toughness term, K_{ict}, is obtained for the filled resin.

The thermal shock resistance in the above work was improved by increasing the volume fraction of a hard filler such as silicon carbide or nitride, but silica and glass beads had only a moderate beneficial effect and alumina had the opposite effect, whether the particles were spherical or angular. Fractography showed that the interface in alumina-filled epoxies was particularly weak.

One definition of thermal shock resistance is the TSRI, thermal shock resistance index, defined by

$$\text{TSRI} = k\sigma/\alpha E \qquad [4.5]$$

where k is the thermal conductivity, σ is the tensile strength, α is the linear coefficient of thermal expansion and E is the Young's modulus. A few values are given in Table 4.3.

4.2.2.1 *Temperature cycling of carbon fibre reinforced resins*

The following research examples illustrate the importance of changing temperatures.

Simpson *et al.* [18] subjected 0°/90° carbon fibre reinforced bismaleimide

(PMR-15) composites to 160 thermal cycles between 300°C and 20°C, and detected microcracks in the 90° ply, using penetrant X-ray radiography. The cracks were said to be caused because the transverse residual strain increased, while the transverse strength decreased. The authors related the cracking to the cure chemistry.

Morris and Cox [19] were concerned with simulating the temperature cycles experienced by space structures in low earth orbit. They subjected carbon/epoxy unidirectional laminates to repeated thermal cycling between hot and cold reservoirs at +138°C and −151°C, respectively. It was deduced from stereoscopic analysis that parts of the circumference of each fibre/resin interface were often microcracked, even after the first cycle. The cracks occurred where the distance between neighbouring fibres was greatest, and were related to large plastic strains in the resin near to the fibres. With further cycles, the cracks appeared where the fibres were closer together. The amount of damage was dependent on the resin/fibre system used, and some were more resistant than others. Motomiya *et al.* [20] have demonstrated that finite element analysis can be used to identify high stress areas in laminates at the design stage and hence improve durability.

4.2.3 Thermal spikes (see also Chapter 3)

A thermal spike is a special case of thermal shock. Sudden, large increases in temperature are followed soon afterwards by a similar decrease, or a decrease is followed by an increase. Spikes are best known in connection with fighter aircraft executing dives from a great height, or with components exposed to the reflected heat from the thrust of a vertical take-off aircraft, but the same effect could be achieved in any ordinary commercial products if they suffer sudden rapid heating for only a few minutes. Conventional dc electrical currents, alternating electric fields, microwave heating, chemical reactions in reactors, friction against a countersurface and damping during forced oscillations can cause rapid heating.

If a reinforced plastics component has absorbed moisture or solvent before the spike occurs, microcracking is possible by other means than straightforward thermal shock, provided that the matrix is brittle. The mechanism of the damage process can be very complex, but one important aspect is the available free volume, that is, the space within the matrix not taken up by the atoms themselves. The free volume is used to accommodate the molecules of absorbed liquid, but it is reduced when the temperature rises, because the resin's own molecules become mobile and expand. Consequently, there is no longer enough space available to accommodate the water or solvent molecules, nor is there time for them to escape by the slow process of diffusion, although they will inevitably move about and possibly cause further problems at the cooling stage.

One significant consequence of microcracking is that the material becomes capable of absorbing more moisture than before, although this will happen to a lesser extent anyway because of subtle changes in the resin morphology. Collings and Stone [21] showed that some laminates (which had already absorbed as much moisture as they normally could) took up even more moisture when stored in wet atmospheres after thermal spiking, that is, more than the equivalent unspiked specimens did. Sometimes matrix cracking occurred. Pritchard and Stansfield [22] compared the behaviour of two epoxy/carbon systems, one being brittle and the other toughened by a special additive. They also evaluated the tough PEEK/carbon system then known as APC-2 (ex ICI, later ex Fiberite). These were the same materials used in the work illustrated by Fig. 4.1. The tough epoxy performed better than the untoughened one, but the behaviour of the thermoplastic composite was far superior, because it had intrinsically a much lower moisture absorption, and also because of its inherent resistance to crack propagation. Figure 4.2 shows the difference between the three systems.

In the absence of spiking there would be little or no increase in moisture uptake with time for any of the three materials, because they were already very close to equilibrium before the spiking programme began. The spikes consisted of trips from ambient temperature to 135°C for 5 min, followed by cooling to ambient again. Between spikes, the samples were kept at 72°C, 96% RH for a week, then removed and cooled. It was reported that much of the damage occurred on cooling from the maximum temperature, when the resin contracted onto the moisture-filled free volume, causing internal stresses. Less damage was caused if the peak temperature of 135°C was held constant for a much longer period. It must be stressed that the tendency to undergo matrix cracking or interfacial weakening is dependent on the detailed nature of the materials and damage is by no means inevitable.

4.2.4 Physical ageing (see also Chapter 12)

Physical ageing [23–25] is a characteristic property of all glasses (a category of materials that includes many plastics – namely the amorphous or noncrystalline ones). It is a feature of all thermosets, and some thermoplastics, and is associated with the redistribution of internal stresses first introduced during processing at high temperature. The process leads to shrinkage and other physical property changes, with the possibility of fracture in extreme cases.

Physical ageing is a distinct process, separate from (but superimposed on) any other effects such as those of heat, hot air or any other environmental influence. It occurs to some extent even in a vacuum at ambient temperature, although the rate of the process is an inverse function of the difference

Weight % moisture absorbed

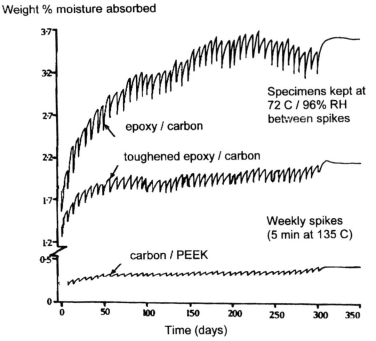

Figure 4.2 Graph showing the increase in moisture absorbed in the same three materials mentioned in Fig. 4.1, during a series of thermal spikes. They were first maintained at 72°C /96% RH (relative humidity) until saturation. They were then periodically removed from that environment, cooled to ambient temperature and then heated rapidly to 135°C, holding the temperature for 5 min. This spiking procedure was carried out every week. Each local peak represents one spike. After each spike, the samples were returned to the stated environment [22]

between the storage temperature and the T_g, so it will be very slow indeed at room temperature for most plastics.

4.2.4.1 Physical ageing in thermoplastics

Physical ageing has been extensively investigated in thermoplastics over the last 30 years. Thermoplastics will often creep, or show an increase in modulus and a reduction in dimensions; they will increase their T_g and reduce their impact strength. This is all as a consequence of being held at elevated temperatures for extended periods of time, during which the structure becomes slowly more organized and compact, and if possible, less amorphous. Similar changes are observed in thermosets. The crosslinking in thermosets means that some of these effects may not be so marked.

Physical ageing in thermoplastics can lead to crazing, a failure process normally produced simply by high tensile stress. Crazing involves the prior development of regions called crazes, resembling cracks bridged by several strands of oriented polymer.

Processing of thermoplastics will usually involve heating the material to such a high temperature that the viscosity of the melt allows free flow. When the melt enters the mould it is quickly cooled, and close-to-the-wall stresses can be generated, associated with shear of the cooling material by the internal molten stream. The whole of the material will be quenched into a thermodynamically non-equilibrium state, and it is the attempt by the system to achieve a new thermodynamic equilibrium at the temperature of operation that results in physical ageing. Release of these internal stresses can in principle lead to separation of the resin from the fibres (debonding), and to the generation of defects, but unless assisted by the ingress of solvent, the changes are not dramatic and only lead to a modest reduction in fatigue life and impact strength, caused by the formation of microdefects.

The closer the storage temperature to the T_g, the faster the ageing process and the greater the observed creep and the release of the internal stress. If an article is clamped or in some other way constrained, the released stress becomes concentrated at the fixing points and failure occurs. In composite materials, this stress redistribution is influenced by the presence of the reinforcing fibres and the main effect observed is debonding of the matrix from the interface. It should be stressed that loss of integrity in this way at a microscopic level does not necessarily lead to failure, but any debonded fibre interfaces reduce the reinforcing action of the fibres and they become more vulnerable to environmental degradation. If a debonded glass fibre composite is kept in an aggressive medium, for instance, attack on the glass by alkaline solvents or by acids can lead to embrittlement and fracture of the fibre more easily than would otherwise be the case. Physical ageing is therefore worth considering in assessing the long term durability of a component, even when thermal and thermo-oxidative degradation is not an issue. The practical implication is that reinforced plastics with glassy or amorphous matrix resins should not be kept indefinitely at temperatures near to their T_g, and certainly not while rigidly clamped.

4.2.4.2 Thermosets

Physical ageing of crosslinked thermosets has not been as extensively investigated as that of thermoplastics. One report [25] describes its effects on a TGDDM/DDS (tetraglycidyl diaminodiphenyl methane/4,4'-diaminodiphenyl sulfone) cured (epoxy) material containing 11% of diglycidyl orthophthalate and catalysed by boron trifluoride/ethylamine.

The resin was given a final cure at 177°C and postcured at 250°C for 16 h. The ultimate tensile strength, strain to break and static toughness of ($\pm45°$) symmetrically reinforced epoxy composites decreased along with increases in the sub $-T_g$ annealing time at 140°C. No weight loss was observed in materials annealed at 140°C and an enthalpy relaxation peak developed and moved progressively to higher temperatures; this was, as in the case of thermoplastics, 'thermoreversible'. A linear relationship was observed between the enthalpy relaxation process, associated with a loss of free volume in the material and the logarithmic ageing time. There was a parallel drop in the ultimate tensile strength.

Physical ageing leads to an increase in the compressive yield strength, which increases along with the glass transition temperature. Systems with a T_g less than about 160°C underwent ductile yielding at a creep rate of 0.002 in min^{-1} (0.05 mm min^{-1}), whereas the cured resin systems with T_g values greater than about 160°C underwent brittle fracture before reaching this creep rate. Similar behaviour was observed for several vinyl ester resin thermosetting systems.

Organic matrix composites have temperature-dependent moduli, and so when they are exposed to high temperatures they can undergo changes in their physical shape under relatively minor stresses. Cases exist of racing yachts, travelling as deck-cargo on freighters, being irreparably damaged as a result of physical ageing at the high deck temperatures experienced in the tropics.

Physical ageing is accompanied by a decrease in the mechanical damping or internal energy dissipation characteristics of the material and these changes correlate well with the observed loss in fatigue properties. The behaviour of resins in fibre reinforced composites is almost identical to that observed in ordinary thermoplastic materials and is most dramatic within the temperature window $(T_g - 50) \leq T \leq T_g$. If physical ageing readily occurs near the use temperature, the most obvious effect is that a tough resin can become brittle. The 'toughest' system should be identified by using specimens aged under actual service conditions. It is believed that the increased yield stress after physical ageing is the main factor contributing to the change in fracture toughness [26].

4.3   Chemical mechanisms

4.3.1   Thermal and thermo-oxidative degradation

4.3.1.1   Overview

The highest temperature at which an unreinforced polymer can safely be used for short and intermediate time periods is often determined by physical and mechanical considerations, that is, its tendency to lose mechanical and dimensional stability at elevated temperatures. Long term use at decidedly high temperatures, in contrast, is quite likely to be constrained by irreversible chemical degradation processes which are so slow that they do not affect applications involving shorter usage.

Chemical decomposition of the matrix resin does not usually begin abruptly at a precise temperature, but proceeds slowly at low temperatures, increasing in rate as the temperature is raised. The onset temperature will be dependent upon the molecular structure of the resin, filler and additives. Decomposition will be either heat triggered (thermal) or dependent on heat-plus-oxygen (thermo-oxidative), and is affected by the surface to volume ratio of the sample.

Good dimensional stability in reinforced polymer composites is no guarantee of chemical stability, so engineers and designers must be careful to ensure that the deleterious effects of thermal or thermo-oxidative degradation are completely understood before recommending the use of a particular material for a specific application.

4.3.1.2 Thermal degradation

Exposure to 'extreme temperatures' is a phrase needing definition. Whether or not degradation is observed depends upon a number of factors. A material may be able to retain its physical properties at elevated temperatures in an inert atmosphere, but can often be observed to fail at lower temperatures and over relatively short periods of time, when exposed to a reactive gas. Hot air can be considered such a gas for present purposes. High velocity gas flows and internal stresses also have profound effects on the effective lifetime of a component. A material may become ineffective in a particular application, long before it shows obvious visible evidence of chemical degradation, such as discoloration or cracking. Much effort has been expended in attempts to understand the detailed chemical processes that occur during degradation.

Polymer degradation is a process controlled by chemical kinetics and often involves a sequence of simple reactions working together, each of which has its own temperature dependence [27,28]. It is possible to correlate the rate at which degradation occurs with the temperature. The same extent of degradation may be observed for a material being exposed at a high temperature for a short time, as for a much longer period at a slightly lower temperature.

Purely thermal degradation of polymer molecules can be thought of in terms of two types of chemical reaction: (i) main chain scissions (i.e. breaking of covalent bonds along the chain backbone) and (ii) side group/ substituent reactions.

(i) *Main chain scission* processes can best be understood by considering two addition polymers: polymethylmethacrylate and polypropylene. Neither of these resins is generally considered to be a matrix for advanced reinforced plastics, but for chemical reasons they are particularly appropriate to illustrate the principles. At temperatures above 200°C, the backbone chain molecules of polymethylmethacrylate undergo scission, leading to the formation of macroradicals [29]. These macroradicals can 'unzip' by splitting off monomer molecules, one by one, in a process which is effectively the reverse of the polymerization reaction. In the case of polymethylmethacrylate, the whole polymer is converted back to monomer in 100% yield by this unzipping (properly termed a 'depropagation') process. The term ceiling temperature can be defined for many polymerization reactions and corresponds to the point at which the polymerization and depolymerization processes are in thermodynamic equilibrium. This temperature also defines the point at which unzipping begins to predominate.

In contrast, polypropylene does not undergo unzipping; instead, chain transfer chemistry dominates [28]. Through this alternative pathway, instead of monomer molecules, the chains break into much larger fragments (which can involve several monomer units). A complex series of degradation products containing fragments of between one to about 70 carbon atoms can be detected. It is important to note that chain scission reactions do not always lead to the production of volatile products, particularly at lower temperatures. The cleavage of molecular chains into non-volatile fragments results in a drop in molecular weight but the process happens at temperatures well below those at which volatile products are observed. This is important to keep in mind when faced with thermal analysis data as an indication of decomposition temperatures, as these techniques are often dependent on their being able to detect the loss of volatile products and can give the impression that nothing is happening if no volatiles are evolved. Thus whilst thermogravimetry – the measurement of weight change on heating – shows no weight loss from polypropylene in an nitrogen atmosphere until above 300°C, mechanically deleterious molecular weight changes can be detected some 50°C earlier [30].

(ii) *Side group reactions* can be grouped into two classes. In the first, neighbouring side groups react with one another; in the second, all or part of the side groups are split off. Side group reactions are exemplified by polyacrylonitrile, PAN [27]. Neighbouring nitrile groups react to form nitrogen-containing cyclic structures, which at higher temperatures evolve ammonia to form a graphite-like material. Industrially, this process is utilized in the production of carbon fibres.

Polyvinylchloride (PVC) provides the classic example of the complete loss of a substituent, namely a chlorine atom. The chlorine atom abstracts a neighbouring hydrogen atom to form hydrogen chloride (HCl). This has the potential to act as a catalyst for further degradation and can attack the fibres in a glass reinforced composite. The elimination of the acid leaves a double bond in the backbone and as sequences of these double bonds build up, discoloration of the material results.

4.3.1.3   Degradation of condensation polymers

All the reactions considered so far have been free radical in nature. However, many polymers, including most of the so-called engineering plastics, undergo chain scissions via non-radical processes. Thermoplastic polyesters, typified by PET (polyethylene terephthalate), invariably undergo backbone scission at the ester links. For PET the chain breakage occurs above about 280°C in the absence of air. The result is an immediate drop in molecular weight and sequential scission will generate short chain fragments that can volatilize from the polymer at this high temperature. Although the primary degradation pathway of PET is via this ester elimination reaction, secondary reactions also occur [29]. The major volatile product, acetaldehyde, is produced by a reaction which effectively stitches some of the chains back together again. Radical reactions will also occur at higher temperatures, generating methane and carbon monoxide gases. Nylon polymers (i.e. the polyamides) can undergo chain scission at the amide links by a similar process to the ester cleavage in polyesters, but in many respects the degradation is rather more complex (see for instance the reviews by Zimmerman [31], by Weber [32] and by Steppan *et al.* [33]). Thus nylon 6,6 shows a decrease in molecular weight at 300°C under nitrogen, but crosslinks at 330°C [34], with the concurrent formation of volatile cyclic products.

4.3.1.4   Thermooxidative degradation (heating in air)

It is rare for polymers to degrade chemically by heat alone. This is simply because in practice, very few polymers are used in the complete absence of air and the deleterious effect of oxygen must therefore be taken into account. Polymers will in general degrade faster by thermo-oxidation than they do in inert atmospheres by purely thermal degradation, because of the highly reactive nature of oxygen. The oxygen molecule is a diradical species and it can efficiently initiate damaging radical reactions. Virtually all hydrocarbon (i.e. carbon and hydrogen containing) polymers will undergo oxidation at a rate that depends on the chemical structure and physical state of the material. Radical chain reactions involving oxidation often autoaccelerate, and whilst oxidation may be slow in the initial stages, it can dramatically accelerate as the reaction proceeds.

Oxidation chemistry is complex, but three reactions tend to predominate [27]:

$$P\cdot + O_2 \rightarrow POO\cdot \qquad \text{(step 1)}$$
$$POO\cdot + PH \rightarrow POOH + P\cdot + O_2 \rightarrow POO\cdot \text{ etc.} \qquad \text{(step 2)}$$
$$POOH \rightarrow PO\cdot + HO\cdot \qquad \text{(step 3)}$$

In the first step, oxygen reacts with a polymer radical $P\cdot$ which has been formed by a purely thermal process, to generate a peroxide radical (step 1) which itself can react to form a hydroperoxide (step 2). Crucially, another polymer radical is generated as a result of this process, and in turn reacts with another oxygen molecule. The situation is exacerbated by the decomposition of the hydroperoxide species to generate alkyl and hydroxyl radicals (step 3), which in turn generate further polymer radicals $P\cdot$. Thus the number of reactive radicals increases steadily and the rate of oxidation increases proportionately.

The susceptibility of a resin to oxidation depends on the presence of a hydrogen atom that is readily removable in the initiation step. Thus polypropylene is prone to oxidation because the hydrogen atom adjacent to the methyl group is easily removed. On the other hand, polytetrafluoroethylene (PTFE), with no hydrogen present at all, is exceedingly oxidation resistant! Of course, if oxidation simply involved the addition of an oxygen molecule to the chain backbone its effects would not necessarily be seen as deleterious. Often however, a consequence of oxidation is scission of the chain backbone and a drop in mechanical properties occurs.

This effect can be dramatic, even at relatively low temperatures. For example, polypropylene shows [35] a drop in molecular weight by a factor of about 4 in only 300h at 90°C whilst the elongation at break falls by a factor of 10.

Chain scission occurs by a fairly unusual process, the so-called α, β-bond scission, for example, for polypropylene:

$$\sim CH_2 - \underset{\underset{CH_3}{|}}{\overset{\overset{O\cdot}{|}}{C}} - CH_2 - \underset{\underset{CH_3}{|}}{CH} \sim \quad \longrightarrow \quad \sim CH_2 \overset{\overset{O}{\|}}{\underset{\underset{CH_3}{|}}{C}}$$

$$+$$

$$\cdot CH_2 \underset{\underset{CH_3}{|}}{CH} \sim$$

Although such scissions are infrequent, their effect is pronounced, as only one scission per chain is enough to affect the mechanical properties of the matrix.

Other polymers with similar, aliphatic, carbon sequences will also be susceptible to oxidation. Thus, the aliphatic polyamides such as nylon 6 and nylon 6,6 will oxidize readily. The oxidation of the nylon polymers is characterized by a drop in molecular weight and by discoloration. Oxidation proceeds in a manner similar to that outlined for polypropylene, with the CH_2 unit nearest the amide nitrogen most susceptible to attack [32]. Oxidation is a serious problem for polyamides; as mentioned earlier, nylon 6,6 will embrittle in two years [1] at 70°C. Almost all commercial polyamide formulations therefore include antioxidants, usually based on copper compounds. Similarly, the aliphatic sequences in PET are readily attacked by oxygen and a drop in molecular weight is again observed [36]. Some gel formation is also observed at higher temperatures.

4.3.2 Thermal characteristics of specific resins

4.3.2.1 Polyester resin property changes on heating

Consider unsaturated polyesters as illustrations of some of the earlier discussed principles. They are not outstanding in their thermal stability, but they are extensively used in composite fabrication and have heat distortion temperatures in the range 55–120°C depending on molecular structure. Above this temperature a significant decrease in the modulus is noticed and accelerated creep is observed. Resins cannot be used for load bearing applications within 20–30°C of the T_g, or even 50°C in situations involving substantial solvent or moisture pickup, and standard polyesters are not usually used above about 100°C, even though they do not decompose chemically until around 150°C.

The dc electrical resistivity and loss tangent of one unreinforced orthophthalic polyester resin [37] over the temperature range 70–170°C [4] are shown in Fig. 4.3, with the dc electrical resistivity changing very gradually when the temperature is held constant at 150°C for several days.

This change may be due to chemical decomposition or evaporation of residual styrene monomer or a combination of factors. The resin degrades at the upper end of the temperature range, but an increase in matrix toughness was observed over the range 10–40°C for a similar polyester [38] (Fig. 4.4). The increase was accompanied by a brittle-to-ductile transition, with changes in the fracture surface characteristics.

4.3.2.2 Epoxy resins

Epoxy resin ceiling temperatures depend crucially on both the epoxy resin and the hardener. The difunctional DGEBA (diglycidyl ether of bisphenol A) epoxy resin system is the most common, producing high heat distortion

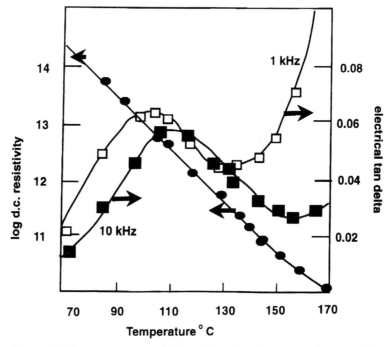

Figure 4.3 Temperature sensitivity of the dc volume resistivity and the dielectric loss tangent of a polyester resin [37]

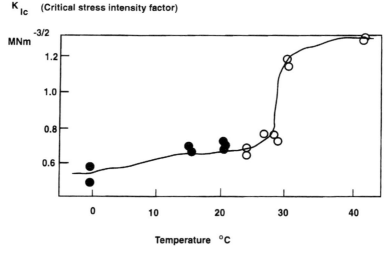

Figure 4.4 Temperature sensitivity of the fracture toughness (K_{Ic}) of a polyester resin [38]. ●, Smooth fracture surface; ○, rough fracture surface. Strain rate, $1\,mm\,min^{-1}$

temperatures with anhydrides [7] such as methyl nadic (218°C) or pyromellitic (290°C) or nadic (305°C).

A high heat distortion temperature can often be achieved at the cost of certain disadvantages, notably brittleness and it is then desirable to use a lower heat distortion temperature system with more ductile characteristics. Some high temperature resins are difficult to process and the choice of fabrication procedures can be limited. Aliphatic amine curing agents with difunctional DGEBA epoxy are restricted to temperatures below 130°C when used in critical load-bearing structures. Epoxy resins based on glycidyl amine resins, such as the tetrafunctional TGDDM can be used at 175°C in military aircraft when cured with an aromatic hardener such as DDS or with an aromatic anhydride. Postcure at 200°C can further increase the T_g to 240°C. Trifunctional epoxy resins have quoted T_g values [39] between 261°C and 307°C and appreciable weight loss is not observed below 370°C, whereas in TGDDM resins the same weight change is observed [40] at 270°C. The cure cycle used has a critical effect on the physical properties. Maximum density of the matrix and hence greatest resistance to water and gas absorption is observed with low cure temperatures for anhydride cured materials [26]. Postcure of epoxy resins will in general improve the mechanical properties, with a few exceptions (sometimes brittleness increases), and there is a close correspondence between the development of the maximum mechanical properties and the onset of degradation of the cured matrix [41,42], signalled by a deepening of the colour.

Thermoplastics are sometimes added to epoxy resins. Thermoplastic-modified epoxy resins [43,44] based on tri- and difunctional epoxy resins cured with DDS and blended with polyethersulfone form the basis for the matrix material in a composite used for the Boeing 777 aircraft. The incorporation of the thermoplastic helps the processing characteristics and also improves the mechanical properties, notably the toughness. The thermoplastic is able to phase separate from the epoxy phase and acts as a reinforcement for the epoxy matrix, enhancing its high temperature properties. The maximum use temperatures of all these resins will typically be 30 to 50 degrees lower than the cited T_g, assuming the same cure schedule.

4.3.2.3 Phenolic resins

Phenolic resins are the oldest fully synthetic polymer matrix materials ever made, being commercialized around 1910. But they have only been taken seriously as conventional laminating resins for structural applications fairly recently, being promoted especially for their favourable fire and heat resistance. The T_g typically ranges from 200°C to over 300°C, which means it is not detectable, being above the chemical decomposition temperature. Loss of rigidity on heating is therefore less important than chemical degradation, as detected by weight changes on heating or by loss of strength at elevated temperatures.

The detailed chemical structure of the resin matrix depends on the synthesis method and altering the catalyst can change the properties significantly [45,46].

Phenolics have good heat resistant properties up to 175°C or higher, but there is still usually a progressive reduction in modulus of typically 30–45% of the ambient temperature values, and permanent degradation can occur at 200°C, although adverse effects take some time to appear. A few high temperature grades can be used at 250°C. It is important that cure is taken to completion, using up the hardener, because if any is left unused (normally it includes hexamethylene tetramine) then ammonia is formed during the cure process [47]. Ammonia is capable of attacking glass fibres and therefore leads to a loss of mechanical properties. Special glasses can be used to minimize this.

4.3.2.4 Polyimides and bismaleimides

Polyimides can be either thermosetting or thermoplastic. They have been developed since the late 1950s, and after a great deal of development they are now in use for a number of applications. The best polyimides have probably the most outstanding overall heat resistance of all the commercially developed matrix resins, although competitors with still higher temperature limits are emerging. Some of the condensation polyimides suffer no weight loss at all below 450°C, whether in air or an inert atmosphere. But weight loss by itself is not an infallible guide to thermal stability, as already mentioned. Another cause of doubt about laboratory data on high temperature resins is the use of predried laboratory specimens. This represents an optimistic scenario, as moisture is thought to accelerate some degradation reactions. Consequently, long term practical use temperatures for some polyimide grades can be well below 300°C.

The relative stability of condensation polyimides depends on the number of aromatic rings in the diamine, decreasing as the number of rings increases. Chemical linkages that are *para* are more stable than *meta* ones. Other categories of polyimide have much lower continuous use temperatures, based on the results of flexural strength retention studies at various constant temperatures. Some addition polyimides, when laminated with glass cloth, lose flexural strength gradually at any temperature over 180°C [48].

Bismaleimides have excellent thermal stability, comparable with the better condensation polyimides mentioned above, although they are also intrinsically brittle. Weight losses frequently start at 350°C or higher. As with many other resins, the use of solvent varnishes to prepare prepregs can result in the tenacious retention of residual solvent in the cured composite, with the result that the compressive modulus at 300°C is less than 40% of its ambient temperature value [49]. This seriously reduces the maximum temperature of use.

A considerable amount of research has been carried out in recent years into the design of matrix materials for the next generation of supersonic civil transports and much of it has focused on amide, imide and related structures. Among resins studied are the polystyrylpyridine (PSP) family, the polyamide-imides, polyetherimides, polyetherketones, and for very high temperatures, the polyphenylquinoxalines, the polybenzimidazoles and the aromatic polyesters [50,51]. Some generalizations emerge [52]:

- Only the strongest chemical bonds should be used in the backbone structure.
- The structure must allow no easy pathways for rearrangement.
- There should be a maximum use of resonance stabilization.
- All ring structures should have normal bond angles.
- Multiple bonding should be utilized as much as possible.

Seldom is the first of these principles the limiting one. Most bonds would be strong enough to give adequate stability, provided that the only possible degradation mechanism required bond cleavage as the first step and was not followed by a chain reaction or elimination of simple fragments. Nearly all systems break down because less energetic mechanisms are operative. Resonance stabilization is advantageous because more energy is needed for bond rupture and if bond angles are normal then once a bond is broken the natural angles of the structure hold the atoms in close proximity, so that bond healing is possible once the excess energy has been dissipated through the molecule. In addition, abnormal bond angles usually imply that the ring is strained and unstable. The multiple bonding principle is designed to ensure that a chain cannot be broken by rupture of a single bond alone – the concept of a ladder polymer is shown in Fig. 4.5.

It should be stressed that these principles all relate to a perfect unit in isolation. In practice, there are interactions within and between molecules that have to be taken into account and a polymer seldom has its perfect idealized structure. In high temperature polymers, aromatic structures are often incorporated into the backbone in order to achieve greater thermal stability.

A number of observations can be made. *para*-Substituted polyphenylene appears to be more stable than the *meta*-substituted equivalent. Similarly *para–para* polyamides are more stable than *meta–meta*, which in turn are more stable than *ortho–ortho*. These polymers are stable at temperatures in excess of 200°C and the *para–para* ones can operate above 500°C.

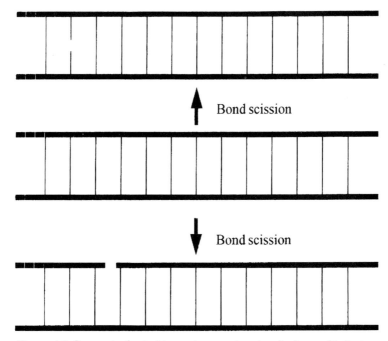

Figure 4.5 Concept of a ladder polymer, showing its insensitivity to bond scission

Fluorination of the ring leads to materials that are at first less stable in oxidizing atmospheres, but at higher temperatures they perform better in inert atmospheres. A change of mechanism appears to take place, involving the elimination of fluorine ions with an increase in the formation of carbonized residue.

The effect of the linking groups in terms of thermal stability indicates that in an inert atmosphere a single C—C bond is more stable than $(>)$ —O— $>$ —CH$_2$— $>$ —CF$_2$CF$_2$— $>$ —C≡C—, —C=C— $>$ —CH$_2$CH$_2$— $>$ —S— $>$ —COO, —CONH—$>$ —CF$_2$— $>$ —NH— $>$ —OCOO—. This order does however not follow for an oxidizing atmosphere. However in the system —[C$_6$H$_4$—X—]$_n$ it is found that —O— $>$ SO$_2$— $>$ CH$_2$— CH$_2$CH$_2$ $>$ C—(CH$_3$)$_2$.

Structure I

Another important structure to be found in high temperature systems is —[(heterocyclic ring—)—C$_6$H$_4$—X—C$_6$H$_4$—] (see Structure I). Isothermal

Table 4.4 Stability of some heterocyclic containing
systems [53]

Compound	Maximum stability temperature (°C)
Dibenzothiophene	545
Quinoline	510–535
Dibenzofuran	518
2,2'-Biquinoline	>482
2,2'-Dipyridine	482
2,2'-Bithiophene	474

weight loss measurements in air at 400°C show that in the case of the polypyromellitimides, the stability is as follows [53]: single bond > —S— > —SO$_2$— > —CH$_2$— > —CO— > —SO—, —O—. On a related system the order observed was single bond > —S— > —O— > —NHCO— > —COO— > —CH$_2$— > —C(CH$_3$)$_2$—. The effect of preparative conditions can play an important part and could account for some of the differences observed [54]. However it is clear that these systems are probably the most stable available for matrix formation. The introduction of heterocyclic groups can also have a profound effect on the thermal stability; see Table 4.4.

It has sometimes been proposed that ladder type structures would enhance the stability of these materials.Their oxidative stability is in fact poorer than that of the linear materials and their processing characteristics make these materials difficult to handle. Incorporation of polymeric anti-oxidants into the material have a significant effect in reducing the stress relaxation rates and improving retention of physical properties. In this context, oligomeric aryl amines are most effective.

4.3.2.5   Other resins

Polycyanurates [55] exhibit service stability at 235°C in air from 800–1200 h while short term exposure to high temperature in air indicates high stability, with degradation onset temperatures greater than 400°C. It appears that long term exposure in air at temperatures in excess of 200°C causes polycyanurates to fail prematurely by outgassing, although this was not observed in ageing tests carried out at temperatures below 200°C. Swelling and blistering of the resin surface was noted at relatively low weight loss values, indicating that a degradation mechanism other than chain scission is a limiting factor in long term service at temperatures within approximately 50°C of the softening temperature. At temperatures above 180°C, thermogravimetric analysis revealed that hydrolysis of the cyanate ester occurs, liberating carbon dioxide.

Furane resins, derived from oats, are worthy of mention because of their outstanding chemical resistance, which makes them good candidates for use as liner cements in chemical vessels and pipes. A combination of chemical and heat resistance is especially useful – silicone resins have thermal stability up to 200–370°C but they are susceptible to steam and strong alkali.

4.3.2.6 Carbon–carbon composites

Although not usually classified as such, carbon is a crosslinked network polymer. Carbon composites are usually produced by the pyrolysis of a resin–carbon fibre composite, and phenolic novolak–carbon fibre systems would be typical examples of a starting material. The resultant materials have thermal properties similar to those of graphite, which makes them attractive as reinforcing materials in high temperature applications. The specific heat is high compared with metals and increases with temperature. A high heat of sublimation and a very high enthalpy are all valuable properties [56]. These properties are valuable where large quantities of heat need to be absorbed without significant loss of structural properties. Consequently, carbon–carbon composites have good thermal shock resistance.

Oxidation in still air occurs above 500°C, whilst steam and carbon dioxide, with threshold temperatures of 700°C and 900°C are other oxidizing agents. Small quantities of metallic impurities such as sodium, potassium, vanadium or copper will increase the rate of oxidation by as much as six-fold, so purifying is beneficial. Probably the biggest volume application of carbon–carbon is as brake discs for aircraft such as Concorde and other high performance civil and military aircraft where these composites have replaced sintered metals by virtue of their low density and superior thermal performance.

4.3.3 Fire resistance

4.3.3.1 Most resins burn readily

In the early years of reinforced plastics, their use in public applications such as buildings and transport engendered much enthusiasm among designers and engineers keen to utilize the unique mechanical and optical properties of these exciting new materials, but their fire performance was strangely neglected. After several disasters it became clear that fire performance was as much an essential design criterion as was modulus, yield stress or clarity.

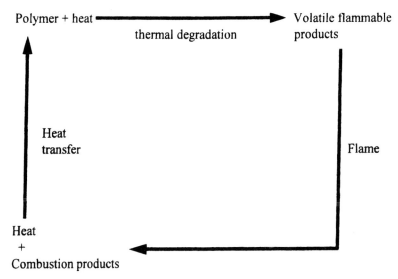

Figure 4.6 The fire cycle

Strictly speaking, the resins themselves do not burn; rather it is the highly volatile products of degradation that diffuse out from the polymer and burn in the gas phase immediately above the polymer surface. The heat generated promotes further degradation of the polymer, and the cycle continues, as shown in Fig. 4.6.

If strong heating leads to the evolution of flammable volatiles, therefore, there will be implications for the acceptability of the material as a fire risk. Evolution of black smoke or highly flammable gases is extremely undesirable. The toxicity of the fumes in plastics fires is said to be less important than smoke density in causing fatalities.

4.3.3.2 Oxygen depletion

The complex chemistry occurring in the combustion zone and in the solid polymer during a fire is far from fully understood. It is clear however that although in the early stages of a fire there is a plentiful supply of oxygen, ambient oxygen levels can quickly fall. This, coupled with the limited diffusion of oxygen into the polymer matrix means that the subsequent degradation is predominantly thermal, rather than thermo-oxidative.

In the oxygen depleted environment of a fire, combustion is often incomplete, and large quantities of carbon monoxide are produced. Carbon monoxide is held responsible for most fire deaths and smoke is considered the other major factor in preventing escape. Toxic gases such as hydrogen cyanide are invariably produced upon the thermal decomposition of nitrogen containing polymers such as nylons and polyurethanes; they are also evolved from burning natural products.

Resins heated above T_g or T_m will rapidly lose dimensional and mechanical integrity. The poor thermal conductivity of most resins (typically 0.1–$0.2\,W\,m^{-1}\,K^{-1}$) and consequently of most reinforced plastics means that under high radiant heat, high surface temperatures are soon reached and degradation becomes rapid. As we have seen, most organic resins will rapidly degrade with the evolution of volatiles at temperatures typically between 300–400°C. These compounds, typically hydrocarbons or oxyhydrocarbons, burn readily in air.

4.3.3.3   Fire retardant grades involving additives

Attempts to reduce or eliminate the tendency to burn can involve careful choice of resin (see later) and of the nature and physical form of the reinforcement, but the main emphasis is usually on adding special additives to the resin composition [57].

Additives come in many types, such as:

1   Mixtures that release hydrogen chloride or bromide during a fire. The HCl and HBr work by interfering with the radical chain reactions in the combustion zone, effectively quenching the flame. The HCl or HBr must be liberated at an appropriate point, if it comes too soon, combustion will be unaffected. It is thus important to ensure that the additives decompose over an appropriate temperature range. Synergists such as antimony compounds are often used with chlorinated or brominated additives as they function by extending the range over which flame quenching occurs. This 'right place – right time' philosophy favours the use of chemically bonded additives introduced during the polymerization. Physically mixed additives can leach from the matrix in use or volatilize at low temperatures in the early stages of a fire without decomposition. The loadings required can be high and can adversely affect mechanical properties. Chlorinated or brominated additives are becoming increasingly unpopular anyway, as the gases evolved are corrosive and will quickly damage electronic components. It must also be remembered that most polymer waste is disposed of by incineration or landfilling. The fate of halogenated additives in such situations has raised environmental concerns.

2   Aluminium, magnesium or other hydroxides, which release water on heating. The presence of these compounds in large enough quantities is very effective, but it places a strict limit on the temperature at which the products can be processed or used without provoking the formation of

water vapour, which in the case of aluminium hydroxide occurs at around 200°C, i.e. low enough to exclude many thermoplastics processed at higher temperatures. The water vapour has the important effect of diluting the combustion gases and its production is an endothermic (cooling) process.

3   Phosphorus compounds.

4   Tin, molybdenum, boron and to a lesser extent iron compounds.

Fire retardant resin systems can perform quite well in terms of fire and smoke resistance, although low flammability does not necessarily imply low smoke. Unfortunately there can be adverse effects on processing, mechanical properties or chemical resistance.

Intrinsically non-flammable polymers are few, but phenolic resins have a good reputation both in fire and smoke performance, which has resulted in their becoming increasingly favoured for reinforced plastics structures, for example, underground transport, where such concerns are greatest. Polyether ether ketone (PEEK) is also a low fire and smoke polymer. Unsaturated polyesters, vinyl esters and epoxy resins burn readily, but modified versions are available with improved behaviour. For example, both bromine and chlorine are used extensively in the form of chlorendic (HET) acid, tetrachlorophthalic anhydride (TCPA) and tetrabromophthalic anhydride (TBPA) which can be reacted into the polyester in small quantities and can act as permanent (non-migrating) flame retardants.

4.3.3.4 Smoke

In addition to toxic gases (and to be fair, practically all organic materials, including naturally occurring ones, generate carbon monoxide), certain resins are associated with the production of large quantities of dark smoke. The smoke is itself rarely toxic, but its choking, disorientating qualities impede escape. It is the denseness of the smoke which most clearly differentiates synthetic resins from natural organic materials such as wood. Smoke is a sign of incomplete combustion and is associated with a high carbon content. For example, polystyrene has a high carbon:hydrogen ratio (14:1 by mass) and decomposes to form monomer and so on, which often burns incompletely and generates large quantities of sooty smoke, consisting largely of unburnt carbon. In contrast polypropylene (carbon:hydrogen ratio = 6:1 by mass) burns very cleanly.

4.3.3.5 Comparing the performance of materials by fire testing

Short of building a full scale structure and setting it alight, it can be difficult to define a material's fire performance. In practice we rely on small-scale

Table 4.5 Some fire and combustion tests for plastics

Test	Description
BS 2782: –	
Method 141	Determination of flammability by oxygen index
Method 143	Determination of flammability temperature of materials
BS 6401	Method for measurement of the specific optical density of smoke generated by materials
BS 6853	Fire precautions in the design and construction of railway passenger rolling stock
NES 711	Determination of the smoke index of products of combustion from small specimens of materials
NES 713	Determination of toxicity index of the products of combustion from small specimens of materials
NES 715	Determination of the temperature index of small specimens of materials
ASTM D635	Rate of burning and/or extent and time of burning of self-supporting plastics in a horizontal position
ASTM E 662	Specific optical density of smoke generated by solid materials
ASTM D 1929	Ignition properties of plastics
ISO 3795	Road vehicles: determination of burning behaviour of interior materials
UL 94	Tests for flammability of plastics materials. Horizontal and vertical burning tests.
ATS 1000.001	Airbus Industry Test Specification for fire, smoke and toxicity

tests, although the flame spread rate and mass loss rates are very different from those performed in special full size rooms, as in ISO 9705, and the data obtained are therefore controversial [58].

There are many small scale tests, often specific to one application. In this section we can only examine a very brief selection. Further details are available in the specialist literature [58–61] and a few standard tests are listed in Table 4.5.

Possibly the best known, most widely used and most criticized test is the limiting oxygen index (LOI) measurement. The LOI is defined as the percentage of oxygen in a nitrogen/oxygen mixture that just sustains the combustion of the polymer. A 'candle' of polymer is ignited in a high oxygen atmosphere and the proportion of oxygen in the test atmosphere progressively decreased until the combustion is extinguished. The significance of this test is that air contains approximately 21% oxygen and thus a polymer that requires more than this value to sustain combustion (i.e. which has an LOI greater than 21) should be self-extinguishing under normal atmospheric conditions. Note that 'self-extinguishing' is not the same as 'non-inflammable' and a polymer with an LOI in excess of 21 may combust readily under the forced draught conditions found in a fire. Graphite carbon, for instance, has an LOI in excess of 40 and thus it is highly self-extinguishing and difficult to ignite, as anyone with experience of barbecue briquettes will know. However, once forced with fire lighters the carbon will burn quite happily. The results from LOI tests are dependent on temperature, and nearby burning materials in a real fire usually ensure that the plastics material being considered is not at ambient temperature. The LOI test is also dependent on specimen geometry.

Whilst LOI measurements are an indication of the ignitability of a material, flame spread tests are a measure of how rapidly a fire will take hold after ignition occurs. Flame spread tests can be performed horizontally or vertically (or any angle in between) depending on the application they are designed to mimic. For example, the flame spread test, ASTM E84-84 measures the flame spread on building materials mounted to simulate the underside of a ceiling and exposed to an 88 kW flame source.

Char yield has often been overlooked in the past, and when considered it has often been measured in a rather *ad hoc* fashion (for example by simply burning a sample in a crucible over a bunsen flame) or as an 'extra' after some other experiment. Combined smoke and char measurements have been made on purpose-built combustion rigs [62]. Controlled pyrolyses under air or an inert atmosphere are more reproducible and are increasingly favoured, although the small sample size is a limiting factor [63].

Char yield is increasingly regarded as an important measure of fire performance. The formation of large quantities of char (from organic resins, invariably carbonaceous) has several advantages. First, conversion of resin to inert char rather than volatile flammable products reduces the fuel available to feed the fire. Furthermore, a carbonaceous char will be self-extinguishing. In addition, a surface char seals the polymer surface, preventing the escape of volatile materials.

Smoke obscuration tests are performed most often by burning a sample in a sealed chamber, with the smoke density measured optically by measuring the amount of obscuration of a light beam. Such tests can be misleading. For example if the light attenuation is determined over a horizontal optical path (ASTM D2843-70) the results can be influenced by the tendency of the smoke to form layers of different density. The NBS smoke chamber (ASTM E662) is designed to overcome this problem by measuring across a vertical light path. Tests designed to measure the density of smoke in a vent-flue are also available. In all cases the amount of light scattering is affected by smoke particle size.

Table 4.6 Limiting oxygen index data for common polymer systems [60]

Polymer	Limiting oxygen index
Polypropylene	17–29
Nylon 6,6	21–38
Polyether sulfone	37–42
Phenolic resin	18–66
Epoxy resin	18–49
Unsaturated polyester resin	20–60
Polytetrafluoroethylene, PTFE	95
Polyethylene terephthalate, PET	20–40

Cone calorimetry is rapidly becoming one of the standard tools of fire testing [64,65]. Originally designed to measure heat release rates, the cone calorimeter is built around a truncated cone-shaped electric heater (hence its name) capable of generating heat fluxes between 0 and $100\,kW\,m^{-2}$. The instrument has been developed beyond the original concept, and can also be used to measure ignition times, volatile material fluxes (a measure of the fuel load) and smoke generation, whilst allowing easy access to the gaseous combustion products for analysis. Cone calorimetry is more reliable than conventional methods [66] of smoke measurement and has been widely used in reinforced plastics research, for example, for the study of reinforced phenolic, polyester and vinyl ester matrices [67]. For example, peaks in the rate of heat release have been correlated with composite structure and the contribution to increased flammability of delamination has been demonstrated.

The problem with all such tests, however, is their interpretation and comparison. Physical form can influence the test (for instance the sample may readily deform out of the flame region), and some tests are more demanding than others. The LOI test has a flaw in that the sample is burning from the top down, and the heat flow into the polymer below the flame is necessarily limited. The data in Table 4.6 also clearly show that the range of measured values can vary dramatically depending on the exact sample form, test protocol and sample chemistry.

Trying to crosscorrelate data from different tests or to use the data to define a 'good' polymer unambiguously is impossible. Is a polymer system with a low LOI (unfavourable) but low smoke production (favourable) better than a system with a high LOI (favourable) but high smoke production (unfavourable)? And what about a material with high smoke production (unfavourable) but a low horizontal flame spread rate (favourable)? The choice can involve a compromise between conflicting requirements and the relevance of many tests to real life remains contentious.

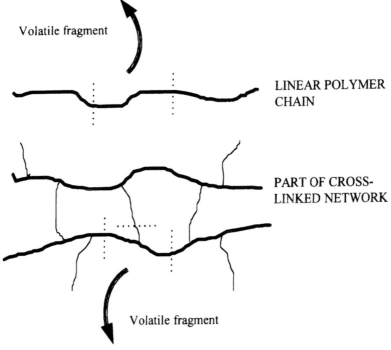

Volatile fragment

LINEAR POLYMER
CHAIN

PART OF CROSS-
LINKED NETWORK

Volatile fragment

Figure 4.7 Minimum number of bond scissions required to release a volatile fragment from a linear polymer is two. Three scissions are required for a highly crosslinked network

4.3.3.6 Choice of resin

Highly crosslinked thermosetting resins in general have a better fire resistance than that of most unreinforced thermoplastics, although some are prone to black smoke evolution. Not only do the thermosets possess superior dimensional stability, since the highly crosslinked nature of the chains does not allow them to flow, but crosslinking also makes the evolution of small, volatile, flammable decomposition products much more difficult. A crosslinked network requires the scission of three appropriately placed bonds to generate a mobile fragment. Only two bond scissions are required in a linear thermoplastic backbone (Fig. 4.7). (This is the same philosophy as underlies the development of the 'ladder' polymers described earlier.)

Furthermore, the ability of small molecules to diffuse out from a crosslinked network is severely restricted because of the very limited chain mobility. In contrast, the mobility in a polymer melt is many times greater. Oxidatively the crosslinked resins are often more stable, as the diffusion of oxygen into the matrix is similarly restricted.

So what can be done to improve the fire performance of the resin matrix? From a mechanical standpoint we must ensure that dimensional stability is maintained, even at elevated temperatures. On this basis thermosets are preferred to thermoplastics and where thermoplastics are used, high T_g and T_m values and fibrous reinforcement are preferable.

The nature of the reinforcement is important, as a three-dimensional reinforcement may prevent delamination and the exposure of fresh resin-rich layers; woven roving is less likely to allow structural collapse than chopped strand mat. Fibre bundles have been reported to act as wicks, bringing fresh resin into the combustion zone, although they can also have the reverse effect [68]. Carbon fibres can conduct heat away from the combustion zone, limiting the decomposition of the polymer.

Chemically, the strategies are clear from consideration of the fire cycle. We can modify the degradation chemistry, reducing volatile production, or quench the combustion process. Alternatively we can restrict the heat flow back into the polymer.

The ideal approach to reducing volatile yield is to use an inherently stable matrix polymer and reference has already been made to the chemical structures that confer maximum thermal stability. Alternatively, there are additives which function in the condensed phase and modify the degradation pathway. Phosphorus-based additives, usually working through a phosphoric acid intermediate, are popular additives, particularly for polyurethanes. They act predominantly in the condensed phase, increasing char yield and reducing volatile production.

4.3.3.7 Char-promoting additives and intumescent foams

Such are the benefits of forming a layer of char at the surface of combusting polymers that additive systems are now available even for resins that are not inherently char forming. These additive systems are complex but have the advantage that they can be tailored to provide a highly blown, foamed char. These intumescent chars have very low thermal conductivities and are thus particularly effective at breaking the fire cycle at the heat transfer step. The degree of intumescence can be remarkable – a 1 mm intumescent fire resistant coating for structural steelwork may produce 50 mm of char. The additive system has to be chosen carefully. Usually three components are used. Two react together to form the char, whilst the decomposition of the third component produces a gas, usually ammonia, which acts as the blowing agent [69]. It is impossible to foam a solid char, so the system must be chosen such as to ensure that the generation of the blowing agent occurs at a point when the initial product of the reaction between the char forming agents is still mobile. Except in the form of coatings, intumescent formulations are not widely applicable in reinforced plastics, because the formation of the char disrupts the matrix/reinforcement interface.

A rather simpler approach has been successfully tried with nylon 6,6. The simple addition of polyvinyl alcohol (a polymer capable of char formation) increased char yield five-fold and decreased the rate of heat release more than two-fold [70].

4.3.3.8   Heat shields

'Black Box' in-flight recorders are usually constructed from a metallic box with an external covering of polymeric material which has to insulate the contents from temperatures of 60°C for polyester and 350°C for metal tape for periods of up to 30 min. In a typical structure the external cover may be a rubber fabric material with the binder based on silicone polymers, inside this may be an intumescing cellulosic material, impregnated with products from the condensation of amines with aliphatic aldehydes [71]. Mixtures of chlorinated and sulfonated polyethylene have been used as heat shield materials. These materials contain antimony oxide as a synergist for halogen-containing flame retardants and also silicon oxide as an active filler. Limited oxygen indices in excess of 40 have been reported and over a 15 min period they are stable to approximately 200°C. The resins are crosslinked, and this increases the char formation, with a consequent increase in fire stability as a carbonaceous layer is formed early on in the fire to act as a heat shield and reduce volatilization. The construction of heat shield materials has been improved over a number of years and a variety of different systems have been formulated, depending on whether they are used with rockets or with shuttle re-entry vehicles. In order to cope with the thermal stresses these systems tend to be based on a crosslinked elastomer base, highly filled with silicon dioxide, carbon black, asbestos, graphite or phenolic microspheres. The use of organic fillers allows the generation of very lightweight materials and there are patents on the use of cork, phenolic resins and rubber chip as fillers. The main feature of all these systems is that the degradation forms a high char yield and this acts as a thermal insulator, traps gas and generates a stable resistant coating. Efforts have been made to use organic-silicon rubbers, which under suitable circumstances can degrade to a crosslinked silicon oxide type of structure [72].

Acknowledgements

The support of the EPSRC and the various commercial and defence organizations with which the authors have been associated in the past is gratefully acknowledged.

References

1 Society of Chemical Industry, *High Temperature Resistance and Thermal Degradation of Polymers*, Monograph No. 13, SCI, London, 1961.

2 A A Hodgson, 'Fibrous silicates', *Lecture Series 4*, London, Royal Institute of Chemistry, 1965.

3 'Properties and uses of Kevlar-49 aramid fibre and of reinforced plastics of Kevlar 49', *DuPont Trade Literature Bulletin K-1*, June 1974.

4 G L Hart, *The Chemical Stability of Carbon Fibres and their Composites*, PhD Thesis, Kingston Polytechnic, 1975. See also G L Hart and G Pritchard, 'Mechanisms of corrosion in carbon fibre-reinforced phenolic resins', in *Developments in Composite Materials–1*, ed. G S Holister, London, Applied Science, 1977.

5 P M Hergenrother, 'Polymeric materials for high speed civil transport', *Trends Polym Sci* 1996 **4** 104.

6 A A Donskoy, 'Elastomeric heat shielding materials for internal surfaces of missile engines', *Internat J Polym Mater* 1996 **31** 215.

7 R G Weatherhead, *FRP Technology*, London, Applied Science, 1980.

8 K E Stansfield, *The Effects of Stress and Thermal Spiking on the Hygrothermal Response of Carbon Fibre Reinforced Plastics*, PhD Thesis, Kingston Polytechnic, 1989.

9 D Short and J Summerscales, in *Fibre Composite Hybrid Materials*, ed. N L Hancox, Barking, UK, Applied Science, 1981.

10 R Taylor, 'Thermophysical properties', in *International Encyclopedia of Composites*, ed. S M Lee, New York, VCH, 1989, Volume 5.

11 K F Rogers, L N Phillips, D M Kingston-Lee, B Yates, M J Overy, J P Sargent and B A McCalla, The thermal expansion of carbon fibre reinforced plastics. Part 1 – the influence of fibre type and orientation, *J Mater Sci* 1977 **12** 713.

12 C Zweben, 'Fibrous composites: thermomechanical properties', in *Encyclopaedia of Materials Science and Engineering*, ed. M B Bever, Oxford, Pergamon, 1986, Volume 3.

13 W J Liou, R H Ni and C I Tseng, 'Effects of moisture content on the thermal expansion strain of carbon/epoxy composites', in *Proceedings ICCM-10*, Whistler, BC, Canada, August 1995. Eds A Poursartip and K Street, Cambridge, UK, Woodhead, 1995, Volume 4, pp 521–528.

14 R B Prime, in *Thermal Characterization of Polymeric Materials*, ed. E A Turi, New York, Academic Press, 1981.

15 R Strümpler, G Maidorn, A Garbin, L Ritzer and F Greuter, 'Electromechanical properties of conductive epoxy composites', *Polym Polym Compos* 1996 **45** 299.

16 G Montes, M Robertson and A Puckett, 'A characterisation of thermal shock failures in cultured marble vanity tops', *Proceedings of ANTEC '88*, Brookfield, CT, USA, Society of Plastics Engineers, 1988.

17 M Kubouchi, K Tsuda, T Nakagaki, K Arai and H Hojo, 'Evaluation of thermal shock fracture toughness of ceramic particulate-filled epoxy resin', in *Proceedings, ICCM-10*, Whistler, BC, Canada. Cambridge, UK, Woodhead, 1995.

18 M Simpson, P M Jacobs and F R Jones, 'Generation of thermal strains in carbon fibre reinforced bismaleimide (PMR-15) composites. Part 3 – a simultaneous thermogravimetric mass spectral study of residual volatiles and thermal microcracking', *Composites* 1991 **22** 105.

19 W L Morris and B N Cox, 'Thermal fatigue of unidirectional graphite/epoxy composites', in *Proceedings, ICCM-VI*, eds F L Matthews, N C R Buskell, J M Hodgkinson and J Morton, London, Elsevier Applied Science, 1987, Volume 4.

20 S Motomiya, T Yamagishi, M Ikezoe, S Walsh and D Walsh, 'Finite element analysis of composite sinks undergoing thermal cycling', Paper 12B in *Proceedings of the 46th Annual Conference, Composites Institute*, Washington, D.C. USA, The Society of the Plastics Industry, 1991.

21 T A Collings and D E W Stone, *Hygrothermal Effects in CFC Laminates. Part 2- Damaging Effects of Temperature, Moisture and Thermal Spiking*, Royal Aircraft Establishment Technical Report 84003, Farnborough, January 1984.

22 G Pritchard and K Stansfield, 'The thermal spike behaviour of carbon fibre reinforced plastics', *Proceedings, ICCM-VI*, eds F L Matthews, N C R Buskell, J M Hodgkinson and J Morton, London, Elsevier Applied Science, 1987, Volume 4, p. 190.

23 L C E Struik, *Physical Ageing in Amorphous Polymers and Other Materials*, Amsterdam, Elsevier Scientific, 1978.

24 J C Bauwens, in *Failure of Plastics*, eds W Brostow and R D Corneliussen, New York, Hanser 1986, p 235.

25 E S W Kong, S M Lee and H G Nelson, 'Physical aging in graphite-epoxy composites', *Polym Compos* 1982 **3** 1, 29.

26 B L Burton and J L Bertram, in *Polymer Toughening*, ed. C B Arends, New York, Marcel Dekker, 1996.

27 N Grassie and G Scott, *Polymer Degradation and Stabilisation*, Cambridge, UK, Cambridge University Press, 1985.

28 I C McNeill, 'Thermal degradation', in *Comprehensive Polymer Science*, eds G C Eastmond, A Ledwith, S Russo and P Sigwalt, Oxford, Pergamon Press, 1989, Volume 6.

29 H Zimmerman, in *Developments in Polymer Degradation*, ed. N Grassie, London, Applied Science, 1984, Volume 4.

30 G Camino and L Costa, 'Thermal degradation of polymer-fire retardant mixtures – 1', *Polym Deg Stab* 1981 **3** 423.

31 J Zimmerman, 'Polyamides', in *Encyclopaedia of Polymer Science and Engineering*, 2nd edn, New York, Wiley-Interscience, 1988.

32 J N Weber, 'Polyamides', in *Kirk-Othmer Encyclopaedia of Chemical Technology*, 4th edn, New York, Wiley-Interscience, 1996.

33 D D Steppan, M F Doherty and M F Malone, 'A simplified degradation model for nylon 6,6 polymerization', *J Appl Polym Sci* 1991 **42** 1009.

34 V V Korshak, G L Slonimskii and E S Krongauz, 'Hetero-chain polyamides. VII – thermal destruction of polyhexamethylene adipamide', *Izvest Akad Nauk SSSR, Otdel Khim Nauk* 1958, 221.

35 N S Allen (ed.), *Degradation and Stabilisation of Polyolefines*, London, Elsevier Applied Science, 1983.

36 L H Buxbaum, 'The degradation of polyethylene terephthalate', *Angew Chem Internat Edn* 1968 **7** 182.

37 G Pritchard, *The Crosslinking of Polyesters: A Study of Network Formation by Physical Methods*, PhD Thesis, University of Aston in Birmingham, 1968.

38 D Ho, *Effect of Chain Length on the Retention of Polyester Mechanical Properties in Water*, MPhil Thesis, Kingston Polytechnic, 1983.

39 Y Saito, K Kamio, A Morii and T Adachi, 'A novel trifunctional epoxy resin with low viscosity and high heat resistance', *Proceedings 32nd International SAMPE Symposium at Anaheim, CA*, SAMPE, Covina, CA, USA, 6–9 April 1987, pp 1119–1125.

40 K L Hawthorne and F C Henson, 'High performance tris(hydroxyphenyl) methane-based epoxy resins', in *Epoxy Resin Chemistry II*, ed. R S Bauer, American Chemical Society Symposium Series 221, 1983.

41 A J Kinloch and R J Young, *Fracture Behaviour of Polymers*, London, Applied Science, 1983.

42 J E Selley, in *Unsaturated Polyester Technology*, ed. P F Bruins, London, Gordon and Breach, 1976.

43 A J MacKinnon, S D Jenkins, P T McGrail and R A Pethrick, 'A dielectric, mechanical, rheological and electron microscopic study of cure and properties of a thermoplastic modified epoxy resin', *Macromolecules* 1992 **25** 3492.

44 A J MacKinnon, S D Jenkins, P T McGrail and R A Pethrick, 'A dielectric, mechanical, rheological and electron microscopic study of cure and properties of a thermoplastic modified epoxy resin. Incorporation of reactively terminated polysulphones', *Polymer* 1993 **34** 3252.

45 R A Pethrick and B Thomson, '^{13}C NMR studies of phenol-formaldehyde resins', *Br Polym J* 1986 **18** 171.

46 R A Pethrick and B Thomson, '^{13}C NMR studies of phenol–formaldehyde resins – analysis of sequence structure in resins', *Br Polym J* 1986 **18** 380.

47 H-G Elias, *Macromolecules*, New York, Wiley, 1977.

48 G J Knight, 'High temperature properties of thermally stable resins', in *Developments in Reinforced Plastics – 1*, ed. G Pritchard, London, Applied Science, 1980, Chapter 6.

49 J A Parker, D A Kourtides and G M Fohlen, 'Bismaleimide and related maleimido polymers as matrix resins for high temperature environments', in *High Temperature Polymer Matrix Composites*, ed. T T Serafini, Park Ridge, NJ, USA, Noyes Data Corporation, 1987.

50 J P Critchley, G J Knight and W W Wright, *Heat Resistant Polymers*, New York, Plenum Press, 1983.

51 W W Wright, in *Degradation and Stabilisation of Polymers*, ed. G Geuskens, London, Applied Science, 1975, Chapter 3.

52 I K Partridge, *Advanced Composites*, London, Elsevier Applied Science, 1989.

53 G M Bower and L W Frost, 'Aromatic polyimides', *J Polym Sci* 1963 **A1** 3135.

54 R A Dine-Hart and W W Wright, 'Effect of structural variations on the thermo-oxidative stability of aromatic polyimides', *Makromol Chem* 1972 **153** 237.

55 I Hamerton, *Chemistry and Technology of Cyanate Ester Resins*, Glasgow, UK, Blackie Academic, 1994.

56 C R Thomas, *Essentials of Carbon–Carbon Composites*, Royal Society of Chemistry Monograph 32, Cambridge, UK, Royal Society of Chemistry, 1993.

57 G Pritchard (ed.), *Plastics Additives–an A–Z Reference*, London, Chapman and Hall, 1998.

58 V Babrauskas, 'Sandwich panels performance in full scale and bench scale fire tests', *Fire and Materials* 1997 **21** 53.

59 *Annual Book of ASTM Standards*, The American Society for Testing and Materials, Philadelphia, PA, ASTM, USA.

60 *Flammability Handbook for Plastics*, ed. C J Hilado, Lancaster, PA, Technomic, 1990.

61 R G Gann, R A Dippert and M J Drews, 'Flammability', in *Encyclopaedia of Polymer Science and Engineering*, 2nd edn, New York, Wiley-Interscience, 1988.

62 W J Kroenke, 'Metal smoke retarder for poly(vinyl chloride)', *J Appl Polym Sci* 1981 **26** 1167.

63 P Carty and S White, 'Char formation in polymer blends', *Polymer* 1994 **35** 343.

64 V Babrauskas and W J Parker, 'Ignitablity measurements with the cone calorimeter', *Fire and Mater* 1987 **11** 31.

65 V Babrauskas, 'The cone calorimeter – a versatile bench-scale tool for the evaluation of fire properties', in *New Technology to Reduce Fire Losses and Costs*, eds S J Grayson and D A Smith, London, Elsevier Applied Science, 1986, pp 78–87.

66 H Morgan and P J Geake, *Smoke Particle Sizes – A Preliminary Comparison Between Dynamic and Cumulative Smoke Production Tests*, Borehamwood, UK, Fire Research Station, 1988.

67 J R Brown and N A St John, 'Fire-retardant low-temperature-cured phenolic resins and composites', *Trends Polym Sci* 1996 **4** 416.

68 D H E Kunz, 'Flame retarding materials for advanced composites', *Makromol Chem Macromol Symp* 1993 **74** 155.

69 A P Taylor and F R Sale, 'Thermoanalytical studies of intumescent systems', *Makromol Chem Macromol Symp* 1993 **74** 85.

70 S M Lomakin, G E Zaikov and M I Artsis, 'Advances in nylon 6,6 flame retardancy', *Internat J Polym Mater* 1996 **32** 173.

71 M A Shashkina, R M Aseeva, A A Donsky and G E Zaikov, *Internat J Polym Mater* 1996 **33** 1.

72 A Mersberg and J Nee, 'Heat resistant protective coatings', *Mater Perform* 1980 **9** 1.

5

Cyclic mechanical loading

KARL SCHULTE

5.1   Scope of chapter

Polymer matrix-based composite materials offer substantial improvements over metals in structural applications. Their light weight, high strength and high stiffness make them candidate materials for primary components especially in the aerospace industry, where they are replacing metallic materials, components and structures. However, they are also finding increasing application in the automation and machine industry.

The characteristic response of polymer matrix-based composites to a tensile or compressive loading, however, is substantially different from that of metals. Whereas in metals, damage development under static loading exhibits only one primary failure mode, which is the initiation and propagation of one single crack (which can be described with simple fracture-mechanics tools), composite materials exhibit a combination of different failure modes. There are the initiation and multiplication (rather than propagation) of cracks, including transverse, longitudinal and angle-ply cracking in the matrix along fibres (intraply cracks), delaminations (interply cracks), fibre fracture and fibre/matrix interface debonding. This chapter reviews the material properties under cyclic tensile, compressive and variable amplitude loading conditions for various types of fibre-reinforced plastics laminates. Special attention will be given to continuous fibre reinforced composites. Temperature and environmental conditions will be discussed as well. A related chapter (Chapter 11) covers interphase aspects of composite fatigue failure.

5.2   Strain criteria

Favourable specific mechanical properties of continuous fibre-reinforced plastics have made them attractive materials for application to many engineering structures.

Components made of carbon fibre-reinforced plastics already achieve weight savings of about 20% compared with conventional constructions of light metals and these savings are often accompanied by greater reliability [1]. Such weight savings are already possible despite the fact that composite materials are in the early stages of their application in industry. Further weight savings will be realized in the future when more knowledge about composite processing design has been developed. For highly loaded primary structural components, the very specific properties of advanced composites cannot yet be efficiently included in the design of the part, because safety considerations have led to extremely conservative design criteria. In the aircraft industry the maximum allowable strain in a structure is limited to about 0.4%, which is calculated from the maximum expected strain ever likely to be reached in the component during the life of an aircraft, multiplied by a safety factor of 1.5. A component is designed such that even at stress-critical locations, the achievable strains never exceed this level. This strain criterion was chosen because of poor hot/wet notched compression performance and compressive strength after impact requirements. Under static tensile loading above 0.4% strain, first-ply-failure (mostly transverse cracks in laminate plies perpendicular to the loading direction) can occur, although the exact value of the critical strain is dependent on the stacking sequence chosen, as well as on the matrix properties. In the design only a small fraction of the potential strength of a composite material can be used. This is shown in the example of a conventional composite material (see Fig. 5.1 [2]).

The tensile stress–strain behaviour of a single fibre and of the unreinforced resin used as the matrix are shown schematically, together with the behaviour of a cross-ply laminate. Despite the fact that the matrix system has a strain to failure of about 2%, the first transverse cracks can be recognized in the cross-ply laminate at strains slightly greater than 0.4%; only a small part of the strength potential of the laminate is used. With only a slight increase in the allowable strain from 0.4% to, for example 0.5%, an enormous further utilization of the strength potential of the composite laminates could be achieved. This can possibly be realized by using toughened matrix systems, allowing a shift of the crack initiation to higher strain levels [3]. In spite of the fact that a great body of literature is available on the durability of composite materials structures, the body of knowledge has not yet matured to a level where an overall satisfactory understanding of their long term performance and fracture behaviour exists.

Fibrous composites are a relatively new technology and there are too many material and structural variables to discuss fully all the details of their mechanical behaviour. Therefore, the present chapter will concentrate on certain categories of fibre reinforced composites, all with polymer matrix systems.

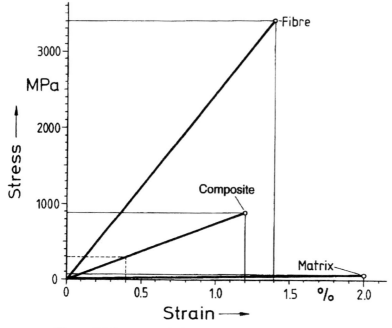

Figure 5.1 Stress–strain behaviour of fibre unreinforced resin matrix and a cross-ply laminate (schematic)

5.3   Mechanical properties of fibre reinforced plastics

5.3.1   Relevant factors

The mechanical properties and the damage mechanisms during loading and fracture of composite laminates are dependent on a wide variety of factors:

1   Geometry:

- stacking sequence of the laminate
- thickness of the specimen
- width of the specimen
- off axis loading
- ply thickness of a lamina
- notched or unnotched specimen
- type of laminate (continuous fibre, short fibre, fabric).

2   Environmental factors:

- temperature
- moisture content
- aggressive environment.

3   Influence of constituents:

- type of fibre
- type of resin.

4   Type of loading:

- static tensile or compressive loading
- cyclic loading.

In this chapter, the dependence of the mechanical properties on these various factors will be shown in a comparative study.

The elastic mechanical properties of carbon fibre-reinforced laminates are highly dependent on the properties of the fibres and the matrix chosen and on the direction of loading relative to the fibre orientation. In a unidirectional laminate, with all fibres orientated in one direction and loaded parallel to the fibres, the properties are mainly dependent on those of the fibres and can be estimated by the rule of mixtures, taking into account the volume fractions of the fibres and the matrix. Equation [5.1] shows that E, the Young's modulus parallel to the fibres is simply given by

$$E = E_F \cdot V_F + E_M \cdot V_M \qquad [5.1]$$

where V_F is the fibre and V_M the matrix volume fraction. E_F is the elastic or Young's modulus of the fibres and E_M is the elastic modulus of the matrix. For a more detailed discussion, the reader is referred to the literature [4].

5.3.2 Classical laminate theory

For a unidirectional laminate the elastic stress–strain relations define an orthotropic material for which the generalized form of Hooke's Law, relating the stress σ to the strain ε,

$$\varepsilon_{ij} = \frac{1}{E_{ijkl}} \sigma_{kl} \qquad [5.2]$$

can be used.

For the case of an orthotropic laminate under conditions of plane stress, which can be assumed when the thickness of a laminate ply is small compared to its length and width, the stress in each constituent ply of a laminate can be determined with the aid of classical laminate plate theory [5–7]. The elastic stress–strain relationship for an orthotropic unidirectional plate then becomes

$$
\begin{bmatrix} \varepsilon_1 \\ \varepsilon_2 \\ \lambda_{12} \end{bmatrix} = \begin{bmatrix} \dfrac{1}{E_1} & -\dfrac{v_{12}}{E_1} & 0 \\ \dfrac{v_{12}}{E_1} & \dfrac{1}{E_2} & 0 \\ 0 & 0 & \dfrac{1}{G_{12}} \end{bmatrix} \cdot \begin{bmatrix} \sigma_1 \\ \sigma_2 \\ \sigma_{12} \end{bmatrix}
$$

[5.3]

which contains four independent constants, the elastic modulus in the fibre direction E_1, the elastic modulus transverse to the fibre E_2, the in-plane Poisson's ratio v_{12} and the in-plane longitudinal/transverse shear modulus, G_{12}. (λ is Poisson's ratio.)

In general, stresses in orthotropic laminates caused by:

- applied stresses or strains in the structure, and
- hygrothermal effects

can be calculated.

For a complete description of stresses and strains according to the laminate theory, the reader is referred to the original literature, as quoted above.

A real laminate used in engineering practice is normally a combination of plies stacked together at various angles, mainly 0°, 90° and ±45° (compare Section 5.3.3). The schematic stress–strain relation of thin 0°, 90° and ±45° laminates produced from graphite epoxy tape prepreg material is shown in Fig. 5. 2 [8]. It shows that in order of decreasing stiffness, the 0° laminates (fibres in loading direction) rank first ($E_x \gg 128\,\mathrm{GPa}$), the ±45° laminates rank second ($E_x \gg 23.2\,\mathrm{GPa}$) and the 90° laminates rank last ($E_x \gg 9.5\,\mathrm{GPa}$). The fracture strain is smaller for the 90° laminates ($\varepsilon_f \gg 0.5\%$) and somewhat larger ($\varepsilon_f \gg 1.2\%$) for the 0° laminates. The fibres dominate the stress–strain behaviour of these laminates. Even if the resin system is ductile, the transverse tensile behaviour tends to be brittle, because of the influence of the fibres. The ±45° laminate is the only laminate with strong matrix dominated properties, which are caused by the relatively unhindered shearing of the matrix. This results in the schematic linear–elastic, ideal–plastic behaviour, as indicated in Fig. 5.2.

The transverse contraction of these laminates is completely different under uniaxial applied strain. The 90° laminate contracts very little ($v_{12} \gg 0.02$), whereas the contraction of the ±45° laminate is very large ($v_{12} \gg 0.773$). This difference in the Poisson's ratio causes additional stresses in orthotropic laminates, where the constituent plies have different ply orientations. The contraction of an angle-ply laminate, for example with 90° plies, is hindered by the small Poisson's ratio of the 90° plies. This leads to transverse stresses in other plies. Near free edges and around unloaded holes, the transverse stresses are equilibrated by interlaminar stresses. For

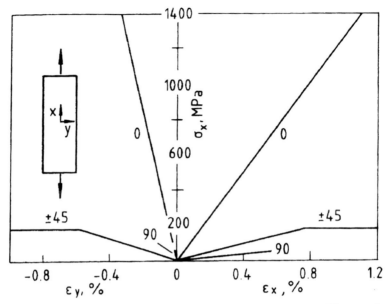

Figure 5.2 Stress–strain curves for laminates with different ply angles under uniaxial loading (CFRP T300/914c) [8]

this reason, two areas of different stresses are recognized in undamaged laminated plates: (i) the specimen free edges and holes with an inhomogeneous stress distribution and (ii) the area of homogeneous stresses, starting at some distance from the free edge.

5.3.3 Edge stresses

Classical laminate theory considers the stresses to be uniform in the different plies. The fact that these stresses vanish at the specimen free edges is not considered in the classical theory. For this reason a different approach is necessary to analyse these stresses. An analytical method (Puppo and Evensen [9]), a finite-difference solution technique (Pipes and Pagano [10]) and finite-element methods (Rohwer [11]) have been applied. As an example, the stresses at the free edge of $\pm 45°$ laminates were analysed. Figure 5.3 summarizes the results of the different investigators.

At a distance of about twice the specimen thickness away from the edge, σ_x and τ_{xy} approach the uniform stress, which can be calculated from laminate theory, whereas the interlaminar shear stresses τ_{xz} vanish. The finite-difference method does not allow the conclusion that the shear stress contains a singularity at the interface, although there are indications that this might be the case.

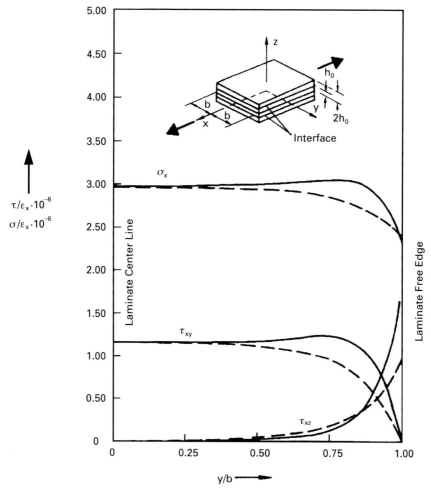

Figure 5.3 Comparison of the most important stresses at the interface in a [±45]$_s$ laminate[10]. ——, Pipes-Pagano; – –, Puppo-Evenson. b = 8h, θ = +45°

In orthotropic laminates with transverse plies (as shown before) the small Poisson's ratio of these plies introduces large stresses perpendicular to the loading direction, especially if other plies with a large Poisson's ratio are present. An example from Rohwer [11], shown in Fig. 5.4, shows the displacements in thickness and width direction of a quasi-isotropic laminate. The deformation of the edge element of the grid in the 90° ply indicates severe tensile stresses in the thickness direction. Thus, for this laminate the tensile stresses in the 90° ply at the specimen edge are very critical, which is also shown in Fig. 5.5.

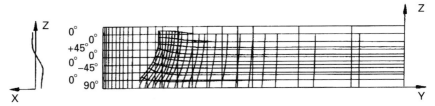

Figure 5.4 Displacements of a [0₂, +45, 0₂, −45, 0, 90]ₛ laminate due to axial tension, $\varepsilon_x = 0.5$ [11]

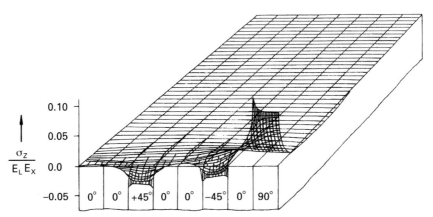

Figure 5.5 Stresses in thickness direction due to axial tension [11]

5.4   Fatigue behaviour

5.4.1   Comparisons between metals and composites

Fatigue failure in metals is known to consist of crack initiation and growth. Crack initiation starts with dislocation movement. Submicrocracks are then formed at slip bands. They subsequently grow and coalesce to form a crack of detectable size (short crack) to complete the crack initiation process [12]. This is then followed by the growth of a single crack until final rupture. The period of crack initiation and submicrocrack growth covers most of the fatigue life.

In neat polymers the fatigue failure process is quite similar, but in certain amorphous thermoplastics, such as polystyrene, phase crazes are formed during the initiation [13] with the subsequent fatigue crack propagation phase to final failure. (Crazes resemble regions with many cracks bridged by oriented fibrils).

In short fibre reinforced composites with aligned or randomly distributed fibres, cracks initiate at flaws, such as pores or in resin-rich areas with local strain inhomogeneities caused by improper fibre alignment or at fibre ends.

The local load transfer from the fibre into the matrix can lead to an overstressing of the matrix or a fibre/matrix debonding. After that a crack propagation can occur, but not as smoothly as in metals [14].

In continuous fibre reinforced composites the fatigue process is characterized by the initiation and multiplication, rather than propagation, of cracks. Crack initiation occurs early in fatigue life, and coincides with the first ply failure in the laminates, that is the first cracking of the weakest ply [15]. While in metals, crack growth accelerates during fatigue, crack multiplication in composites decelerates, resulting in uncontrolled final rupture of the composite, also called 'sudden death' [16].

The similarities between fatigue mechanisms suggests that one might borrow from the well developed metals technology, mainly to describe fatigue damage progression. However, this is only possible for a few special cases, as with mode I or mode II delamination growth. The differences in general point out the need to develop a specific methodology suitable for polymer composites.

Because of the virtually unlimited variations in fibre arrangements (short and continuous fibres, aligned and randomly distributed, laminated, woven, knitted or braided) there are too many materials available to discuss fully all the details of their fatigue behaviour. This chapter will therefore concentrate on some selected composites for demonstration purposes. We shall discuss both tension fatigue, which is the most important load situation in metals, and compression fatigue. The influence of harsh environments will be mentioned as well.

5.4.2 Fatigue of neat and short fibre reinforced polymers

Fatigue failure in neat polymers or short fibre reinforced polymers involves a phase in which a defect zone initiates, with the development of crazes or microcracks, followed by crack propagation to final rupture [17]. This statement means that an analysis of fatigue behaviour can use the tools already developed for metals. This will be demonstrated in the cases that follow.

Figure 5.6 shows the fatigue behaviour of two polymers. One is a brittle bismaleimide (BMI) and the other a ductile polyether sulfone (PES) both reinforced with short fibres, unidirectionally aligned. The loading is parallel to the fibre direction with an R-value $R = 0.1$ [18]. In the $S–N$ plot both materials exhibit a practically linear range up to 10^6 load cycles, with only a slight drop in the upper stress level. However, this level ascertained under fatigue loading is considerably lower than the appropriate static tensile test values, which amount to a fracture stress of 910 MPa for the BMI and 1000 MPa for the PES, materials, respectively. Here the absolute values are lower for the PES than for the BMI composite (in contrast to the

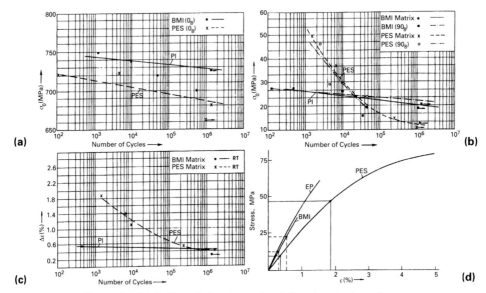

Figure 5.6 (a) Fatigue behaviour of unidirectional aligned bismaleimide (BMI) and polyether sulfone (PES) short fibre composites. [0_8], loading parallel to fibre direction. (b) Fatigue behaviour of BMI and PES neat resins and of corresponding [90_8] laminate. (c) Total strain amplitude plotted against fatigue life for the BMI and PES neat resins. (d) Stress–strain behaviour of neat polymer BMI , PES and an epoxy (EP)

tensile test data). It is not possible to ascertain unambiguously whether this drop in fatigue strength is due to the material itself or to defects in the composite.

Figure 5.6 shows the fatigue behaviour of neat BMI and PES resins and of the appropriate [90_8] transverse test pieces. Both the BMI neat resin and the [90_8] pieces exhibit a very low fatigue strength in their short term fatigue life, compared with the PES material, reflecting the poor static values already mentioned. However, within the range of the endurance limit BMI is now considerably better than PES. This is especially true for the neat resins. Plotting the cyclic strain versus fatigue life (Fig. 5.6(c)) it turns out that the endurance limit for both resin systems is reached at about the same cyclic strain values. A slight increase in strain (load) level for the BMI resin leads to a dramatic decrease in fatigue life, while the PES can withstand a higher strain (load) level for some thousand load cycles. This is understandable if the different static stress–strain behaviour of the two materials is taken into consideration (Fig. 5.6(d)). Up to 0.3% strain, which is about the endurance limit, both polymers behave linearly (totally elastically) in their stress–strain response, but the stress in BMI is higher than in PES due to its

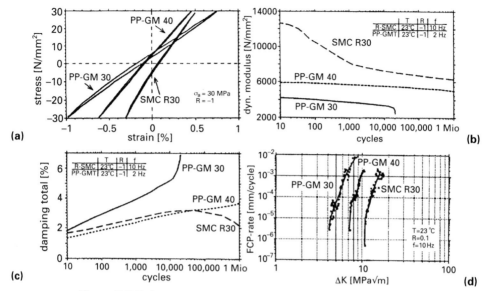

Figure 5.7 Fatigue behaviour of SMC-R 30 (30 wt%–23 vol%), PP-GMT 30 (30 wt%–13 vol%) and PP-GMT 40 (40 wt%–19 vol%) glass fibres. (a) Hysteresis loop, (b) stiffness reduction (change in dynamic modulus), (c) change in mechanical damping, (d) fatigue crack propagation diagram

higher elastic modulus. At higher strain levels the onset of plastic deformation begins. PES is an extremely ductile material and it can resist higher strain levels during short term fatigue cycling.

The stress–strain response and its variation during fatigue cycling is important for the fatigue behaviour of a material. Its continuous recording therefore provides valuable information. These hysteresis measurements are feasible, independent of the *R*-values, in specimens with and without cracks, and the initiation phase for crazing, shear yielding, debonding and crack magnification can be investigated. The recording of the hysteresis loops allows us to determine energy related characteristics (i.e. lost energy, stored energy and damping) for non-linear viscoelastic materials [19], including the variation of the dynamic modulus (stiffness reduction).

Figure 5.7(a) shows the hysteresis loops for GMT (glass mat-reinforced polypropylene) and SMC (sheet moulding compound) materials reinforced with different amounts of glass fibre after 10 000 load cycles at $R = -1$, with a $\Delta\sigma$ of 60 MPa [13]. The different fibre volume fractions are reflected in the different slopes of the hysteresis loops, while the shift in the hysteresis loop indicates microcracking for the SMC and plastic deformations in the compressive loading phase for the GMT. The variation in the stiffness is summarized in Fig. 5.7(b) and at the damping in Fig. 5.7(c), respectively.

Microcracking in the SMC results in a strong stiffness reduction, while plastic deformation leads only to a modest reduction in the dynamic modulus. However, the material damping clearly shows plastic deformation. The chopped glass fibre reinforced SMC and GMT materials fail by crack propagation. This allows us to analyse these materials with the well developed fracture mechanics tools. Figure 5.7(d) compares the fatigue crack propagation rates while plotting them against the ΔK values. With a higher number of glass fibres, the curves shift to higher ΔK values. This is especially true for the ΔK_{th} (threshold).

5.5   Fatigue of continuous fibre reinforced polymers

5.5.1 Tension fatigue test results

Most carbon fibre reinforced plastics (CFRP) used and investigated to date are produced from preimpregnated continuous carbon fibre prepregs. Polymers reinforced with aligned short carbon fibres have certain advantages as materials for structural components, because they can easily be formed into complicated shapes with satisfactory mechanical properties. Woven fabrics produced from carbon fibres find increasing application in the aerospace and many other industries, because they are easy to handle, they have the ability to conform to complicated shapes and the in-plane properties are more isotropic than those of equivalent unidirectional materials.

The fatigue behaviour of the different cross-ply laminates is summarized in a S–N plot in Fig. 5.8. The laminate with the continuous fibres could sustain the highest number of load cycles at the highest stress levels with a very slight slope of the Wöhler curve [15]. The endurance limit was reached at about $\sigma_e = 670 \text{MPa}$. The Wöhler curve of the fabric had a comparatively steep slope, reaching the endurance limit at about 420 MPa [20], while the cross-ply laminate with the aligned short fibres had an endurance limit of about 380 MPa, again with a very small slope of the Wöhler curve [14].

During fatigue cycling, stiffness reduction can be used as an analogue to monitor damage development [21]. Its change can be related directly to stress redistributions, which can be expected if internal damage occurs in composite laminates. A typical stiffness reduction curve for the cross-ply laminate with continuous fibres is shown in Fig. 5.9(a). The shape of the curve suggests three regions of interest: an initiation region (stage I) with a rapid stiffness reduction of 2–5%, an intermediate region (stage II), in which additional 1–5% stiffness reduction occurs in an approximately linear fashion and a final region (stage III), in which stiffness reduction occurs in abrupt steps ending in specimen fracture. A similar behaviour in stiffness reduction due to tension fatigue loading can be observed for the cross-ply

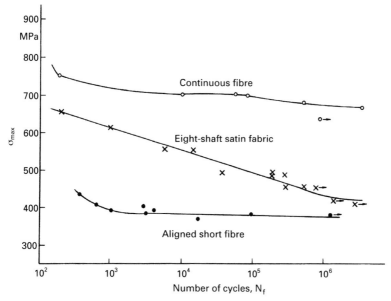

Figure 5.8 Variation of peak stress level for failure with number of cycles to failure for three laminates investigated

laminate with aligned short carbon fibres (Fig. 5.9(c)). In the woven fabric (Fig. 5.9(b)) the initial stiffness reduction (stage I) in early fatigue life is followed by a region with a gradual reduction (stage II). In contrast to the cross-ply laminates with continuous and short carbon fibres, final failure did not initiate abruptly (sudden death behaviour) within the last 10% of fatigue life, but after about 50% of the fatigue life a further pronounced decrease in stiffness occurred, loading by a stepwise reduction to fracture of the test coupons (stage III).

The differences in fatigue behaviour of the three laminates investigated give some interesting insights, especially when they are discussed together with the different damage mechanisms typically observed, as follows.

Under fatigue loading in stage I, the formation of transverse cracks can be observed very early, depending on the cyclic strain level. The formation of transverse cracks dominates the stiffness reduction ascertained in early fatigue life in stage I and is proportional to the transverse crack density [22].

During fatigue in stage II a constant decrease in the elastic modulus can be observed, visible in the linear part of the curve for all three laminates investigated. For the fabric this linear part is comparatively small, leading after about 50% of the fatigue life to an accelerated stepwise stiffness reduction until final failure, whereas in the other two laminates the sudden

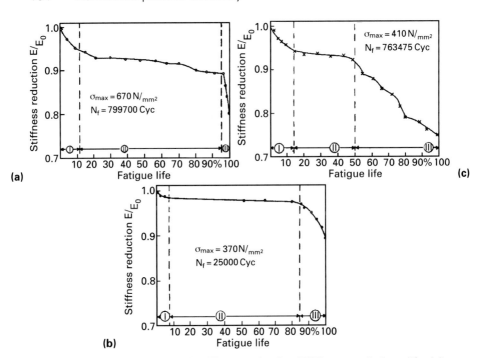

Figure 5.9 Normalized stiffness reduction (E/E_0) versus fatigue life. (a) Continuous fibres, (b) aligned short fibres, (c) eight-shaft satin fabric

stiffness reduction initiating final rupture is limited to the last 5–10% of the fatigue life.

The damage mechanisms occurring in stage II seem to be typical fatigue mechanisms. However, some essential differences exist between the three laminate types, as described below. The typical damage mechanism observed during stage II is the development of delaminations starting from the edge of the test coupon, which are the main mechanism for the reduction of stiffness during this stage. In addition to delamination, longitudinal cracks along the 0° fibres have developed, initiated at transverse cracks. The longitudinal cracks increase in length and number with increasing number of fatigue cycles. The formation of longitudinal cracks and their development can be regarded as a typical fatigue mechanism; the constraint stresses might not be high enough to develop cracks under static loading, but they are high enough to develop cracks under fatigue loading.

In the interior of a cross-ply laminate, local delaminations develop at intersections of longitudinal and transverse cracks [23]. Local tensile stresses are provided by the longitudinal cracks. The tensile nature of the stress produces an out-of-plane tensile stress component at the interface. In addition, an out-of-plane tensile stress component is provided by the transverse cracks. At intersections of these cracks the stresses are additive and the propensity for the nucleation and growth of delaminations will be highest.

Failure of a composite cross-ply laminate containing fibres in the loading direction is always associated with fracture of 0° fibres. The pattern of fibre breaks suggests the involvement of transverse cracks, whose tips reach the 0° ply. The longitudinal stress in a 0° ply is increased by about 8% near the line at the transverse crack in the adjacent 90s [22]. Therefore fibre fractures are concentrated along a small band along the crack tip line [23]. This mechanism appears to supersede any tendency of the fibres to break in a spatially random array determined by the expected strength distribution of single fibres, and causes the fibre breaks to occur in a more localized configuration than would otherwise be predicted from statistical strength arguments. It seems that failure of the 0° fibres, at least at certain locations as described above, has some influence on the decrease in the stiffness during fatigue loading.

Final failure of a composite laminate is preceded by a sudden and rapid drop in stiffness. Formation of longitudinal cracks and coalescence of delaminations between adjacent longitudinal cracks leads to the isolation of strands of 0° fibres from the remaining material. This total separation occurs occasionally. The strand failure now leads to a sudden, rapid and stepwise drop in stiffness.

In the laminate with the aligned short carbon fibres, fatigue during stage II is primarily characterized by the growth of existing cracks and only to a small amount by the initiation of new cracks. An important difference in failure at the very end of fatigue life (stage III) was detectable in this type of laminate when compared to the laminate with continuous fibres.

Starting from the specimen edges, longitudinal cracks are formed, which propagate at an angle of about 2° to the load direction in the 0° layers until the specimen ends are reached. A realistic explanation for this particular angle of crack direction does not exist at present; however, it appears that one reason for this angle arises from the alignment of the short fibres in bundles, a length similar to the fibre length of 3 mm. In Fig. 5.10 a model of the prospective crack path is given. It can be assumed that the cracks will normally follow along these bundles (A) (which in turn are not perfectly oriented) until they reach the bundle ends. Here, they can take the more favourable direction perpendicular to the applied load, that is, through the resin-enriched regions (B). But this process is stopped soon by the interface with the next bundle, along which the crack is now forced to propagate. In fact such a propagation mode results in an angle of crack direction which is very close to 2°.

During fatigue in stage I, the fabric composite transverse cracks initiate and propagate along the transverse threads in the warp direction. The

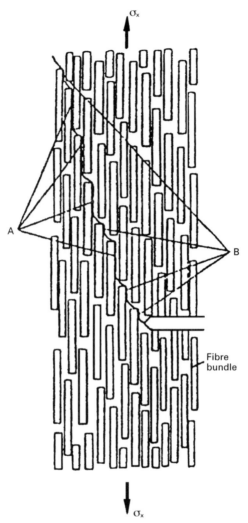

Figure 5.10 Model of crack propagation in a laminate with aligned short fibres

material in the undulated region, where a fill thread encircles a warp thread, has a lower stiffness than in the surrounding regions with the straight threads. This lower stiffness causes a local softening and the maximum strain appears at the centre of undulation and initiates the transverse cracks.

During fatigue stage II, longitudinal cracks along the fill threads develop and increase both in length and density. A similar mechanism to that already described for cross-ply laminates causes this longitudinal crack growth and the fill threads contract because of tensile loading. This contraction is hindered by the warp threads. Stresses of a tensile nature appear between the fill threads, whereas the warp threads are under a compressive load in the fibre direction.

At the undulation of warp and fill threads, delaminations develop in the interior, being initiated at intersections of the transverse and longitudinal cracks, where local tensile stresses are provided by the transverse warp threads. As soon as delaminations occur, the stiffness of a test coupon decreases considerably and enters stage III, that is, the last 50% of fatigue life, producing final failure.

During stage III the first fibre fracture occurs in the fill threads at the centre of undulation. Later, total failure of a fill thread results in a stepwise stiffness reduction, as is documented in Fig. 5.9(c).

The fatigue behaviour of the woven fabric shows that this textile structure is sensitive to fatigue loading. This sensitivity is confirmed by testing thermoplastic matrix systems (polyamide 12, PA 12) reinforced with a glass fabric. A strong stiffness reduction, with a pronounced creep deformation was observed under tension–tension fatigue [24].

The fatigue behaviour is an important parameter to investigate for composites with a textile reinforcement structure. Warp knitted, braided and woven reinforcements will find increasing attention in future.

5.5.2 Compression fatigue test results

With increasing use of fibre reinforced composites it becomes more and more important to investigate the compression behaviour. This is in contrast to metals, which have an equivalent (but reverse) behaviour under static compression load to that under static tensile load. Polymer composites are more susceptible to compressive loads. It is therefore important to study the behaviour of fibre reinforced composites under compression fatigue loading [25]. Figure 5.11 shows the fatigue behaviour of a carbon fibre reinforced polymer under tension ($R = 0.1$) and compression ($R = 10$) fatigue. The very flat $\Delta\sigma$ versus N curve under compression fatigue is surprising, because while under tension fatigue the curve is relatively steep. The result is that under compression fatigue at 80% of the compression strength, the endurance limit is still reached, while under tension fatigue, composite failure is expected at a maximum load in each cycle, which is in the range of about 60% of the tensile fracture strength. Both the cross-ply and the quasi-isotropic laminates show equivalent behaviour. If additional test results are given for tension compression loading ($R = -1$ and $R = -0.55$), it becomes obvious that to understand the test results, it is not sufficient just to plot a Wöhler diagram. Therefore in Fig. 5.11 the maximum stress in each consecutive load cycle, σ_{max}, is plotted

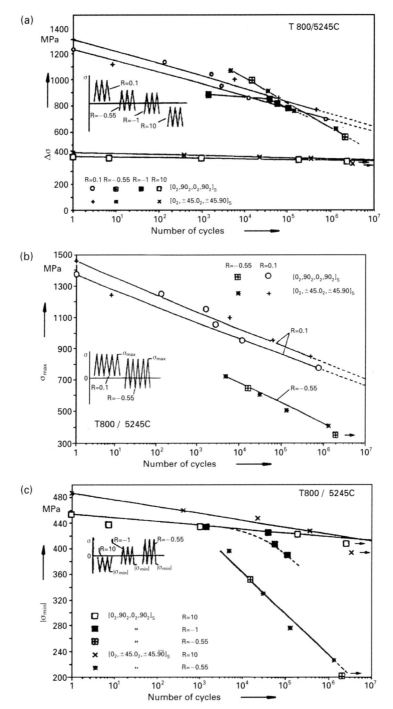

Figure 5.11 S–N diagram for tension, compression and tension–compression fatigue of various laminates. (a) $\Delta\sigma$ versus number of load cycles N, (b) σ_{max} versus number of load cycles N, (c) $|\sigma_{min}|$ versus number of load cycles N

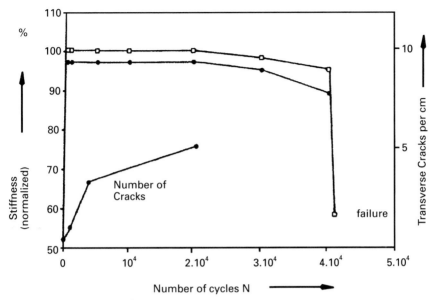

Figure 5.12 Variation of stiffness during the tensile and compressive part, respectively, of a fatigue loaded cross-ply laminate ($R = -1$) and development of transverse cracks. T300/5245C, $[0_2, 90_2, 0_2, 90_2]_s$. $R = -1$, $f = 10\,$Hz, □, compression; •, tension

against N (fatigue life). It clearly shows that if in addition to a tensile load, a compressive load is added in each load cycle, the fatigue life is dramatically reduced.

In Fig. 5.11(c) the results of the fatigue tests, all containing compressive load cycles, are summarized, where the maximum compressive stress in each load cycle (σ_{min}) is plotted against the number of cycles, N, for two different laminates. It is evident that pure compressive cycling results in an extremely flat σ_{min} versus N curve, as already shown in Fig. 5.11. Introducing an additional portion of tensile loading significantly reduces the fatigue life ($R = -1$). This is more pronounced, the higher the tensile portion in each load cycle ($R = -0.55$). A specimen with the laminate stacking sequence of $[0^2, 90^2, 0^2, 90^2]$, would not fail before 10^6 load cycles under tension–tension fatigue when loaded with a maximum stress of about 750 MPa. However, it will fail far earlier, even when fatigued at a lower stress level, by introducing additional compressive stresses. This is especially important at long fatigue life; at short fatigue life, failure occurs in accordance with the compressive–compression mode.

The results can be further corroborated if the normalized secant modulus, separated for tension and compression, is plotted against the number of load cycles (Fig. 5.12). The strong decrease in tensile modulus after the first load cycles is due to the initiation and growth of transverse cracks. It is followed by a region with a nearly constant modulus. In contrast, the compressive modulus remains constant at the beginning, and only at the end of the fatigue lifetime is it dramatically reduced.

During tension–compression fatigue ($R = -1$) damage develops mainly under tensile loading. As a result of this damage, the specimen in the compressive range loses stability and stiffness earlier, as in the case of pure compressive loading. Failure occurs in the compressive part and in essentially less load cycles than under pure compressive loading.

5.6   Effect of heating

Cyclic loading of a polymer, whether reinforced or not, leads in general to a heat buildup in a sample. The extent of the temperature rise depends mainly on the applied load and frequency and on the kind of polymer. In the case of a reinforced polymer, the volume fraction of fibres, their thermal conductivity and the stacking sequence influence the heat generation under cyclic loading. Fatigue-induced damage additionally influences the temperature increase. Stiffness reduction and temperature change give comparable results and both can lead to a similar interpretation of the damage process [26].

Under cyclic loading conditions, when performed at high frequencies of about 10 Hz, an internal generation of heat leads to an increase in specimen temperature. Polymer materials show internal heating due to cyclic loading. Therefore a temperature variation can be used to monitor damage [27,28]. During fatigue loading of a composite material the heat generation can be related mainly to the matrix material used, its viscoelastic behaviour, internal friction and the amount of crosslinking. The type of fibre, its volume fraction and orientation and the stacking sequence in a composite laminate additionally affect matrix loading, and therefore heat generation. In addition to the materials aspects, such conditions as the shape of the sample, the test frequency and the load level influence the amount of heating.

However, a temperature variation of a material can also occur under static loading conditions [29]. In 1851 Thomson (Lord Kelvin) [30,31] showed the proportionality between the load change applied and the resulting temperature variation for an isotropic material under adiabatic elastic deformation and uniaxial stress

$$\Delta T = -\alpha_1 \cdot T_0 \cdot \frac{\Delta \sigma}{c \cdot \rho} \qquad [5.4]$$

where α_1 (K^{-1}) = linear coefficient of thermal expansion, T_0 (K) = ambient temperature, c ($kJ\,K^{-1}m^{-3}$) = specific heat capacity, ρ ($g\,cm^{-3}$) = specific density, $\Delta \sigma$ ($N\,mm^{-2}$) = stress change and ΔT (K) = temperature change.

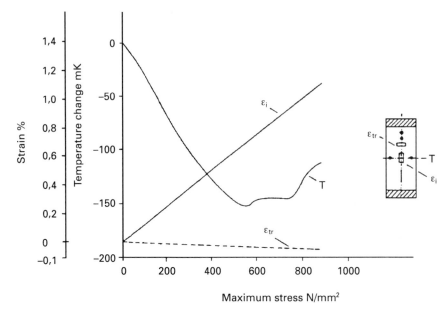

Figure 5.13 Tensile test of a cross-ply laminate. Variation of strain and specimen temperature. Fibre T300, resin LY 556, MY 720 (50:50), lay up $[0_2, 90_2, 0_2, 90_2]_s$. Load rate $= 6.6 \, \text{N} \, \text{mm}^{-2} \text{s}^{-1}$

The above equation describes the so-called 'thermoelastic effect'. The amount of temperature change in a certain volume element depends on the change in the sum of the principal stresses, on its material factors and on the ambient temperature. For an isotropic material with a positive coefficient of thermal expansion, uniaxial tensile stresses lead to a decrease in temperature.

Figure 5.13 shows the results for a cross-ply laminate, containing 0° and 90° plies. The fibres in the 90° plies hinder transverse deformation. They are under compressive load and their temperature change is negative. In addition, the matrix in the 90° plies is under tensile loading. It has a positive thermal expansion coefficient and cools down.

Both these effects are superimposed on the temperature increase in the 0° plies and the result is a cooling down of the entire specimen. At higher stresses, with the initiation of transverse cracks [32] in the 90° plies, the stresses in the 90° plies are locally released and a load transfer from the 90° plies into the neighbouring 0° plies occurs. With an increasing number of cracks, the temperature increase due to loading of the 0° plies becomes more and more dominant (the curve of temperature change goes through a minimum) and at high stresses the temperature increase can be measured. Viscoelastic deformations at the transverse crack tips could contribute to the temperature increase.

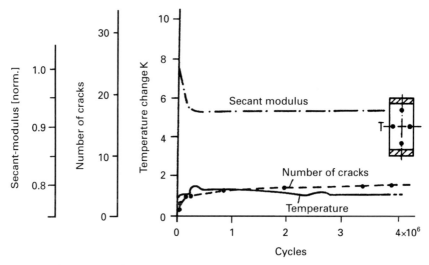

Figure 5.14 Fatigue test of a cross-ply laminate at a low load level. Stiffness reduction, development of transverse cracks and temperature change are plotted versus the number of load cycles. Specimen did not fail. Number of transverse cracks taken from X-ray radiographs. Fibre T300, resin LY 556, MY 720 (50:50), lay up $[0_2, 90_2, 0_2, 90_2]_s$. $R = -1$, $f = 10\,Hz$, $\sigma_{max} = 340\,N\,mm^{-2}$

Figure 5.14 shows the temperature development in a cross-ply test piece. A temperature increase associated with dissipation can be observed at the beginning of the test. Later the temperature remains constant. There are two reasons for this:

- Each specimen has a stable temperature level, depending on load level, frequency, stacking sequence and constituents [27].
- A damaged specimen has a higher stable temperature level than an undamaged one, if a constant damage state is reached [27].

For the present test, where the result is shown in Fig. 5.14, a constant damage state means that transverse cracks initiate in the 90° plies, early in the load history (during the first load cycles). The same plies are also responsible for the slight stiffness reduction observed at the beginning of the test. Later the number of transverse cracks, the stiffness reduction and the specimen temperature remain constant.

In the case of continuous damage growth, which will be observed at higher fatigue load levels, a continuous increase in specimen temperature is also measured. Figure 5.15(a) shows the temperature change for four different tests. All tests were performed until specimen rupture occurred. A continuous temperature increase can be observed throughout the tests. This corresponds to the parallel observation of transverse crack development

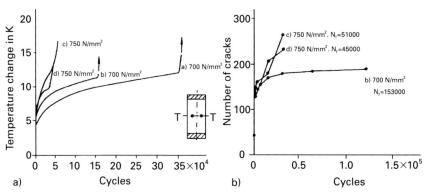

Figure 5.15 Fatigue test of cross-ply laminates at high load levels. All specimens run to failure. Lay up $[0_2, 90_2, 0_2, 90_2]_s$, resin LY 556, MY 720 (50:50), $R = 0.1$, $f = 10\,\text{Hz}$. (a) Temperature change versus number of load cycles, (b) development of transverse cracks versus number of load cycles. Number of transverse cracks taken from X-ray radiographs

(Fig. 5.15), where the number of cracks increases with increasing number of load cycles and the higher the cyclic load level, the more transverse cracks are formed.

The variation in specimen temperature can be used as well as a damage analogue for composite materials. Increasing specimen temperature is a result of increasing (matrix) damage.

High performance polyethylene fibres such as Dyneema® (a reinforcing polyethylene fibre from DSM) show a pronounced time-dependent behaviour under static loading conditions. An increase in strain rate and/or decrease in temperature results in an increase in fibre modulus and strength, but a decrease in work of fracture [33]. It is also known that creep can be observed even in unidirectional PE-fibre reinforced laminates. How far this specific behaviour influences the fatigue behaviour is of great interest and has to be investigated in order to find the appropriate applications for PE-composites.

Figure 5.16 shows the results from a fatigue test performed at a frequency of 10 Hz and a maximum stress in each consecutive load cycle which amounts to 34.4% of the ultimate tensile strength ($\sigma_{max} = 304\,\text{MPa}$) [34]. Figure 5.16(a) plots the variation of the maximum and minimum strain in each load cycle as a function of the number of load cycles. The temperature rise during fatigue life is incorporated into the figure. From the figure it can be seen that during tensile fatigue loading, permanent creep of the composite occurs, resulting in a permanent increase in maximum and minimum strain. The variation in elastic modulus can be calculated from the strain variation and the temperature variation, as summarized in Fig. 5.16(b).

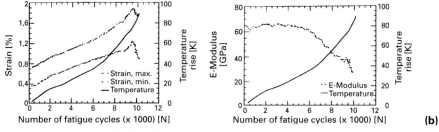

Figure 5.16 (a) Variation of maximum and minimum strain and temperature due to fatigue loading ($R = 0.1$), $\sigma_{max} = 304\,MPa$, $N_B = 10\,050$ cycles, 10 Hz, UD laminate (Dyneema-epoxy 0/0/0). (b)Variation of Young's modulus and temperature due to fatigue loading ($R = 0.1$), $\sigma_{max} = 304\,MPa$, $N_B = 10\,050$ cycles, 10 Hz, UD laminate (Dyneema-epoxy 0/0/0)

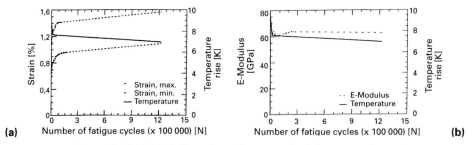

Figure 5.17 (a) Variation of maximum and minimum strain and temperature due to fatigue loading ($R = 0.1$), $\sigma_{max} = 332\,MPa$, $N_B = 1\,243\,150$ cycles, 3 Hz, UD laminate (Dyneema-epoxy 0/0/0). (b) Variation of Young's modulus and temperature due to fatigue loading ($R = 0.1$), $\sigma_{max} = 332\,MPa$, $N_B = 1\,243\,150$ cycles, 3 Hz, UD laminate (Dyneema-epoxy 0/0/0)

Stiffness reduction and temperature rise can be chosen as a damage analogue, when continuously studying the fatigue dependent specimen degradation. The results show that in contrast to composite materials reinforced with glass, aramid or carbon fibres, a pronounced stiffness reduction occurs, even for unidirectional laminates. This stiffness reduction is accompanied by an increase in temperature of more than 80 K. It can further be observed that for a temperature increase in the range between 40 K and 50 K an accelerated stiffness reduction occurs, and ΔT progressively increases as well.

Tests performed at low frequencies, for example, 3 Hz, show that no significant damage occurred even at higher fatigue levels (compare Fig. 5.16). The temperature rose at first by about 8 K and then it decreased to a stable value of $\Delta T = 7\,K$ and remained on this level until more than 1.2 million load cycles, when the fatigue test was stopped. However, the maxi-

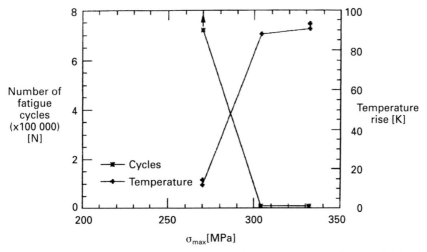

Figure 5.18 Total temperature and fatigue lifetime variation with load level, 10 Hz, unidirectional laminate (Dyneema-epoxy 0/0/0)

mum and the minimum strain in each load cycle increased at the beginning quite rapidly (Fig. 5.17a). During further fatigue testing, the strain variation remained constant but a continuous increase in the medium strain could be observed, which gave rise to the assumption of a cyclically induced steady creep process. No variation in stiffness could be observed during the whole test time (Fig. 5.17b).

It is not only the frequency, but also the maximum fatigue load level that influences the degradation behaviour of a polyethylene composite. Figure 5.18 summarizes the results of various fatigue tests, all performed at 10 Hz but with various load levels. When only 30% of the ultimate tensile strength (σ_{max} = 270 MPa) has been reached in a fatigue loading situation, the fatigue life is close to infinity. However, with only a slight increase in the load level to about σ_{max} = 304 MPa a dramatic reduction in fatigue life occurs. This also results in a rapid increase in specimen temperature.

The results of all fatigue tests are summarized in Table 5.1. They show that there is a strong dependence on the test frequency, independent of the laminate type. At a frequency of 30 Hz, even on the lowest maximum stress level, rupture occurs within a few load cycles, with a temperature rise of more than 80 K, which means at more than 100°C. At a frequency of 3 Hz, even at the highest maximum stress level, only a minor temperature increase can be observed, and there is no rupture of the test coupons.

The results for the fatigue behaviour of the PE-epoxy composites show three phenomena of interest:

Table 5.1 Results of the fatigue tests on the [0,0,0] laminates

Sample no.	σ_{max} (MPa)	Frequency (Hz)	Cycle no. (N)	Temperature rise (K)
2–5	270	30	9 390	95.4
2–9	270	30	7 930	98.2
2–8	270	10	(>) 720 000	14.4
2–3	270	10	(>) 720 000	11.8
2–12	304	10	10 050	87.9
2–4	332	10	8 860	93.1
2–7	332	10	7 930	90.7
2–6	332	3	(>) 1 243 150	8.6

1 Creep induced elongation – it is well known for PE-composites that during long term static loading a temperature-dependent creep behaviour can be observed, which is similar to that of neat PE-fibres [35].

2 Under cyclic deformation there is a continuous increase in the irreversibly dissipated energy, which results in an associated self-heating of the test coupons, while the cyclic modulus decreases (stiffness reduction).

3 In cyclic loading, damage accumulates as matrix cracks initiate. This internal damage development has an effect on both the stiffness (due to load transfer with local strain concentration) and internal heating (due to higher local cyclic deformations, which means locally higher shear stresses).

During fatigue loading there is a balance between the temperature resulting from the dissipated energy per unit volume in each cycle and the heat flux into the environment in the time *t*. At low frequencies there is enough time for the energy to dissipate into the environment. Therefore the endurance limit under fatigue is at low frequencies on a higher load level than at high frequencies. In this last case, the time for a heat flux is too short. With each load cycle additional energy dissipates, leading to accelerated internal heating with the degradation of properties. Cyclic creep and energy dissipation can be made directly visible from the dynamic hysteresis loops. Stiffness reduction and cyclic creep result in shifting and opening of the loops.

5.7   Environmental influences

The effect of adverse environments on the fatigue behaviour of glass fibre reinforced polymers will be discussed, taking polyetherimide (PEI) as an example [36]. PEI is an amorphous thermoplastic, so a low resistance to solvents and similar harsh environments can be expected. Because of its potential applications in aircraft and railways, the hydraulic fluid (Skydrol) and the solvent methyl ethyl ketone (MEK) were chosen.

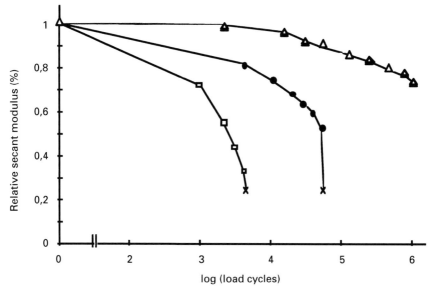

Figure 5.19 Influence of aggressive fluids on the secant modulus of a PEI composite during fatigue (45° orientation, $\sigma_{max} = 95\,\mathrm{MPa}$). △, in air; ●, in hydraulic fluid; □, in MEK

Advanced thermoplastic matrix composites offer a number of advantages over thermosetting matrix-based composites, which include improved toughness, damage tolerance, ease of repair, moisture resistance and easy and fast processability. Cost reduction in relation to processing operations is the main motivation for their use [37]. Thermoplastic polymers such as polyetheretherketone (PEEK), polyphenylenesulfide (PPS) and PEI belong to the new generation of polymers. Amorphous polyetherimides above all meet the most stringent flammability, smoke and toxicity requirements and seem to be excellent candidates for ground transportation or aerospace applications. With increasing interest in PEI composites, however, there has also been increased concern about their environmental stability.

The secant modulus variation of the composite during a flexure test is shown in Fig. 5.19. Specimens with the 45°-orientation and loaded with a maximum stress per cycle of 95 MPa exhibited a significant loss in secant modulus. The specimens exposed to Skydrol hydraulic fluid and to methyl ethyl ketone showed a higher initial deflection, with a more pronounced loss in secant modulus. The specimens tested in air exceeded 10^6 load cycles and the secant modulus decreased to 80% of the initial value. The exposed specimens broke before 10^6 load cycles were reached.

Figure 5.20 shows the *S–N* curves of specimens tested in air and in the two aggressive fluids. Immersion in hydraulic fluid reduced the fatigue life by a factor of about 100. The failure mode in air was rupture in the middle of the

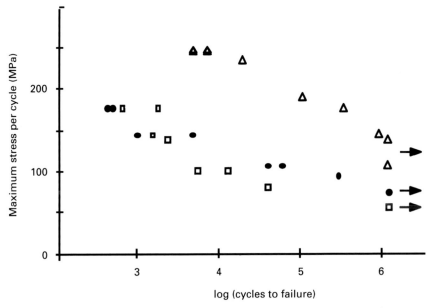

Figure 5.20 Influence of aggressive fluids on the *S–N* curve of PEI composite during fatigue (45° flexural; fabric; *R* = 0.1, *f* = 10 Hz). △, in air; ●, in hydraulic fluid; □, in MEK

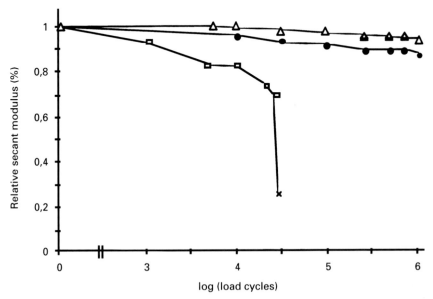

Figure 5.21 Influence of aggressive fluids on the secant modulus of a PEI composite during fatigue (0° orientation, σ_{max} = 120 MPa). △, in air; ●, in hydraulic fluid; □, in MEK

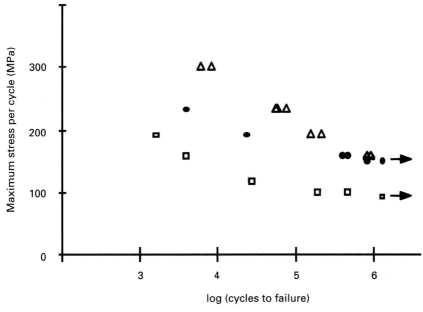

Figure 5.22 Influence of aggressive fluids on the *S–N* curve of PEI composite during fatigue (0° flexural; fabric; R = 0.1, f = 10 Hz). △, in air; ●, in hydraulic fluid; □, in MEK

specimen. The exposed specimens in contrast showed pronounced delamination.

The influence of aggressive fluids on the secant modulus of 0°-oriented specimens is shown in Fig. 5.21. The maximum stress per cycle during fatigue was 120 MPa. As the glass fabric was oriented in the direction of principal stress, the loss in secant modulus was less pronounced than for the 45°-orientation of the fabric. In air and hydraulic fluid the secant modulus decreased only slightly, while in MEK the composite failed after 30 000 load cycles.

The maximum loss in secant modulus of those specimens which did not fail was dependent on the medium: in air the elastic modulus dropped to 90%, in hydraulic fluid to 80% and in methyl ethyl ketone down to 60% of the initial value.

Figure 5.22 shows *S–N* curves for the 0°-orientation of the fabric. All curves descended. At about a maximum stress per cycle of 170 MPa, neither the tests in air nor those in hydraulic fluid showed any effect on the fatigue life. At higher stresses the fatigue life in hydraulic fluid was reduced by a factor 10 compared to the air test. The life of specimens immersed in methyl ethyl ketone decreased by a factor of 100 when compared to the tests in air.

Specimens which were not exposed to the liquids did not show any delaminations. The combination of fatigue and exposure to aggressive fluids produced delaminations in the upper layer of the composite subjected to tensile stress. In specimens immersed in methyl ethyl ketone, moreover, buckling occurred at the compressive side.

These results can be summarized as follows. The more the measured mechanical properties are matrix dominated, the higher is the effect of an aggressive environment. Flexural fatigue behaviour deteriorated as well. In the presence of aggressive fluids the secant modulus measured during cyclic loading dropped more than in air. Exposure to hydraulic fluid decreased the fatigue life by a factor of 10, and immersion in methyl ethyl ketone decreased it by a factor of 100. In contrast to failure in air, failure in aggressive fluids was accompanied by delaminations.

5.8   Prediction of fatigue damage development

A lot of experimental work has been performed since the late 1960s to try to understand the damage degradation mechanisms developing during fatigue loading. An early model to predict strength degradation was proposed by Halpin *et al.* [38]. However, it was based on the classical crack growth process in metallic materials. The interpretation of a fatigue damage progression based on the initiation and growth of only one dominant crack does not reflect the real damage situation in a composite, where various damage mechanisms, such as matrix cracking, fibre/matrix debonding, delamination and fibre rupture occur simultaneously and at various locations.

Therefore, describing the fatigue behaviour of continuous fibre reinforced polymer matrix composites with the tools of the fracture mechanics approach is not yet possible. Only crack propagation behaviour under certain special load situations can be adequately described. This applies to interlaminar and intralaminar crack propagation under both mode I and mode II loading conditions (compare Fig. 5.23(a)). Mode I and II fatigue crack propagation can be measured with specially designed specimens, mode I with the double cantilever beam (DCB) and mode II with the end notch flexure (ENF) test (Fig. 5.23(b)). With both techniques it is possible to simulate the delamination fatigue crack propagation behaviour [39]. Figure 5.23(c) shows the interlaminar fatigue crack propagation rate plotted against the stress intensity range [40] at different R-values.

This means that only certain isolated events can be described. Because of the complicated damage situation, other damage parameters have to be chosen, rather than a single crack propagation. Hahn and Kim [41] proposed that the time rate of residual strength reduction was inversely proportional to the residual strength. An approach to predicting the remaining residual strength of a composite after fatigue loading was devel-

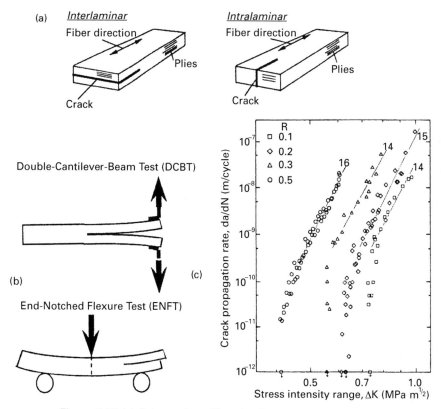

Figure 5.23 (a) Schematics of interlaminar and intralaminar fracture. (b) DCB and ENF specimens. (c) Relationship between mode I crack propagation rate and cyclic stress intensity range for a T300/914c laminate

oped by Yang [42]. None of these models attribute the reduction in residual strength to any specific damage mechanism or explain how they cause final failure.

Reifsnider and Stinchcomb [43] proposed a model for predicting residual strength and fatigue life. They defined a 'critical element' in a composite laminate, which was identified as a local material element whose failure defines global failure for a distinct failure mode. The subcritical elements are defined by the damage events that occur during cyclic loading, leading to local stress redistributions but not to composite laminate rupture. Statistically a composite contains many critical elements. A single critical element approach is therefore not sufficient. Diao *et al.* [44] proposed a statistical model based on the critical element model. Just recently Schaff and Davidson [45] proposed a new residual strength-based model which could be extended to predict the fatigue life of randomly ordered loading spectra [46].

Other cumulative damage models are not based on residual strength, but on other parameters, of which the most recent one is stiffness degradation. Stiffness degradation can be determined non-destructively, *in situ*, by measuring the variation in the elongation of a specimen during fatigue loading, but also indirectly, by measuring the electrical conductivity or the change in dielectric dissipation factor, in the X-ray absorption, or in ultrasonic attenuation, damping coefficient or temperature increase. Measurements of stiffness reduction seem to show less scatter than residual strength measurements [47]. A first approach to describing the damage accumulation in composites based on stiffness degradation was proposed by Wang and Chim [48] for random short-fibre composites. Schulte *et al.* [49] tried to formulate a damage parameter in the form of

$$D = \frac{E_0 - E_i}{E_0 - E_D} \qquad [5.5]$$

with E_i as the instantaneous stiffness in the ith cycle and E_0 as the initial stiffness at the onset of fatigue cycling. E_D is the stiffness at the end of the linear part of the stiffness reduction curve. Plumtree and Shen [50] extended this approach, however, using much more complicated equations.

Residual stiffness degradation models have proved capable of estimating the cumulative damage in composite laminates under constant amplitude loading. In a recent publication Lee *et al.* [51] extended such a model, so that service load spectra could also be predicted.

Although the stiffness reduction curves offer a reasonable basis on which to establish damage curves for constant amplitude loading at different stress levels, these damage curves may not reflect all physical effects which are important for the development of fatigue damage. That, however, would mean that the development of realistic cumulative damage laws for composite materials would require considerable further research and development work.

5.9   Conclusions

Composite materials have matured to become reliable materials in the design of structural parts. Their wide structural variability and individual response to mechanical and environmental loading, however, complicates the business of predicting their long term properties.

References

1 D Schulz, 'Structural certification of Airbus fin box in composite fibre construction', *Proceedings 14th Conference, International Council of the Aeronautical Sciences (ICAS-84)*, Toulouse (France), 1984, pp 427–438.

2 C H Baron, K Schulte and H Harig, 'Influences of fibre and matrix failure strain on static and fatigue properties of carbon fibre-reinforced plastics', *Compos Sci Technol* 1987 **29** 257–272.

3 G Hellard, J Hognat and J Cuny, 'Mechanical characteristics of resin/carbon composites incorporating very high performance fibres', *Proceedings International Conference on Composite Materials ICCM-V*, San Diego. London, Elsevier Applied Science, 1985, pp 1633–1639.

4 A Kelly and N H Macmillan, *Strong Solids*, 3rd edn, Oxford, Clarendon Press, 1986, 242ff.

5 A Puck, 'Zur Beanspruchung und Verformung von GFK-Mehrschichtverbund-Bauelementen', *Kunststoffe* 1967 **57** 284–293.

6 R M Jones, *Mechanics of Composite Materials*, New York, McGraw-Hill, 1975.

7 S T Tsai and H T Hahn, *Introduction to Composite Materials*, Lancaster, Technomic, 1980.

8 P W M Peters, 'Damage development in laminated plates under monotonic tensile loading', *Proceedings European Symposium on Damage Development and Failure Processes in Composite Materials*, eds I Verpoest and M Wevers, Leuven, KU-Leuven, 1987, pp 21–30.

9 A H Puppo and H A Evensen, 'Interlaminar shear in laminated composites under generalized plane stress', *J Compos Mater* 1970 **4** 204–220.

10 R B Pipes and N J Pagano, 'Interlaminar stresses in composite laminates under uniform axial extension', *J Compos Mater* 1970 **4** 538–548.

11 K Rohwer, *Stresses and Deformations in Laminated Test Specimens of Carbon Fibre Reinforced Composites*, Braunschweig, DFVLR-FB 82–15, 1982.

12 H Nowak, K Schulte and G Lütjering, 'Microscopical and mechanical contributions to the cycle by cycle crack growth under variable amplitude loading', *Fatigue Crack Growth Under Variable Amplitude Loading*, London, Elsevier Applied Science, 1988, pp 109–133.

13 V Altstädt, W Loth and A Schlarb, 'Comparison of fatigue test methods for research and development of polymers and polymer composites', *Progress in Durability Analysis of Composite Systems*, Rotterdam, A A Balkema, 1996, pp 75–80.

14 K Friedrich, K Schulte, G Horstenkamp and T W Chou, 'Fatigue behaviour of aligned short carbon-fibre reinforced PI and PES composites', *J Mater Sci* 1985 **20** 3353–3364.

15 K Schulte, 'Damage development under cyclic loading', *Damage Development and Failure Processes in Composite Materials*, eds I Verpoest and M Wevers, Leuven, KU-Leuven, 1987, pp 39–54.

16 P C Chou and R Croman, 'Degradation and sudden-death models of fatigue of graphite/epoxy components', *ASTM-STP 674*, 1979, pp 431–454.

17 J A Sauer and M Hara, 'Effect of molecular variables on crazing and fatigue of polymers', *Crazing in Polymers*, Berlin, Springer Verlag, 1990, Volume 2, pp 69–118.

18 K Schulte, K Friedrich and G Hostenkamp, 'Temperature dependent mechanical behaviour of PEI and PES resins used as matrices for short fibre reinforced laminates', *J Mater Sci* 1986 **21** 3561–3570.

19 R Renz, V Altstädt and G W Ehrenstein, 'Hysteresis measurement for characterising the dynamic fatigue of R-SMC', *J Reinforced Plastics Compos* 1988 **7** 413.

20 K Schulte, E Reese and T W Chou, 'Fatigue behaviour and damage development in woven fabric and hybrid fabric composites', *Proceedings ICCM VI*,
London, Elsevier Applied Science, 1987, pp 4.89–4.99.

21 H T Hahn and R Y Kim, 'Fatigue behaviour of composite laminates', *J Compos
Mater* 1976 **10** 156–180.

22 A L Highsmith and K L Reifsnider, 'Internal load distribution effects
during fatigue loading of composite laminates', *ASTM-STP 907*, 1986, pp 233–
251.

23 R D Jamison, K Schulte, K L Reifsnider and W W Stinchcomb, 'Characterisation and analysis of damage mechanisms in fatigue of graphite/epoxy laminates',
ASTM-STP 836, 1984, pp 36–51.

24 J Wendorff, Ch Baron and K Schulte, 'The fatigue behaviour of glass fabric
reinforced thermoplastics' (in German), *Kunststoffe* 1995 **12** 2058–2060.

25 K Schulte, 'Compressive static and fatigue loading of continuous fibre-
reinforced composites', *ASTM-STP 1185*, 1994, pp 278–305.

26 H Neubert, K Schulte and H Harig, 'Evaluation of the fatigue behaviour by
monitoring continuously temperature development in CFRP-laminates', *ASTM-
STP 1059*, 1990, pp 435–453.

27 H Neubert, H Harig and K Schulte, 'Monitoring of fatigue induced damage
processes in CFRP by means of thermometric methods', *ICCM-V I/ ECCM-2*,
Imperial College, London, Elsevier Applied Science, 1987.

28 H Neubert, *Anwendung thermometrischer Methoden bei der Prüfung
kohlenstoffaser-verstärkter Kunststoffe*, Dr.-Ing. Dissertation, Universität-GH
Essen, 1989.

29 C E Bakis and K L Reifsnider, 'The adiabatic thermoelastic effect in laminated
fibre composites', *J Compos Mater* 1991 **25** 809–830.

30 W Thomson, *The Quarterly Journal of Pure and Applied Mathematics 1*, 1857 **1**
57–77.

31 W Thomson, 'Mathematical and Physical Papers 1', *Collected Works*, 1882 **1**
174–332, 882, Cambridge, UK, Cambridge University Press.

32 K Schulte and W W Stinchcomb, 'Damage mechanisms – including edge effects
– in carbon fibre-reinforced composite materials', *Application of Fracture Mechanics to Composite Materials*, Amsterdam, Elsevier Science, 1989, pp 273–325.

33 A A J M Peijs, *High Performance Polyethylene Fibres in Structural Composites*,
Doctoral Thesis, Technical University, Eindhoven, 1993.

34 K Schulte, H Omloo, R Marissen and K-H Trautmann, 'The fatigue and temperature dependent properties of PE-fibre reinforced polymers', *Progress in
Durability Analysis of Composite Systems*, Rotterdam, A A Balkema, 1996, pp
107–113.

35 A A J M Peijs, P Catsman, L E Govaert and P J Lemstra, 'Hybrid composites
based on polyethylene and carbon fibres. Part 2: Influence of composition and
adhesion level of polyethylene fibres and mechanical properties', *Composites*
1990 **21** 513.

36 K Schulte, A Mulkers, H D Berg and H Schoke, 'Environmental influence on the
fatigue behaviour of amorphous glass/thermoplastic matrix composites', *International Conference on Fatigue of Composites*, Paris, eds S Degallaix, C Bathias and
R Fougeres, 1997, pp 339–346.

37 D Leeser and B Banister, *21st International SAMPE Technical Conference*, 25–
28 Sept 1989, Anaheim, CA, p 507.

38 J C Halpin, K L Jerina and T S Johnson, 'Characterisation of composites for purpose of reliability evaluation', *ASTM-STP 521*, 1973, pp 5–64.

39 M Hojo, S Ochiai, C G Gustafson and K Tanaka, 'Effect of matrix resin on delamination fatigue crack growth in CFRP laminates', *Eng. Fract. Mechan.* 1994 **49** 35–47.

40 M Hojo and K Tanaka, *Advances in Fibre Composite Materials*, London, Elsevier Science, 1994, pp 123–155.

41 H T Hahn and R Y Kim, 'Proof testing of composite materials', *J Compos Mater* 1975 **9** 297–311.

42 J N Yang, 'Fatigue and residual strength degradation for graphite/epoxy composites under tension-compression cyclic loading', *J Compos Mater* 1978 **12** 19–39.

43 K L Reifsnider and W W Stinchcomb, 'A critical-element model of the residual strength and life of fatigue-loaded composite coupons', *ASTM-STP 907*, 1986, pp 298–313.

44 X Diao, L Ye and Y-W Mai, 'A statistical model of residual strength and fatigue life of composite laminates', *Compos Sci Technol* 1995 **54** 329–336.

45 J R Schaff and B D Davidson, 'Life prediction methodology for composite structures. Part I-Constant amplitude and two-stress level fatigue', *J Compos Mater* 1997 **2** 128–157.

46 J R Schaff and B D Davidson, 'Life prediction methodology for composite structures. Part II-Spectrum fatigue', *J Compos Mater* 1997 **2** 158–181.

47 Z Hashin, 'Cumulative damage theory for composite materials: residual life and residual strength methods', *Compos Sci Technol* 1985 **23** 1–19.

48 S S Wang and E S M Chim, 'Fatigue damage and degradation in random short-fibre SMC composites', *J Compos Mater* 1983 **17** 114–134.

49 K Schulte, H Nowak, K-H Trautmann and Ch Baron, 'Estimation of the durability of composite materials by means of stiffness reduction', *ECCM-1*, Bordeaux, eds A R Bunsell, P Lamicq and A Massiah, 1985, pp 51–55.

50 A Plumtree and G Shen, 'Prediction of fatigue damage development in unidirectional long fibre composites', *Polym Polym Compos* 1994 **2** 83–90.

51 L J Lee, K E Fu and J N Young, 'Prediction of fatigue damage and life for composite laminates under service loading spectra', *Compos Sci Technol* 1996 **56** 635–648.

6
Weathering

JOHN LAYTON

6.1 Introduction

All materials of construction change in appearance on extended exposure to the weather. Many 'natural' materials such as stone, brick, slate and copper are judged to improve aesthetically on weathering, whereas few plastics become more pleasing to the eye. This is because the bright colours and glossy surfaces that are often specified initially, with the passage of time become faded and dull. Although these changes are difficult to quantify, this chapter attempts to show what it is reasonable to expect from reinforced plastics and how performance can be optimized by use of various ingredients, fabrication techniques and post-treatments.

The term 'reinforced plastics' can be applied to a large number of different materials but the widest coverage has been given here to glass fibre reinforced unsaturated polyester resin, since this has by far the largest share of the market for exterior applications. The term GRP can be assumed to refer to glass reinforced polyester, unlike the rest of the book where several matrix resins are discussed and the term then applies to glass reinforced plastics in general. The major effects of weathering on GRP are loss of gloss and change in colour and so most attention has been paid to these and to other surface phenomena.

Typical applications that require resistance to weathering are boat hulls and superstructures, car and railway carriage bodies, playground equipment and cladding panels for buildings. Whilst loss of gloss or fading of a colour is cosmetic and does not affect physical performance, it is clearly undesirable in such applications.

Other thermoset resins such as epoxides, phenolics and polyurethanes are also used in reinforced form, as are some thermoplastic polymers. Many need to be protected by surface coatings if retention of properties after extended weathering is important.

6.2   Natural weathering

The changes that take place when polymers are exposed to weathering arise mainly from the effects of a combination of ultraviolet radiation and heat from the sun, together with the effects of moisture (precipitation and condensation) and oxygen from the atmosphere.

6.2.1   Ultraviolet light

UV light is the prime cause of breakdown and produces effects which are similar to thermal degradation. These involve the breaking of chemical bonds, giving rise to free radicals which result in permanent chain scission or in crosslinking, depending on the polymer involved, the wavelength of the UV and other factors [1]. Some of the groups formed may be chromophores. Unlike thermal degradation, UV degradation does not occur uniformly throughout the polymer, but particularly with opaque materials, the effects are felt on or near the surface. The two most commonly observed effects are loss of gloss and change in colour.

UV light can be divided into three types according to wavelength. The so-called UV-A portion of the spectrum (400–315 nm) is the least harmful to organic polymers and forms about 6% of the sun's total radiation reaching the earth; UV-B (315–280 nm) is a more damaging part and forms about 0.1% of total; and UV-C (<280 nm) is the most harmful of all to polymers. UV-C is, however, filtered out by the earth's atmosphere. A spectrum of Florida sunlight is shown in Fig. 6.1.

UV radiation below about 350 nm is absorbed by window glass; UV-B is therefore effectively eliminated indoors apart from small amounts generated by artificial lighting.

Atmospheric heat and moisture accelerate and change the nature of damage done by UV radiation to most exterior grade polymers, rather than act as primary causative agents themselves, unless the polymer is under high stress or has been badly fabricated. (For example, air occlusions beneath a polyester gelcoat which pit and eventually break through.)

6.2.2   Climate and geography

Climates can be classified into a number of types such as temperate, subtropical, desert, arctic and Mediterranean [2]. Additionally, they can be industrial, rural or marine in nature. The effects of weathering will vary in each of these, as well as from season to season and from year to year. How, then, can the weathering performance of one material be defined and compared with that of another? The short answer is that it is extremely difficult to do so, because there are so many possible variables. As examples, the same set of test panels can give different rankings when exposed on different sites; moreover, commencement of one twelve month exposure test in the winter and another in the summer, at the same site, can give different end results, and panels made of the same material but in different colours exposed under the same conditions can give different results again.

In order to achieve some degree of uniformity, Florida, USA, is often taken as a reference location for such comparisons [3] because the subtropical climate of that region combines high and fairly consistent UV irradiation with high rain and humidity levels. Desert test stations, such as those in Arizona and in some parts of Australia, are also popular; they receive more UV radiation but less moisture. At the test stations, it is usual to expose flat panel samples, facing the equator, at 5° or 45° to the horizontal, for twelve month periods or multiples of twelve months. Solar energy levels, UV radiation, humidity and rainfall records are kept so that different periods can be compared.

Natural weathering in Florida is often used as a means of forecasting more rapidly effects that will occur in temperate climates. For example, one year's exposure in Florida has been said to be roughly equivalent to 4–5 years in central Europe. Such comparisons are necessarily highly approximate and could cause materials to be rejected that would, in fact, be satisfactory. Nevertheless, Florida exposure often forms part of the specifications for weather resistance throughout the rest of the world.

For results that are more applicable to particular regional conditions, test stations can be used in the areas concerned, exposing panels facing south in the northern hemisphere and north in the southern hemisphere. Anomalous results can still be obtained if, for instance, the test station is in a residential area and the material is used in an industrial area with high pollution levels, and vice versa.

6.3   Accelerated weathering

6.3.1 Outdoor testing

The effects of weathering at locations such as Florida and Arizona can be accelerated by using panel mountings that track the sun, combined with metal Fresnel reflectors to concentrate the rays of the sun, water sprays and cooling devices. This method is known by the initials EMMAQUA which stand for equatorial mounted mirrors with water spray (AQUA). By such means, one year of normal Florida weathering can be achieved in approximately 40 to 45 days [4] and this is therefore a very cost-effective method for materials needing a long service life.

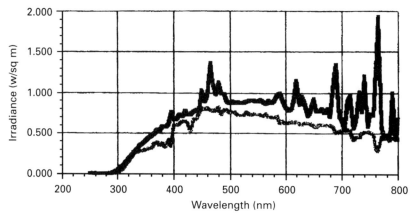

Figure 6.1 Spectrum of Miami sunlight compared with that of a filtered xenon arc lamp [4]. ——, xenon spectrum; ——. sunlight spectrum

6.3.2 Laboratory testing

As an alternative to accelerated outdoor methods, laboratory equipment can be used which aims to reproduce, within a cabinet, the spectrum or part of the spectrum of natural sunlight combined with heating and/or moisture and/or darkness cycles. Test panels are exposed in such equipment which can then be run continuously.

Early accelerated weathering apparatus utilized carbon arc lamps with various filters to generate a spectrum that reproduced some of the harmful effects of sunlight. Improved versions of these are still in use and are specified for certain applications. However, many users and specifiers have turned to xenon arc lamp-based machines as these produce a spectrum that more closely matches that of sunlight (Fig. 6.1). Additionally, they are generally less expensive to purchase and to run than the carbon arc machines. Both types nowadays have facilities for the provision of controlled humidity, water spray, periodic water immersion and/or variable temperature cycles.

In another type of instrument, UV energy is generated by special fluorescent tubes that produce, for example, UV-A or UV-B type spectra only, but not the visible or infrared components of sunlight. The UV-B lamps generate some UV radiation of lower wavelengths than normally reaches the earth's surface and so give rapid results which are not, however, necessarily related to real-life performance. The UV-A lamps reproduce the most relevant part of the sun's spectrum, between 290–350nm, fairly accurately but take longer to produce results. Exposure cycles alternate with heating and condensation cycles. Since other parts of the spectrum do have some effect on weathering behaviour, (for example, the different heating effects of infrared on dark and on light coloured surfaces), fluorescent tube equipment has limitations. However, the low initial costs and running costs of such machines, combined with results that give meaningful comparisons for similar materials, ensure that they maintain popularity.

6.3.3 Comparisons between accelerated and natural weathering

Crump [5] has examined the correlation between six commonly used accelerated weathering devices and natural exposure in Florida, using ten different polyester gelcoats in three different colours: white, red and blue. Changes in gloss and colour were measured after sufficient exposure to cause appreciable, but not excessive, degradation. The best correlation with the Florida results was found using a xenon arc machine and by EMMAQUA, but even with these methods a considerable spread in correlation factors was found, depending on which colour was tested and which property was used as the criterion. The changes could be described in terms of changes in any of the following: ΔL (change in black/white) Δa (red/green), Δb (blue/yellow) or ΔE (total colour change) or in terms of loss of gloss. With the best methods of correlation, acceleration factors of 7–10 times over the Florida weathering rate were obtained. (It should be remembered that there is no guarantee of correlation between Florida results and those from other locations.)

Little other work comparing natural with accelerated weathering using reinforced plastics has been reported, most papers being concerned with unreinforced polymers, surface coatings, fabrics, and so on. For example, results obtained using two types of xenon arc machine with four coloured automotive paints have been compared with results from Florida exposure [6]. This work also compares different instruments of the same type, different sample positions within these instruments, different conditions and different start times, measuring changes in gloss and colour (ΔE) after set periods. (Unrelated difficulties in the actual preparation of identical panels and in overcoming the effect of loss of gloss on colour change – to polish or not to polish – are also discussed.) Surprising discrepancies are reported from machine to machine of the same type. It was, however, concluded from this work that, '. . . If the test method is optimised, the colour difference caused by exposure correlates well with that caused by outdoor exposure in Florida in about 90% of the

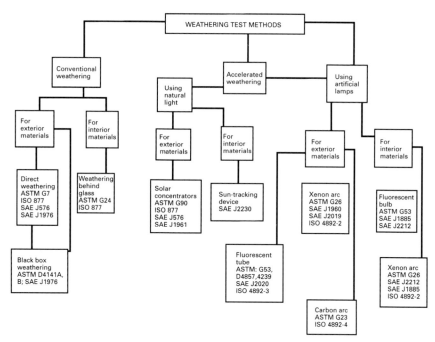

Figure 6.2 Some US and ISO weathering test specifications (after Crump) [5]

cases . . . The results from gloss retention measurements did not correlate quite so well . . .'

Other authors have been less content, commenting in one paper [7] that 'This amount of variability [*between machines of the same type*] makes the use of specifications that require a certain performance level after a specified exposure time highly questionable'. The same authors have also reviewed variations that can occur in superficially identical outdoor weathering tests [8].

It needs to be stated that instruments have improved considerably since these papers were published and that the key to obtaining reliable and reproducible results is in the control of the environmental factors which modern equipment allows and which is required by various standard test methods.

The accelerated weathering of polymeric materials in laboratory instruments is covered by ISO Standard 4892. In the US, ASTM standards G23, G26 and G53 cover such testing, and other ASTM standards are derived from these, as shown in Fig. 6.2. Many car manufacturers have their own acceptance specifications. Reviews of these standards and of factors affecting results have been published [4, 9–11].

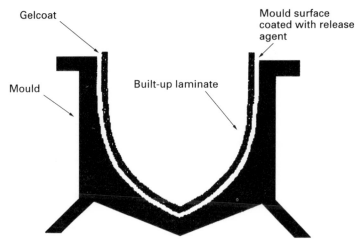

Gelcoat

Mould surface
coated with release
agent

Mould

Built-up laminate

Figure 6.3 Hand lay-up process. Cross-section of GRP boat hull being removed from the mould

6.4   Unsaturated polyester resins

6.4.1 Introduction

Styrenated unsaturated polyesters are the resins most widely used in reinforced plastics for structural and cladding applications. They consist of an unsaturated polyester resin dissolved in styrene monomer, the monomer serving the dual function of acting as a solvent for the resin (making it liquid and hence convenient for impregnating the reinforcement) and as a crosslinking agent to react with the resin to solidify the assemblage in the mould during curing.

6.4.2 Gelcoats and gelcoated mouldings

6.4.2.1 Gelcoats

A widely used method of laminate construction is by the 'contact moulding' process or 'hand lay-up'. In this, a layer of catalysed unreinforced resin, known as the 'gelcoat', is brushed or sprayed onto the mould and allowed to cure and then layers of reinforcing-fibre impregnated with resin are built up behind (Fig. 6.3). Variations of this include resin injection and vacuum-assisted resin injection in closed moulds. After removal of the cured moulding from the mould, the gelcoat forms the outer 'skin' and as such performs the two functions of decoration and protection:

- Decoration – Since the gelcoat is applied directly onto the mould surface, that mould surface is duplicated on the moulding, a high gloss finish resulting from the use of a high gloss mould surface and a textured and/or semi-matt finish from the use of that type of mould surface or release film. Most gelcoats are coloured and opaque and this adds to their aesthetic appeal as well as hiding the reinforcement and any imperfections in the layers beneath.

- Protection – The ingress of water and chemicals into the body of the laminate is substantially prevented by a properly formulated and fully cured gelcoat. Without such a barrier, moisture can be carried in by the reinforcing fibres due to capillary action. The bond between the resin and the reinforcement then becomes weakened, so that the load carrying capacity of the fibres is not fully utilized and strength is lost. The gelcoat is, therefore, a key to protecting the laminate from the environment.

6.4.2.2 Basic gelcoat composition

Polyester gelcoats have traditionally been made by taking a polyester resin and incorporating sufficient thixotropic agent (usually fumed silica) to enable a thick film to be deposited on vertical surfaces without sagging. With the further addition of some accelerator (a cobalt compound) and, if the material is to be a sprayed extra styrene monomer, a basic gelcoat results. In commercial practice, gelcoat manufacturers incorporate other ingredients such as pigments, fillers (to improve flow, abrasion resistance, pigment compatibility, etc.), and additives (to help the release of air bubbles, improve substrate wetting, thixotropy, resistance to UV, etc.).

Gelcoats can be applied to the mould surface by brushing or by spraying. Spraying is usual in most parts of the world but brushing is still popular in some countries such as the UK, especially for marine applications. A brushing gelcoat is normally superior to an equivalent spray gelcoat in its barrier properties (and hence weathering), because extra monomer and/or solvent is required to achieve sprayability, and because additives are needed to overcome a greater tendency towards air entrapment, pigment separation and other problems caused by the spray process. However, application by spray is more convenient than by brush and there is continuing activity by gelcoat producers to improve the properties of spray gelcoats by various means.

6.4.2.3 Polyester resins for gelcoats

The constitution of the unsaturated polyester resin used to make a gelcoat is obviously a prime factor in determining its weathering performance. Base polyesters (which are subsequently dissolved in the monomer) are made by reacting together saturated and unsaturated diacids and diols, and variation of these ingredients enables properties such as flexibility, heat distortion

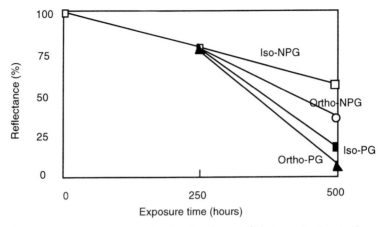

Figure 6.4 Gloss change of gelcoats after artificial weathering [12]

temperature and chemical resistance to be manipulated. The industry-standard resins for gelcoats are isophthalic polyesters, these being made with isophthalic acid as the saturated acid component rather than the less expensive orthophthalic anhydride that is used for general purpose resins. This contributes towards improved toughness, water resistance and weathering properties (Fig. 6.4 and 6.5). Use of neopentyl glycol (NPG) to replace some or all of the commonly used propylene glycol brings further improvements. Such NPG gelcoats (even orthophthalic versions) retain their gloss and colour for longer than the standard isophthalics (Fig. 6.4)

Other diacids and diols have recently been finding favour as ingredients of gelcoat resins. Examples of these are CHDA (1,4-cyclohexane dicarboxylic acid); MP diol (2-methyl-1,3-propanediol); CHDM (cyclohexane dimethanol); 1,5-propanediol; and 1,6-hexanediol. Use of these components can bring about further improvements in weathering properties as well as other desirable features.

Conventional vinyl ester resins are increasingly used in anticorrosion applications and in marine applications. They are essentially epoxy resins, but chemically modified to enable them to cure in just the same way, and with virtually the same chemicals and fabrication procedures as polyesters. They have excellent water resistance and chemical resistance, but poor resistance to UV light, although modifications can be made to make them light stable and suitable for use in weatherable gelcoats [13].

6.4.2.4 Monomers

Styrene is practically the universal monomer used in unsaturated polyesters but it is a known cause of yellowing on outdoor exposure. Complete re-

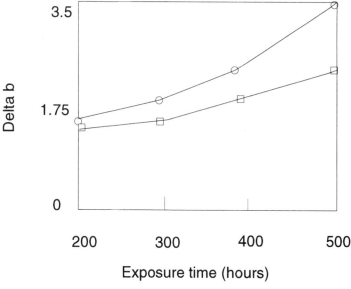

Figure 6.5 Yellowing of gelcoats after artificial weathering [12]. ○, ortho-NPG; □, iso-NPG

placement of styrene is rarely possible because of cure problems with other monomers, but a partial replacement with methyl methacrylate can be shown to decrease the tendency of the cured gelcoat to yellow (Table 6.1), if the lower flashpoint and pungent odour it imparts can be tolerated. Such mixtures are therefore commonly used. Alternatives to methyl metha-crylate are aliphatic diacrylates and dimethacrylates, since these have much lower flammability and odour although they are more expensive.

Table 6.1 Comparison of gelcoats with and without methyl methacrylate after 1000 h accelerated weathering [14]

	Gelcoat with styrene monomer (only)	Gelcoat with styrene: methyl methacrylate (80:20 mix)
ΔE (colour change)	5.31	2.46
Gloss retention (%)	58.02	93.80

6.4.2.5 Accelerators

Accelerators are the ingredients that enable the resin to cure at room temperature after the addition of an appropriate organic peroxide catalyst. Cobalt soaps are commonly used. Cure rates can be increased by certain coaccelerators but aromatic amines such as dimethyl aniline should be avoided in gelcoats as they cause yellowing, both initially and on weathering.

6.4.2.6   UV absorbers

UV absorbers are commonly used in gelcoats to absorb UV light and dissipate the absorbed energy. Although, in the first few years of life in temperate climates, the use of a UV agent makes little difference to the weathering properties of a good quality pigmented gelcoat, experience shows that patchy yellowing of white gelcoats which sometimes occurs in these climates can be overcome by such means. It is thought that one of the causes of this type of yellowing is areas of the gelcoat that have a lower pigment content for various reasons. The incorporation of UV agents helps overcome the yellowing of the resin which shows up in such areas.

In the longer term and under conditions of higher levels of radiation, such as subtropical and accelerated artificial weathering, UV agents can more easily be shown to improve gloss retention and colour stability. Traditional UV absorbers that have been used successfully for many years are benzotriazole and hydroxybenzophenone derivatives. It can be advantageous also to include another type of UV agent, known as a hindered amine light stabilizer (HALS), which does not absorb UV radiation but which acts to absorb any free radicals that have been formed [14, 15]. Selection of the right type of HALS is important because some react adversely with acidic, metallic and brominated compounds and with certain organic pigments. Figure 6.6 shows the effect of various UV agents on light stabilisation of a flame-retardant unsaturated polyester [15].

6.4.2.7   Fillers

In general, the use of fillers in gelcoats increases the rate of gloss loss on weathering. Small amounts of a filler such as powdered clay or talc may, however, be needed to improve application properties.

6.4.2.8   Fire retardant additives

The use of fire retardant (FR) additives, and/or halogenated FR resins in gelcoats detracts from their weatherability and for external applications it is good practice to achieve fire retardancy by using a highly FR laminating resin behind a non-FR gelcoat.

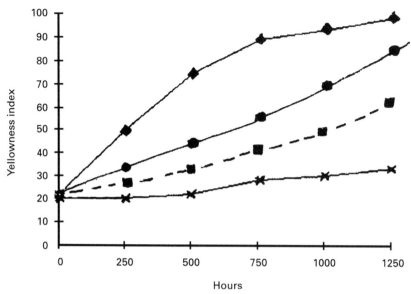

Figure 6.6 Effect of various UV agents on light stability of a flame retardant unsaturated polyester [15]. ◆, Benzophenone type; ●, benzotriazole A; ■, benzotriazole B; ×, benzotriazole + HALS

6.4.2.9 Pigmentation

Most gelcoats are used in a pigmented form, so that they hide the substrate and provide a choice of colour. Careful selection of pigments is of importance since this can affect not only the durability of the finished product, but also other important gelcoat attributes such as rheology, cure, toxicity, bleed (i.e. the tendency for a colour to migrate out of one article, such as a GRP mould, into an adjacent one, such as the product being moulded) and cost. The pigments listed below satisfy the prime requirements for the coloration of gelcoats, and some properties which affect weatherability are discussed.

Rutile (rather than anatase) titanium dioxide is the preferred pigment for making white and pastel gelcoats and this can be incorporated in powder form by high shear stirring on specialized machinery. For full hiding power, 15–20 phr (parts per hundred resin) is required in the cured gelcoat film. Unsuitable titanium dioxide grades can hasten the onset of chalking by catalysing photolytic oxidation of the binder. Chalk-resistant grades reduce this tendency by having deposited on their particle surfaces traces of other chemicals, such as silica and/or alumina [16].

Unlike titanium dioxide, most other pigments need to be finely pre-dispersed in a carrier resin, or 'medium', by grinding equipment, before adding to gelcoats. Such dispersions are known as pigment pastes and the carrier resin used is often a monomer-free, liquid, unsaturated polyester resin, because this gives a long shelf life to the pastes. Although this carrier will crosslink with the gelcoat resin on curing, it usually has a slightly deleterious effect on physical and chemical properties of the final gelcoat, including weather resistance. In fact, the medium can have a more detrimental effect than the pigments themselves and it is preferable for the moulder who requires the best weather resistance properties to purchase prepigmented gelcoats, rather than to colour his own by adding bought-in pigment pastes. This is because the gelcoat manufacturers are often able to avoid or reduce the usage of a pigment dispersing medium in their formulations.

Many furnace- or channel-type carbon blacks produce gelcoats with good durability of colour and gloss since they act as excellent UV stabilizers. However, if an unsuitable carbon black is chosen, it can retard cure and hence cause unsatisfactory durability.

Phthalocyanine blues and greens are the workhorse pigments for these colours but they are invariably used in conjunction with others that improve their opacifying powers. Such colours usually exhibit good permanence.

Ultramarine blues can be used in temperate climates if high quality grades are chosen but tend to fade under more extreme conditions.

Chrome oxide greens are pigments of excellent colour stability, although they lack brightness.

Iron oxide blacks, reds and yellows confer good durability but they are also dull colours. They can affect the durability of certain other pigments, notably some metal toners and some organics.

Lead chromes and molybdates are available in bright yellow, through orange to red. Provided that light fast and sulfur-resistant grades are chosen, they will perform satisfactorily, although they do darken on extended exposure (especially the reds). The darkening of these colours is usually to be preferred to the fading of others, since they remain essentially red, orange or yellow, and as this is a uniform change it often goes unnoticed unless an area has been covered, for instance by a label. There is now a general move away from these pigments on ecological and health grounds.

Cadmium pigments are also available in yellow through to red and these have superior light stability to the lead pigments previously described. (Cadmium yellows, however, fade rapidly if used with large amounts of titanium dioxide.) Cadmium pigments are currently no longer permitted for use in many polymers in several parts of the world, unless they are necessary for safety or related purposes and unless it can be proven that there is no alternative.

In order to replace cadmium and lead pigments with others of similar durability and brightness, it has been necessary to look to 'automotive-grade' pigments free from toxic heavy metals. These, in many cases, have caused other problems, not least of which has been a huge increase in cost. In addition to this, some organic types have poor opacifying properties and large amounts therefore need to be used, further increasing cost and detrimental effect. Typical of the inorganic automotive pigments that can be used successfully in gelcoats are the bismuth vanadate and nickel titanate yellows. Cerium sulfides are also contenders. Appropriate organic automotive pigments include the quinacridone and diketopyrrolo pyrrol reds and the benzimidazolone and isoindolinone yellows.

6.4.2.10 Metallic flakes and powders

Metallic flakes are often used for decorative effect in transparent spray gelcoats. These are sometimes unsuitable for exterior exposure applications and advice should be sought from manufacturers. An effective UV absorber is required because of the potential for yellowing of the clear gelcoat matrix.

When gelcoats are highly filled with metallic powders, an appearance similar to the metal itself can be obtained, especially if they are abraded after curing. The appearance after weathering of such a finish is usually similar to that which would be obtained with the metal itself. For example, a gelcoat highly filled with copper powder can be used architecturally for a copper sheet effect. This will eventually attain a convincing patina, the formation of which can be chemically accelerated with mineral acid.

6.4.2.11 Dyes

Dyes can be used to obtain transparent colours in polyester resins. Considerable care in their selection needs to be taken as most will quickly fade on the addition of catalyst to the liquid resin or on exposure of the cured resin outdoors. However, certain quinone-type aromatic-soluble dyes in the yellow-red range are extremely lightfast and transparent blues and greens can be satisfactorily produced using small quantities of the corresponding phthalocyanine pigments.

6.4.2.12 How long will it last?

Frequently, fabricators or their customers will ask a manufacturer how long a coloured gelcoat will last in an exterior exposure application. Such a question is, of course, unanswerable because all materials change in time, and who is to say when a user will find a change unacceptable? Additionally, apart from climatic effects, there are numerous other factors that will affect the rate of change and which are not in the hands of the gelcoat manufacturer, some being controlled by the fabricator or by the final user.

The two effects of weathering upon gelcoats which are soonest seen are change in colour and loss of gloss. In respect of these changes, many premium gelcoats, if properly fabricated, will have durability comparable to modern automotive finishes. However, products are occasionally not up to standard and not all moulders and users are paragons of virtue. It is necessary, therefore, to consider possible defects and their causes.

6.4.2.13 Fading and darkening

Colour fading or darkening without loss of gloss can be caused by the use of unstable pigments or pigment combinations which change colour after exterior exposure. For instance, many organic pigments which are otherwise stable will fade when they are used in combination with white to make pastel shades, due to the photolytic activity of titanium dioxide. If a judicious choice of pigments is made, then little colour change should be seen before the onset of gloss reduction or chalking, apart from surface dirt which can be removed by cleaning. (However, fading/darkening changes may also be due to factors such as unsuitable additions to the gelcoat before application to the mould, undercure, use of harmful cleaning agents, and so on.)

6.4.2.14 Yellowing

A colour change may sometimes be seen which is due to the darkening of the base gelcoat resin, especially in whites, where it usually causes a yellowing effect. This can be overcome by using a more UV-resistant resin and better UV screening agents, and by ensuring good cure. Other causes of yellowing are moulding faults such as failure to clean buildup of polystyrene from the mould surface, pre-release of the gelcoat, high exotherm of the backing laminate during cure and use of an amine-containing resin behind the gelcoat.

6.4.2.15 Blooming

Another phenomenon that can occasionally occur is the migration, after exposure, of an incompatible pigment or additive, to the surface of the gelcoat to give a matt, faded appearance. This is called 'blooming' and it is due to the incompatibility of that component, or of its degradation products. Certain organic pigments can be a cause of this. Although bloom can easily be removed by polishing, this is often only a short term solution as it can quickly reappear on further exposure.

6.4.2.16 Loss of gloss and chalking

Loss of gloss is normally brought about by erosion of the surface layer of the gelcoat due to chemical and/or physical damage. The colour of the gelcoat then appears to whiten, due to the diffused reflection of light from the matt surface. This is most serious with strong, bright colours because it is on mouldings in these colours that the phenomenon is more easily observed; they look old and drab as the gloss starts to disappear. On paler colours the effect is less readily noticeable and for some applications appearance remains acceptable, especially from a distance, long after the onset of loss of gloss. Indeed, the whiteness of a white structure can even improve by this means because surface dirt is shed, leaving a fresh exposure of white pigment. This phenomenon is termed 'chalking'. A small degree of maintenance such as appropriate cleaning and polishing can retard or prevent the onset of gloss loss and chalking.

It is, of course, erosion of the gelcoat after many years of service that can bring about the eventual mechanical failure of the laminate, if it is not otherwise damaged, by exposing the reinforcement underneath. However, the onset of gloss loss or chalking does not presage the immediate disappearance of the gelcoat which normally lasts for many years longer.

6.4.2.17 Effect of fabrication quality

The time taken for the onset of gloss loss and colour change is affected by the composition of the gelcoat and by climate, as discussed earlier. However, the moulder also has a part to play. If he or she over- or under-catalyses the gelcoat, adds extra monomer, or solvent, applies at too low a workshop temperature, applies too thinly or otherwise departs from the manufacturer's recommendations, the moulder will reduce the durability of that gelcoat. Conversely, good working practices will maximize durability. The use of a textured finish can also prolong aesthetic life.

The resin behind the gelcoat does not usually affect weathering properties, although the choice of a vinyl ester for this purpose has been said [17] to cause and/or accelerate yellowing, possibly because of amine accelerators used for cure.

6.4.2.18 Customer care

The final user of the moulding can also deleteriously affect the weathering properties by using the wrong cleaning agents based upon strong alkalis, solvents or abrasives. Conversely, a beneficial effect can be obtained by the occasional use of a mild cleaning agent and the application of a wax specially formulated for the upkeep of GRP. A breathable fabric covering will also extend life.

6.4.2.19 Angle and direction of exposure

Of major importance for architectural applications is the effect of direction and angle of exposure on weathering. Thus GRP on one side of a building will deteriorate at a different rate from that on another, and horizontal panels will deteriorate at a different rate to areas which are set at other angles, such as roofs and window surrounds. Shading effects from nearby buildings, trees, and so on will also be of relevance.

6.4.2.20 Atmospheric pollution

A properly formulated and cured gelcoat will withstand most chemicals in the concentrations likely to be found in the atmosphere for many years, although sulfur dioxide and related chemicals can darken gelcoats containing lead-containing pigments; this may be seen near engine exhaust outlets. It has been known for complaints to be made of deterioration of gelcoats which have later been found to be purely due to the deposition of dirt from the atmosphere, for example, at bus and railway stations. Such deposits can usually be removed without damaging the gelcoat, by mild detergents or special cleaners designed for this purpose. As previously stated, harm can result from the use of abrasives, solvents, strong acids or alkalis and of cleaners based on these materials.

6.4.2.21 Fungal growth

Cured polyester resin is not subject to fungal attack, but when the surface has eroded sufficiently, it will trap spores carried in the air which may grow and become unsightly. These can be removed by cleaning. Polishing or the application of a surface coating will delay their reappearance.

6.4.2.22 General recommendations

Manufacturers' recommendations should be followed for the selection of gelcoats with good weathering properties. A standard isophthalic marine gelcoat will prove satisfactory for many purposes, but more expensive premium products may be required for some. Caution is, however, recommended in accepting some of the claims concerning comparative weatherability test results. The difficulty of relating test data to real-life performance has been pointed out earlier in this chapter, and if procedures such as hot postcuring the panels before testing, or excessively cleaning them after testing have been used, then unduly favourable results can be presented!

The best colours to use for long life architectural applications are buffs, creams, off-whites, whites, greys and pastel shades; the least suitable are the bright, full shades of red, green, yellow, blue, brown, and so on. This is because eventual deterioration will occur by loss of gloss and chalking, which is far less visible on the first mentioned colours than on the second. Lighter colours reflect more heat and this also slows deterioration. A degree of texturing, ribbing or patterning is to be preferred to an unbroken high gloss surface.

Different considerations may apply to other applications. A car body, for instance, would continually face different directions and could reasonably be expected to be cleaned and polished, so that bright colours and a gloss finish are more appropriate.

6.4.2.23 Practical expectations

As previously described, natural weathering exposure tests for general applications are normally carried out by exposure of panels facing the equator, at an angle of 5° or 45° to the horizontal. Under these test conditions in a temperate climate, a typical general purpose isophthalic pigmented gelcoat will last about 4–5 years before noticeable loss of gloss; premium 'superior weathering' gelcoats can be expected to take up to twice as long. In subtropical test stations such as Florida, these times are considerably shortened and some loss of gloss from general purpose gelcoats will often be noticed after only one year's exposure. It must be emphasized that under practical conditions much longer times than those indicated will be experienced before the onset of loss of gloss in these various climates, because of less severe exposure angles, directions and periods. Also because mouldings are often cleaned, polished or otherwise maintained, boats and buildings exist which are still in an acceptable condition after more than 30 years.

The rate of deterioration of gelcoats in cold climates is generally reduced when compared with that in temperate or subtropical ones, provided that embrittlement does not result in stress cracking or impact damage that would allow penetration of moisture into the substrate.

6.4.2.24 Case histories – buildings

A 50 ft (17 m) GRP church spire was erected in Smethwick, near Birmingham, UK, in 1961 and when the church was demolished 35 years later, the spire was taken to the Avoncroft historic buildings museum at nearby Bromsgrove. A crude orthophthalic gelcoat, to which phthalocyanine green

Figure 6.7 Modular GRP stores building

pigment paste had been added, was used on the exterior. On recent (1996) close-up examination of the spire, the gelcoat surface was seen to have entirely disappeared leaving a fibrous surface of chopped strand glass mat embedded in green resin. The approximately 3 mm thick laminate however appeared to be structurally sound and good for many more years, since the glass and a certain amount of ingrained dirt were protecting the under-lying structure and the steep angle allowed all water to drain off im-mediately. The intention at that time, however, was to recoat the surface with an isophthalic flowcoat (gelcoat plus wax to overcome air inhibition of cure) in the original colour and to re-erect the spire on a brick base at the museum.

A number of houses and apartment blocks were constructed in Harrow, north of London, UK, in 1966, featuring 'Resiform' preformed load-bearing panels made from a mixture of polyester resin, sand and alum, reinforced with chopped strand glass mat. Behind this was a mineral wool insulation layer, covered with plasterboard, and integral metal fixing points were provided. The surface was textured in a number of variations. When in-spected in 1996, they were in exceptionally good condition, although some had been painted. Local authority housing officials who recently refur-bished one of the sites, expressed their satisfaction with the performance of the panels.

A modular GRP stores building (Fig. 6.7) was constructed at Wollaston (a semi-rural location in central England) in 1968, to demonstrate the suitability of this material for construction purposes. The walls were of a folded plate design made with filled, FR resin and a green (chrome oxide) pigmented FR gelcoat; the roof utilized a translucent FR resin and a translucent FR gelcoat. When examined 28 years later, the green gelcoat on the outer walls had some general grime and algae growth, but when cleaned was found to be smooth, uniformly green but completely matt (it was originally semi-gloss). There had been some impact damage from low-level traffic. Vertical, corrugated infill-panels using a bright green (phthalocyanine) pigment were also in good condition in respect of both colour and surface uniformity, again apart from some vehicular damage at edges. However, on the translucent roof, only traces of the original FR gelcoat remained and it had badly discoloured. Some areas where the gelcoat had been lost were smooth, but others had patches of exposed reinforcement. However, the structure was mechanically sound and light transmission inside the building was still sufficient to avoid the need for artificial lighting during the day. Fire-retardant gelcoats would not now be recommended for exterior exposure and today's materials would undoubtedly have a much improved service life.

Mondial House is a multistorey building in central London, built in 1974. It is clad with GRP panels faced in white isophthalic gelcoat, with a filled, FR resin behind. These extend from above the windows on one floor to form a parapet on the floor above. The top and underside of each panel have a smooth surface, whilst the vertical faces are ribbed. The original finish was semi-gloss. This building was inspected 19 years after construction and the panels were found to be in good condition. The gelcoat had not yellowed, chalked or stained although the surface was completely matt. Except for one or two areas of porosity on the north facing side, there was no serious surface degradation. Aesthetically, the building looked good but for close inspection would have benefited from cleaning. Structurally the panels were sound although there was water ingress in a few areas of unrepaired damage.

Some GRP dome structures supported by metal frames were constructed in the mid-1970s for the Sharjah International Airport Terminal Building, United Arab Emirates and these were recently inspected. The cream, isophthalic gelcoat was complete but appeared dull when viewed close-up and a few blisters were seen. However, the laminate structure was sound with no signs of delamination.

A less sanguine view of the architectural use of GRP is taken by Fidler who reviewed various projects, mainly in connection with building restoration work [18]. Several incidences of unacceptable deterioration of newer replacement items when compared with the longevity of more conventional constructional materials are cited, although much of this appears to have been due to poor workmanship.

Figure 6.8 The *Neptune*, a 9m GRP vessel

6.4.2.25 Case histories – marine

In 1956, a GRP motor cruiser, the *Bebe Grand*, was constructed in the UK, and was taken to South Africa in 1958, where it has remained since. Despite being involved in two serious accidents, the boat is still in frequent use. The gelcoat was painted above the waterline in 1985. Another vessel built around the same time was the 8 m cruiser *Norwood*, which utilized a foam sandwich construction. This vessel has also since been painted but a review by our editor [19] reports that it was still in good seagoing condition in 1996.

There are many examples of elderly GRP boats in various parts of the world that are still in their original gelcoats. *Neptune*, a 26 ft (9 m) Creighton Inlander built in 1968 is a typical example recently (1996) inspected at random in the UK (see Fig. 6.8). When examined, its white gelcoat was seen to be in good condition, although it was somewhat dirty; there was one star crack on the deck, probably where a heavy weight had been dropped and a few isolated craters on the front of the superstructure. On polishing a small area of gelcoat, a glossy surface could be obtained which was uniformly white and there was every indication that such a finish could have been obtained all over with a little effort. Nearby were three more examples of boats built before 1976, one of them having recently been cut back and polished to good effect.

6.4.2.26 Surface coatings other than gelcoats

For a few applications (e.g. some automotive parts) it is the practice to apply an additional surface coating onto the gelcoat after fabrication of the part; this is often a two-pot polyurethane or an epoxy type. A smooth surface suitable for receiving such a coating may be obtained by using a sandable gelcoat based on an orthophthalic resin plus a soft filler such as talc. This is lightly abraded and any traces of wax or silicone release agents are removed before application of the coating.

Similar two-pot surface coatings are also used to refurbish weathered articles, such as boat hulls, fairground decorative mouldings and kiosks. The durability of these coatings is well suited for further extended exposure.

6.4.3 Transparent and translucent sheeting

Flat and corrugated transparent and translucent glass-reinforced polyester sheets are used in areas of buildings requiring light transmission. For glass-like transparency, the refractive index of the resin in the cured state is matched to that of the glass fibres, usually by substitution of some of the styrene monomer with methyl methacrylate. As with gelcoats, this step also improves gloss retention and resistance to yellowing on exposure [20].

6.4.3.1 Laboratory studies

Some studies on unpigmented unreinforced polyester resins [21,22] have shown that rapid yellowing is caused by a narrow band of the UV spectrum around 330 nm and that this effect varies little with resin composition. UV of other wavelengths can reverse yellowing by bleaching and so, in practice, this effect is lessened. The tensile strength of resin castings increases initially on exposure, due to postcuring effects which temporarily override degradation but strength is reduced in the longer term [23].

The internal structural breakdown of reinforced sheeting has been investigated by acoustic emission and by scanning electron microscope observations [24], and low crosslink density resins are recommended for best durability. Other authors have also investigated the effect of varying resin ingredients and have shown, for example, the advantages of NPG over propylene glycol and of isophthalic acid over orthophthalic anhydride in strength retention properties [20,25].

6.4.3.2 Practical effects

The installation of a panel onto a sloping roof approximates to standard outdoor weathering test conditions, which are severe. The sheeting may

Table 6.2 Gloss retention of translucent sheeting with various surfaces after exterior exposure in UK [26]

Type of surface	Initially (%)	2 years (%)	4 years (%)	5 years (%)
Gelcoat	100	98	97	97
Polyester tissue	100	95	92	92
Glass tissue	100	93	82	78
No tissue or gelcoat	100	75	30	12

have no gelcoat to protect it, nor any opacifying pigments to absorb or prevent the penetration of UV radiation. Hence it is not surprising that early examples suffered from premature erosion of the surface resin, leaving exposed glass reinforcing mat that was unsightly and mechanically weakening. Breakdown starts at points where the fibre bundles are located near the surface and it can be assumed that cyclic loading due to temperature fluctuations causes much of this because of the widely differing thermal expansion coefficients of the glass reinforcement and polyester resin; also because of differential swelling of the two components when moisture is absorbed. Improvements are brought about if fibres are kept away from the surface by the use of tissue-induced resin-rich surfaces, gelcoats, or the incorporation of a film of polyethylene terephthalate or polyvinyl fluoride on the exterior surface. Table 6.2 illustrates the effect of some of these steps on gloss retention [26].

With good quality modern polyester sheeting, loss of light transmission after 5 years in the UK is said to be less than 10% (after cleaning) and a 'service life' up to 30 years is achieved. A study of physical property retention after 7 years exposure has reported bend strength falling by less than 30% and tensile strength by less than 20% [27].

Weathered GRP sheeting, even with exposed fibres, can be renovated using a polyester flowcoat, or an acrylic lacquer containing UV agents specially formulated for this purpose, so that its mechanical and aesthetic life is considerably prolonged. Much of the market for transparent and translucent GRP sheeting has now been lost to unreinforced polycarbonate which has excellent light transmission retention properties when exposed and which can be extruded in double- and triple-glazing form.

An area of the market which has been well retained by polyesters is fire-retardant translucent sheeting. To obtain a good degree of fire-retardancy with an unfilled polyester resin, the use of a halogenated resin and/or halogenated additives is required; most of these detract from good weathering properties. However, polyesters based on dibromoneopentyl glycol (DBNPG), when combined with efficient UV absorbers, can be used to produce glass-reinforced sheeting with the necessary degree of fire and weather resistance. Such products, although they still yellow more rapidly than the non-FR versions, are superior in longevity to their unreinforced PVC equivalents.

6.4.4 Hot-moulded compositions

Sheet moulding compounds (SMC) and related polyester bulk and dough hot-moulding materials such as BMC and DMC, retain their physical properties well enough on exposure but, since they are heavily filled, their appearance tends to deteriorate rapidly [28]. Uncoated hot-moulded compositions have, however, been utilized for applications requiring exterior exposure, such as motor car bumpers (fenders) and protective panels, as well as meter boxes and water tanks. The use of neutral colours such as white, cream or grey, and of surface textures such as ribbing, and the avoidance of a high gloss finish helps to retain an acceptable appearance.

Other applications have utilized a paint finish after moulding. The performance of such a finish is considerably improved by use of an in-mould coating as a primer; this is applied as a liquid or powder onto the mould surface before or during the moulding cycle. Recent developments have lead to pigmented in-mould coatings with sufficient gloss and colour stability to enable them to be used as the final finish, without subsequent painting [29].

Many automotive applications require the part to be coated at the same time and with the same finish as the rest of the assembled metal components and so such finishes will have the durability normally associated with the modern motor car, provided that they continue to adhere.

Some low pressure/low temperature moulding materials can be used with a conventional or modified conventional polyester gelcoat which is applied to the warm tool, prior to moulding.

6.4.5 Miscellaneous polyester applications

Pultruded sections are used for handrails, cable trays, ladders and even bridge construction. The parallel orientation of the continuous fibres ensures that they are protected with a covering of resin, which enhances weathering properties; a polyester veil can be used to improve protection. Despite this, some deterioration in surface gloss is experienced fairly rapidly on standard pultrusions if a surface coating is not used, but mechanical properties are retained for many years.

A good example of a natural product combined with a synthetic product to maximize the properties of both is the stone-chipping faced polyester building panel. The chippings, which can be of various sizes and colours, are embedded in the surface layer of a heavily filled, glass fibre reinforced sheet, produced on a continuous machine and cut to standard sizes. Such panels are used extensively in building construction and very little change after 13 years of service has been observed. From accelerated tests, a service life exceeding 30 years is forecast [30].

6.5   Phenolic resins

Phenolic resin-based composites have the virtue of being inherently fire retardant and give off little smoke when they do burn. Applications include tunnel linings and train components including whole front ends. Offshore oil industry structures such as gratings and firewalls are also candidates for phenolics. Composites based on these resins are available only in dark colours and are not suitable for long term external weathering applications without protection because of erosion and high moisture absorption, leading to distortion. Conventional two-pot polyurethane and epoxy surface coatings can be postapplied to phenolic composites but will usually compromise the desirable fire properties. Until recently, gelcoats for phenolics have not been available. However, it has been announced [31] that one has been developed based on a filled FR acrylic system which is pigmentable and resistant to chalking. Other similar developments are expected.

Phenolic laminates are also used in rocketry and space travel applications as heat shields during exit and re-entry into the earth's atmosphere. At very high temperatures, the surface burns away to leave a carbonaceous layer that glows and which is gradually sacrificed by ablation, protecting and insulating the under-surface for a considerable period.

6.6   Epoxide resins

Epoxide resins are known for their toughness and chemical resistance properties. Because of this, they are widely used for composites in critical applications such as aircraft and aerospace, oil rigs, chemical storage tanks and high-speed boats, often in conjunction with carbon or aramid fibres.

For such critical applications, moisture uptake at high humidities can lead to changes in physical properties that are vitally important, especially as service temperatures varying between $-30°C$ and $+110°C$ can be encountered. Much study has been devoted to this (e.g. Wright [32] and Eckstein [33]). Water diffuses into even the most resistant resin, especially if the surface is damaged by weathering. It both plasticizes and swells the resin and reduces the bond between it and the reinforcement. For conventional epoxy resins, for each 1% water pickup, T_g decreases by about 20°C [32,35]. Eckstein [33] examined 70 different epoxy resin formulations and found that water absorption could vary by a factor of 10 for different resin types and by a factor of three for different hardeners used with the same resin. In the aircraft industry, exterior composites are invariably protected by surface coatings; in critical areas, polyvinyl fluoride film is applied and on leading edges that are subject to rain erosion, a polyurethane rubber coating or stainless steel sheathing may be used. A long term study of the resistance to tropical weathering of the graphite fibre/epoxy composite used in F/A 18 aircraft is still in progress in Australia and Malaysia but interim reports [34] suggest that this material, using Hercules AS4/3501–6 material, will be very durable.

For the less critical applications, standard (epichlorhydrin/bisphenol A) epoxy resins retain their physical properties well, but are particularly subject to the adverse effects of UV radiation on their appearance. This causes rapid yellowing and chalking due to their aromatic structure and UV agents are ineffective. Therefore, all mouldings made with these resins which are to be subjected to extended outdoor exposure will benefit from protection by surface coating or shading.

Aliphatic and cycloaliphatic epoxy resins have less tendency to yellow and can even be used as a basis for epoxy gelcoats. Such gelcoats are not widely available due to difficulties in application and adhesion to the substrate, as well as high costs. Use of non-aromatic anhydride curing agents such as hexahydrophthalic anhydride will produce further improvements in colour stability and these systems are suitable for high temperature curing applications.

Conventional polyester gelcoats are sometimes used on epoxy laminates but care needs to be taken to ensure that a sufficient permanent bond between them has been obtained. Tie-coats are available for this purpose.

Weather-stable epoxy surface coatings are widely available and are sometimes used to protect polymer composites; however, the resins used for these coatings are not suitable for use as gelcoats or matrices for reinforced plastics.

6.7   Other thermosetting polymers

6.7.1 Silicone resins

Silicone resins based on polyorganosiloxanes have excellent inherent resistance to UV and weathering and are used for making reinforced compounds and laminates which also have valuable electrical insulation properties and heat resistance. (Silicone modified acrylate surface coatings containing UV agents are used to protect various other polymers, e.g. transparent polycarbonates for glazing and car headlights.)

6.7.2 Polyurethanes

Polyurethanes are used in reinforced resin injection moulding (RRIM) for parts such as car bumpers. These materials are not normally stable to UV light and discolour rapidly when exposed. However, if they are pigmented with carbon black, and/or stabilized with UV agents [36] they can be used for exterior applications without protection other than a layer of unreinforced material which is sprayed onto the mould surface (in the manner of a polyester gelcoat) before injection to provide a protective barrier against the exposure of the reinforcement. Aliphatic polyurethanes generally show better colour and strength retention properties after weathering than aromatic-based polyurethanes. Surface coatings, both in-mould and post-applied, can also be used on urethane RRIM mouldings to achieve improved weatherability.

6.8   Thermoplastics

6.8.1 General purpose thermoplastic polymers

The weathering resistance of thermoplastic polymers can vary between excellent and poor. Many of the pioneer thermoplastics, such as polystyrene, cellulose nitrate and polyolefins proved to have poor resistance to weathering. Polystyrene homopolymer, for example, is susceptible to yellowing and crazing on exposure to both UV and visible light.

It was soon found that many polymers with poor weathering properties could be transformed by compounding with stabilizers such as antioxidants, UV agents, carbon black, and so on, into materials which were suitable for external use, well-known examples being synthetic rubbers, unplasticized polyvinyl chloride (UPVC) and polypropylene.

The use of carbon black pigmentation/stabilization in thermoplastics for external exposure is widely applicable but has to be undertaken with care since it increases radiant heat absorption and hence the possibility of distortion. The type, particle size and degree of dispersion of the carbon are important for optimum efficacy. Additions in the order of 1–3% are generally sufficient for this purpose but in rubbers, where the carbon black also has a reinforcing function, considerably more is used. Other pigments have varying effects, an example being iron oxide which protects polyolefins yet catalyses the decomposition of PVC.

Polymethyl methacrylate (PMMA), exceptionally amongst the early thermoplastics, was found to be very weather resistant, even in non-stabilized, unpigmented form. Pure PMMA sheet, after 17 years exposure in New Mexico, suffered only 10% loss of light transmission, due to slight surface roughness [37]. This can be further improved with the aid of UV agents. Coating or laminating an acrylic film onto the surface of other thermoplastics can be used to improve their outdoor performance.

Polyvinyl chloride is an example of a polymer that is intrinsically unstable to heat and to UV exposure but which can be compounded with stabilizers into a material widely used for external plumbing and glazing application. In this instance, stabilization involves provision for the absorption of hydrogen chloride decomposition product, as well as stabilization with UV agents and/or pigments such as carbon black. White PVC compounds are extensively used for cladding buildings, an application for which they prove quite durable, and large areas of the material are exposed to the weather. Investigations of specific aspects of the weathering of the outer layer after 11 years in south-facing Canadian (Ottawa) locations have recently been reported [38].

6.8.2 Engineering thermoplastics

Many of the so-called 'engineering thermoplastics' retain their outstanding physical properties well after external exposure, if properly compounded with stabilizers, carbon black, UV agents, and so on [39]. Retention of colour and surface gloss is more problematical and protection with a surface coating is often required, film laminating and metallization being alternatives to conventional types.

Polyamides such as nylon 6 and 6:6 absorb up to 10% moisture after moulding as well as being susceptible to UV and visible light degradation. Unstabilized polymer therefore exhibits rapid change in appearance on exposure, involving yellowing, dirt retention, microbiological growth, cracking and crazing. Other types such as nylons 6:10, 11 and 12 have lower moisture absorptions, but all types, when stabilized with carbon black and/ or UV agents, show much improved outdoor properties.

Polyether ether ketone (PEEK) has poor resistance to UV. Reinforced grades are available and, where some degree of UV resistance is required, carbon black may be added. Polyphenylene oxide (PPO) and polysulfone are susceptible to photodegradation due to their aromatic content, although PPO in its usual form as a stabilized polymer alloy is somewhat better. It is inadvisable to use any of these polymers for prolonged exposure without protection.

Polyacetals are also unstable in unmodified form but are often copolymerized to overcome this. Incorporation of carbon black and UV agents further improves matters.

Polycarbonates have good stability, being well-known glazing materials; however, they also need UV agents and a barrier coating to prevent premature yellowing. Carbon black is an alternative if transparency is not required.

Polyethylene terephthalate (PET) is susceptible to deterioration by UV at wavelengths below 315 nm, causing embrittlement, crazing and yellowing of thin films and sheeting [40]. Both PET and polybutylene terephthalate (PBT) bulk polymers, however, are used in stabilized form for exterior mouldings, sometimes in admixture with other polymers such as acrylo-nitrile styrene acrylate.

Acrylonitrile-butadiene-styrene copolymer (ABS) has generally poor weathering properties; styrene-acrylonitrile (SAN) is better if properly stabilized.

Many fluorinated polymers have excellent weathering resistance. Ethyl-ene/tetrafluoroethylene copolymers are particularly appropriate for film manufacture and developments have included examples reinforced with woven glass and perforated metal foil which can be used outdoors [41]. Fluorinated polymers reinforced with glass fibre are used for roofing applications.

Publications dealing with the weathering properties of polymers *per se* in more detail are available (e.g. Rabek [1], Davis and Sims [35], Plastics Design Library [39], Halliwell [42]).

6.8.3 Reinforced thermoplastics without fibres

The reinforcing effect of carbon black on rubbers has been previously mentioned. Thermoplastics are also sometimes classed as 'reinforced' when they contain non-fibrous mineral fillers. These fillers are used to increase the stiffness of an otherwise low modulus polymer, perhaps one that has been elastomer modified. A typical example of this is filled, elastomer-modified polypropylene which is widely used for car body components such as sill covers and side mouldings. Fillers adversely affect the surface appear-ance of such materials so components are often moulded in black or dark grey with a grained finish. There is a tendency for even these colours to fade on weathering, which can be exaggerated by surface contamination with polishes designed to preserve paintwork. (Proprietary products are avail-able for restoring their colour.) Their mechanical properties remain satis-factory for the life of the vehicle.

6.8.4 Fibre-reinforced thermoplastics

Like the thermoset resins, thermoplastics can also be fibre-reinforced with, for example, short glass fibres to improve stiffness and strength, but at the expense of ductility. Such materials are usually fabricated by injection moulding although some interesting techniques are under development for pultrusion, filament winding, stamping, and so on. Surface appearance of injection moulded items is generally affected detrimentally by the reinforcement, moulding discontinuities such as knit lines and wall thickness transitions being emphasized. Texturing is therefore often used to improve surface appearance but, unless the reinforcement can be protected by a surface coating, it is still prone to be exposed by UV attack or erosion. Such exposure is aesthetically unacceptable for many applications and appearance is not improved by the dirt retention and microbiological growth which can accumulate on the uneven surface. However, the presence of fibrous reinforcement does mean that any surface fissures, which would otherwise propagate through the mass of the moulding, are initially arrested, and so strength and integrity are maintained until the matrix/fibre bond is weakened by moisture ingress as discussed earlier in this chapter.

Some studies of unreinforced polystyrene mouldings that had been irradiated with UV while under an applied tensile stress showed that molecular scission is accelerated by the stress [43]. This could have interesting implications for weathering testing conditions which are generally unstressed.

Similar studies were made on polypropylene [44], both with and without glass fibre reinforcement. Surface effects were again seen with both materials but UV penetration was considerably reduced by the presence of glass.

A study of polystyrene and polyether sulfone, with and without 20% glass reinforcement [45], found that, after one year's exposure in Saudi Arabia, strength and ductility were retained better in the reinforced polymers, although the values in all cases fell to 50–77% of their original values. Other studies have shown that short glass fibres cause a significant increase in as-moulded residual stress in such polymers as well as in semi-crystalline polymers [46,47]. This can lead to stress cracking in service.

To summarize, many fibre-reinforced thermoplastics are not suitable for extended exterior exposure where the maintenance of appearance is important unless their surface is protected. Some unprotected applications that are found include car door handles, rear-view mirror housings and body panels (for all of which polyamides, polyethylene terephthalate and polybutylene terephthalate are used), as well as parts for caravans, boats and snowmobiles.

6.9   Reinforcements

The choice of fibrous reinforcements can have as important an effect on moisture and hydrolysis resistance as the resin matrix, especially when polyester resins are used. Low-alkali content E-glass is almost universally used for glass reinforcements and is coated with an organosilane size to protect it, as well as to improve the bond between it and the resin. If chopped strand glass mat is utilized as a reinforcement, the type of binder used in its manufacture can affect hydrolysis resistance, a powder binder generally being preferable to a polyvinyl acetate emulsion binder (see also osmosis and blister resistance, in Chapters 3 and 7).

Frequent attempts to introduce natural fibres (such as sisal and hemp) as reinforcements have not been particularly successful due to poor strength and high water uptake.

As previously described, surfaces of composites intended for external weathering are usually protected by a gelcoat and/or a conventional surface coating, and maintenance of gloss and colour of these is not generally affected by the choice of reinforcement underneath. However, in the final stages of attack, when the outer layer has been destroyed, the exposure of the fibres themselves leads to rapid deterioration by 'wicking' of moisture along the interface between the resin and the fibre, and in the case of natural fibres, via the fibres themselves. Such water ingress destroys the interfacial bond between resin and fibre and can also attack some resins.

It should be noted that certain organic fibres such as the polyaramids (e.g. Kevlar, Nomex) are affected by UV light. This is of importance on thin (single ply) unpigmented sheeting reinforced with these materials, where properties can deteriorate markedly on exposure [48].

Non-fibrous fillers usually detract from the weathering performance of polymers, although mica, glass flake and other platelet particles are used for added protection in surface coatings. Silica flour, polyaramid powders and glass spheres are also used to improve abrasion resistance. Gelcoats containing these materials have been commercially available. Organosilane and other reactive coatings can improve the bond between resin and filler, and hence improve various properties, including resistance to weathering.

References

1 J F Rabek, *Polymer Photodegradation*, London, Chapman and Hall, 1995.
2 M R Kamal, 'Cause and effect in the weathering of plastics', *Polym Eng Sci* 1970 **10**(2) 108–121.
3 ASTM G7-89, *Recommended Practice for Atmospheric Environmental Exposure Testing*, The American Society for Testing Materials, Washington, DC, 1989.
4 A B Wootton, 'Accelerated weathering specifications used in the polymer industry', *Polymer Testing '96*, Rubber & Plastics Research Association, Shawbury, Shropshire, UK, Sept 1996, Paper 12.
5 L S Crump, 'Evaluating the durability of gel coats using outdoor and accelerated weathering techniques: a correlation study', *Proceedings 51st Annual Conference*, Composites Inst, SPI, Feb 1996, 22-B.
6 Association of Automobile Industries Working Group on Test Methods for Paints, 'Comparison of outdoor and accelerated exposure methods', *J Coatings Technol* 1986 **58**(734) 57–65.
7 R M Fischer, W D Ketola and W P Murray, 'Inherent variability in accelerated weathering devices', *Prog Org Coatings* 1991 **19** 165–179.

8 R M Fischer, W P Murray and W D Ketola, 'Thermal variability in outdoor exposure tests', *Prog Org Coatings* 1991 **19** 151–163.

9 R P Brown, 'Test procedures for artificial weathering', *Polym Testing* 1993 **12** 459–466.

10 L Crewdson, 'Correlating outdoor and lab weathering', *European Coatings J* 1993 (1–2) 34–44.

11 J W Martin, 'Methodologies for predicting the service lives of coating systems', *NIST Building Science Series*, 1994, 172, US Dept of Commerce.

12 B R Bogner and H R Edwards, 'The cost/performance advantages of iosphthalic acid based polyesters in gelcoated laminates', Paper 12-F, *Proceedings 51st Annual Conference, SPI Composites Institute*, SPI, New York, USA, February 1996.

13 G D Wigington, 'Light stable vinyl ester resins designed for UV resistant gel coats', Paper 15-C, *Proceedings 46th Annual Conference, Composites Institute*, SPI, New York, USA, February 1991.

14 J Layton, unpublished work, Scott Bader Company Ltd., Wollaston, Northants, UK.

15 Information from Ciba Ltd.

16 M P Diebold, 'The causes and prevention of titanium dioxide induced photodegradation of paints', *Surface Coatings Internat* 1995 **6** 250–256.

17 H R Edwards, 'The influence of gel coat and skin coat resins on yellowing', *Amoco Chemical Company TS Report*, Sept 6, 1994.

18 J Fidler, 'Glassfibre-reinforced plastic and cement facsimiles in building restoration', *Transactions Association for Studies in the Conservation of Historic Buildings* 1987 **12** 17–25.

19 G Pritchard, 'Neither steel nor wood!', *Practical Boat Owner* March 1996 (351) 88–90.

20 R S Yamasaki, 'Surface weatherability of glass fibre reinforced polyester sheeting. II – Effect of constituents', *Compos Technol Rev* 1982 **4**(4) 125–129.

21 R C Hirt, R G Schmitt and W L Dutton, 'Solarisation studies on polyester resins using a heliostat-spectrometer', *J Solar Energy Sci Eng* 1959 **3**(2) 19–22.

22 N D Searle, 'The activation spectrum and its application to stabilisation and weatherability tests', *Proceedings ANTEC '85*, 1985, pp 248–251.

23 M R Kamal, 'Effect of variables in artificial weathering on the degradation of selected plastics', *Polym Eng Sci* 1966 **6** 333–340.

24 K Tanaka, 'Weatherability of glass fibre reinforced polyester sheets', *Proceedings 42nd Annual Conference SPI*, 1987, Session 10-A.

25 R S Yamasaki, 'Surface weatherability of glass fibre reinforced polyester sheeting. I – Effect of modification of resin chemical composition', *Compos Technol Rev*, 1982 **4**(3) 84–87.

26 A A K Whitehouse and D Wildman, Paper 28, 'Surface weathering characteristics of reinforced polyesters' *4th International Conference, Reinforced Plastics (BPF)*, 1964, British Plastics Federation (after A Davis and D Sims, see Ref [35]).

27 B Alt, 'Erfahrungen beim Bewittern von Ungesaettigten Polyesterharzen', *Kunststoffe* 1968 **58**(12) 961–966.

28 D Roylance and P McElroy, 'Weathering of polymer composites', *Composites Asia Pacific 89 Conference Proceedings Adelaide*, 19–21st June 1989, pp 29–39.

29 J Humberstone, 'New In-mould coatings for RTM', *BPF Composites '96 Conference Papers*, British Plastics Federation Sept 1996.

30 Stenni Cladding Panels, *British Board of Agrement*, Certificate 92/2838.

31 Anon, 'Coloured phenolic FRP breakthrough', *Plastics and Rubber Weekly* 1996 (March 1st) 8.

32 W W Wright, 'The effect of diffusion of water into epoxy resins and their carbon fibre reinforced composites', *Composites* 1981 **12** 201.

33 B H Eckstein, *ACS Org Coatings Div Plastics Chem Pre-prints* 1978 **38** 503.

34 R J Chester and A A Baker, 'Environmental durability of F/A – 18 graphite/epoxy composite', *Polym Polym Compos* 1996 **4**(5) 315–323.

35 A Davis and D Sims, *Weathering of Polymers*, Applied Science Publishers, London, 1983.

36 G Capocci, 'Advances in the light stabilisation of polyurethanes', *30th Annual Polyurethane Tech/Marketing Conference*, 1986, pp 220–227.

37 L G Rainhart and J Schimmel, 'Effect of ageing on acrylic sheet', *Internat Solar Energy Soc Paper (US Section)*, 1974.

38 D J Carlsson, M Krzymien, D J Worsfield and M Day, 'Volatiles released during the weathering of PVC', *J Vinyl Additive Technol* 1997 **3** 100–106.

39 Plastics Design Library, *The Effect of UV Light and Weather on Plastics and Elastomers*, New York, Plastics Design Library, 1994.

40 M Day and D M Wiles, 'Photochemical degradation of poly(ethylene terephthalate), I–III, *J Appl Polym Sci* 1972 **16** 175–215.

41 H Fitz, 'Witterungsbestaendige Fluorkunstoffe im Ausseneinsatz', *Kunstoffe* 1989 **79**(6) 519–524.

42 S M Halliwell, 'Weathering of polymers', *RAPRA Rev Reports* 1992 **5**(53) Rapra Technology Ltd, Shawbury, UK.

43 B O'Donnell and J R White, 'Photo-oxidation of polystyrene under load', *J Mater Sci* 1994 **29** 3955–3963.

44 B O'Donnell and J R White, 'Stress-accelerated photo-oxidation of polypropylene and glass-fibre-reinforced polypropylene', *Polym Degradation and Stability* 1994 **44** 211–222.

45 M M Qayyum and J R White, 'Weathering of glass fibre reinforced polystyrene and poly(ethersulfone)', *Polym Compos* 1990 **11**(1) 24–31.

46 C S Hindle, J R White, D Dawson, W J Greenwood and K Thomas, 'Characterisation of injection mouldings: residual stresses, orientation and distortion', *Proceedings SPE 39th ANTEC*, Boston, 1981, pp 783–785.

47 M Thompson and J R White, 'The effect of a temperature gradient on residual stresses and distortion in injection mouldings', *Polym Eng Sci* 1984 **24**(4) 227–241.

48 F Larsson, 'The effect of ultraviolet light on mechanical properties of Kevlar 49 composites', *J Reinf Plastics Compos* 1986 **5** 19–22.

7

Review of the durability of marine laminates

TIM J SEARLE AND JOHN SUMMERSCALES

7.1   Introduction

This chapter is about the structural integrity in ocean conditions of fibre–resin composites, so that it is not mainly concerned with cosmetic changes, such as gelcoat deterioration, although osmotic hull blistering is discussed. Those concerned with the weathering and durability of unsaturated polyester gelcoats will find these topics discussed more extensively elsewhere in this book, for example in Chapter 6, which concentrates on weathering. Emphasis will be placed here on the conclusions from recent and significant investigations of the behaviour of fibre reinforced plastics (FRP) in sea water. When a research report is summarized, details will generally be given of the materials systems used.

7.1.1   Overall status of reinforced plastics in marine applications

Reinforced plastics possess an extraordinary combination of properties, which make them uniquely appropriate for marine structural use. High strength and low weight are combined with corrosion resistance, good dielectric properties and non-magnetic properties. There is a long history of using glass fibre reinforced plastics (GRP) in the marine environment, for general purpose boat hulls, naval vessels such as minehunters, offshore oil structures, and more recently, components such as propellers.

Boat hulls were among the earliest applications of modern reinforced plastics. The United States Navy introduced their first 8.5 m (28 foot) fibreglass personnel boat in 1947, and was contracted to pass 3000 vessels by 1966. The orders for the first two vessels specified construction by the pressure bag method and by the vacuum injection system in order to achieve higher quality than was obtained by the then current hand-lay-up technique [1]. One of the earliest large GRP structures for submerged operation was the fairwater of the submarine *USS Halfbeak* (SS-352) which entered service in 1954 [2].

Subsequent experience has shown that fibre reinforced laminates have the potential to provide extended service in the marine environment. Early losses in mechanical strength have tended to stabilize after a year or so. However, it is essential that all the constituents of the material are selected carefully to avoid the inclusion of hygroscopic/hydrophilic components. Careful control of the manufacturing process is likely to improve the durability of the product considerably.

Glass is overwhelmingly the preferred reinforcing fibre in marine applications although aramid (e.g. Kevlar®) and carbon are used where cost permits. Where carbon is used, there is a theoretical possibility of dangerous corrosion cells being set up with exposed carbon fibres electrically connected to metals. This problem is easily avoided.

7.1.2 Design principles

Many early FRP marine structures were modelled on those made with traditional materials, but with substantial overdesign. Thick laminates were employed because of caution arising from unfamiliarity with the long term behaviour of the materials. Useful experience has been gained since the late 1950s which makes this caution less necessary nowadays.

A classic book on the structural design of marine GRP is the manual by Gibbs & Cox [3]. The late Charles Smith's book [4] is some 30 years younger and is an excellent introduction to the design of marine structures in composite materials. There are various published classification rules from the different national bodies [5–14].

7.1.3 Recent developments

Some idea of progress in FRP marine developments is given by the following list [15]:

- mine warfare vessels
- submarine structures (sonar domes and fairings)
- submersibles
- navigational aids (buoys and towers)
- offshore oil exploration and exploitation (drilling risers and life rafts)
- hydrofoil flaps
- hovercraft blades and skirts
- passenger ferry hulls
- powerboat structures (see Fig. 7.1)
- racing and luxury yachts
- work and pleasure boat hulls (see Fig. 7.2)

Figure 7.1 The plethora of GRP yachts and powerboats occupying a Plymouth marina

Figure 7.2 Severn class lifeboat in composite materials (reproduced by kind permission of the Royal National Lifeboat Institution, Poole, UK)

Figure 7.3 Ninety foot (27.4 m) carbon fibre composite mast on the racing yacht *Victoria Group*

- composite masts (see Fig. 7.3) and
- composite sailcloths.

Since the late 1980s there have been further developments in the components area, for example:

Figure 7.4 Prototype composite propeller built by the Advanced Composites Manufacturing Centre at the University of Plymouth

- high pressure sea water pump for reverse osmosis desalination [16]
- marine propellers [17–23] (See Fig. 7.4)
- sea water circulation pumps for large aquaria [24]
- riser guide tube bellmouths for floating oil production platforms [25]
- blind-flange bulkheads for ocean thermal energy conversion [26]
- floating piers for marinas [27]
- cladding and guttering for the Second Severn Crossing, (cable-stayed bridge over the Bristol Channel) [28]
- walkway gratings for offshore platforms [29]
- strengthening of primary steel on existing offshore structures [30].

7.1.4 Estimates of the durability of FRP in marine environments

There have been few formal assessments of actual performance, but it is widely considered that FRP hulls should last considerably longer than 20 years, and when well made, they have been known to last more than 40 years. Considering that the products in question were made in the early

Figure 7.5 The launch of the GRP composite minesweeper, HMS Sandown, in April 1988. (Reproduced from *Advanced Composites Engineering*, Sept 1989, p 11, with the kind permission of *Engineering* magazine)

days of the reinforced plastics industry, we can reasonably expect further improvements in lifetime for those vessels being made today. Some naval vessels already in service are expected to last for much longer (see next paragraph).

In one recent investigation, companies in the Finnish FRP industry were asked to estimate the lifetime of their products [31]. The average estimated lifetime across all product segments was greater than the 15 years they estimated for process equipment, tanks and containers (actually, process equipment has proved more durable than this estimate anyway, see Chapter 9 for the results from the Dow survey). The marine sector was given the longest average estimated lifetime, at 28 years, typically 25–30 years. Mableson *et al.* [32] predicted a hull life of 60 years for *HMS Wilton*, the world's first GRP mine counter measure vessel, 15 years after the vessel had been launched (see Fig. 7.5).

Attempts to support such estimates simply by using laboratory data and accelerated testing methods, with high temperatures and so on, have not been particularly successful. To quote Nagae and Otsuka [33], '. . . many researchers have investigated the corrosion behaviour of GRP [glass reinforced plastics] in hot water and strong acids . . . However, little is known about corrosion in milder or more practical conditions'. The reason is of course the time required to monitor durability over many decades, that is, over periods longer than the lifetime of many manufacturing companies and their clients.

The widespread use of GRP for marine vessels of less than 60 m length suggests that there is a high level of confidence in the ability of laminates to perform in service. The materials are nevertheless not without problems. This chapter will consider their response to the marine environment and will note any features of design and manufacture which may compromise the performance. The resulting emphasis on potential difficulties should be balanced by considering the obvious and numerous successes of reinforced plastics vessels in marine applications.

7.2   Water absorption

7.2.1   Resins (and those laminates with negligible water absorption in the fibres)

All organic resins absorb water to some extent, whereas glass and carbon fibres (unlike aramid) do not. A great deal has been discovered about the long term effects of absorbed water on the properties of resins and reinforced plastics. Hot water is often quick to cause visible damage, affecting most thermosetting resins and E-glass fibres adversely, but cold water is usually very slow to produce any signs of deterioration in well manufactured laminates, especially postcured ones. There are some exceptions to this statement, and osmotic blistering is fairly common, but water damage does not usually threaten structural integrity without the assistance of mechanical stress or some other additional influence.

7.2.1.1   Fick's laws

The rate of deterioration appears to be dependent on the absorption rate and on the amount of moisture absorbed. Springer [34] has outlined analytical expressions for determining moisture content and moisture distribution in laminates, as a function of time, assuming that the material obeys Fick's laws of diffusion.

In Fickian diffusion, the absorption behaviour is as shown in Fig. 7.6. The absorption (weight gain) and desorption (weight loss) curves can be plotted against $t^{1/2}$ (square root of time) and the resulting graphs are linear until 60% of the maximum absorption, M_m has been reached, but later it is concave towards the $t^{1/2}$ axis and asymptotically approaches the final equilibrium value. M_m is a constant when the material is fully submerged in a

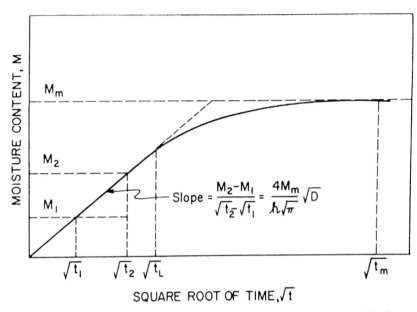

Figure 7.6 Illustration of the change of moisture content with the square root of time, for Fickian diffusion. For $t < t_L$, the slope is constant. (Reproduced from Springer [34], p 52, with kind permission of the author and Chapman & Hall)

liquid, but it varies with the relative humidity when the material is exposed to moist air. M_m can be predicted in that case using $M_m = a\Phi^b$, where Φ is the relative humidity, and a and b are constants. The value of b was once believed to be in the range 1 to 2, but recent experimental evidence suggests that it is near 1.

The 'moisture problem' can be solved by analytical means when the following conditions are met:

- Heat transfer in the material is by conduction only and described by Fourier's law.
- Moisture diffusion is described by a concentration dependent form of Fick's law.
- Temperature within the material approaches equilibrium much faster than the moisture concentration, and hence energy (Fourier) and mass transfer (Fick) equations are decoupled.
- Thermal conductivity and mass diffusivity depend only on temperature. They are independent of moisture content and of stress levels inside the material.

If the above assumptions are valid, then the temperature and moisture distributions within the material can be calculated using the procedures for analysing the 'Fickian' diffusion process. These procedures [34] require a knowledge of the following material properties:

- density (ρ)
- specific heat (C)
- mass diffusivity (D: normally a second rank tensor for fibre reinforced polymers)
- thermal conductivity (K: normally a second rank tensor for fibre reinforced polymers)
- maximum moisture content (M_m).

Other factors that must be known are the geometry, boundary conditions, initial conditions and a relationship between the maximum moisture content and the ambient conditions. It is commonly assumed that ρ, C, D and K are not significantly affected by the moisture content and that values for dry materials can be used without much error in the calculations. There is some experimental evidence that this assumption is reasonable.

7.2.1.2   Failure of Fickian predictions

Calculations performed on the basis of Fick's law for laminates will fail if any of the following apply:

- The matrix contains voids.
- The matrix itself exhibits non-Fickian behaviour.
- Cracks, delaminations or other damage features develop in the material.
- Moisture propagates along the fibre–matrix interface.

Deviations from Fick's law become more pronounced at high temperatures and are often found for materials immersed in certain liquids, notably solvents, hydrocarbons and sea water. However, in many laminates, even where moisture transport is by a non-Fickian process, the results obtained using Fick's equations together with appropriate apparent D and apparent M_m values are not very different from the observed behaviour.

7.2.1.3   Resin softening

Resin softening is sometimes noticed after several years' immersion of laminates in water, and can be quantified by using a Barcol hardness meter. Softening can be a consequence of several processes, the most important being plasticization, which must be taken into account in the design of marine laminates. The softening process will generally lower the glass transition temperature (T_g) and therefore affect the mechanical properties and will also accelerate subsequent diffusion behaviour. This topic is discussed in Chapter 4 and some equations designed to predict the extent of lowering of the T_g are mentioned.

Most of the experimental evidence for T_g lowering in resins by water comes from studies of aerospace materials, where the T_g must remain very high throughout the service life of a component, rather than marine laminates. One representative study of relatively low T_g materials was by Apicella et al. [35], who determined the extent of the depression of the glass transition point (ΔT_g) experimentally for epoxy resins. A TGDDM/DDS (tetraglycidyldiaminodiphenyl methane/diaminodiphenyl sulfone) epoxy showed ΔT_g values of 50–80°C, while DGEBA (diglycidyl ether of bisphenol A) cured with more reactive linear amines had ΔT_g values of 20–40°C at equivalent levels of absorbed water. Carbon fibre/TGDDM epoxy prepreg materials equilibrated in water at low (near 0°C) temperatures with 5% water uptake exhibited ΔT_g values of 100°C, measured calorimetrically and mechanically.

Amorphous PEEK (polyetheretherketone) thermoplastic resins had much lower water uptakes and the ΔT_g was limited to just 2°C. The more usual semi-crystalline PEEK, that is, the form commercially available, has an even greater resistance to water absorption and therefore an even greater stability in wet conditions, but it suffers from a high cost disadvantage.

Nechvolodnova et al. [36] studied the effect of moisture absorption on some Russian epoxy resin films (EDT-10 resin and a resin based on a diglycidyl ether of resorcinol (DGER) and meta-phenylene diamine). Equilibrium moisture absorption at 20°C reached 4–5%. The ΔT_g values were in the range 50–80°C, with consequent reductions in static modulus, yield stress and tensile strength at room temperature.

7.2.1.4   Alternatives to Fick's laws

Fick's laws assume a single phase diffusion model, that is, all the water molecules absorbed are assumed to behave in the same way and are all equally free to move throughout the resin, regardless of the strong attractions of polar groups in the resin, or of hindrances to movement resulting from the geometry of the resin structure. The Langmuir model, a more general theory of two-phase diffusion, introduces the concept that water molecules can become trapped by the laminate. The model is discussed by Bunsell [37] in the course of an account of how the resin's electrical properties are affected by moisture. It is assumed that at any given moment only a fraction of the water molecules can diffuse freely, but this fraction increases with temperature. The two models become almost indistinguishable as laminate thickness increases.

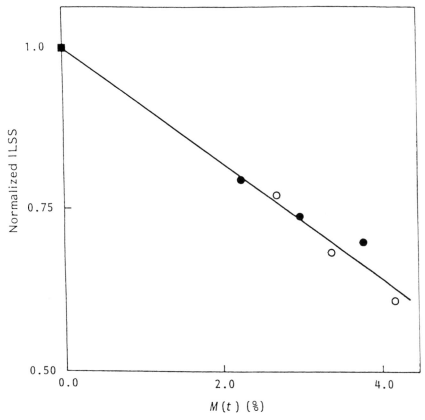

Figure 7.7 Normalized interlaminar shear strength versus absorbed water content. Least squares adjusted equation; correlation coefficient, $r = 0.988$ ■, as fabricated; ○, distilled water; ●, salt water. (Reproduced from d'Almeida [40], p 449, with kind permission of the author and Elsevier Science)

7.2.1.5 *Measuring moisture content in laminates*

Electrical resistance measuring devices together with dielectric and infrared instruments are available as non-destructive commercial products for field use, notably by small boat surveyors [38]. Clegg [39] has reported comparative tests on three of the most popular UK/Ireland moisture meters which use capacitance measurements.

More advanced techniques are also available [38]. Microwave, nuclear magnetic resonance (NMR) and positron annihilation techniques may be useful, but they tend to be complex and expensive techniques for the measurement of specimens which can be transported to the laboratory.

7.2.2 Aramid fibres

Glass and carbon/graphite fibres do not absorb significant quantities of moisture, although glass fibres are eventually attacked, the rate of attack being significant with hot water. Aramid fibres (e.g. Kevlar, Nomex, Twaron) in contrast are polymers, just like the matrix resins and therefore they absorb water. The amount is typically about 5% w/w. The fibre/matrix interface is weak, even in the dry condition.

D'Almeida [40] has reported interlaminar shear strength (ILSS) results for 65 v/o Kevlar 49 fibres in DER 383 epoxy resin cured with 27 phr DEH 50 aromatic polyamine with a void content below 0.5%. The ILSS values decreased as a function of immersion time, with specimens immersed in distilled water declining faster than those in saline solution, which is absorbed more slowly. When the results were normalized to water content the results were coincident, suggesting that the degradation mechanism was the same in both cases (Fig. 7.7).

7.3   Osmosis and blistering

The occurrence of a rash of tiny blisters on the external surfaces of GRP boat hulls is the best known, but not the only manifestation of osmosis in laminates. Surveys differ widely in their estimates of the scale of the problem, ranging from 2% or 3% of boats up to more than a third. The frequency of osmotic problems seems to depend on the prevailing climate as well as on local fabrication practices and the materials used.

Blister formation is not simply a matter of the age of the hull. Some laminates never blister, while others start to give trouble less than a year after manufacture. The harm done by gelcoat blistering is usually only cosmetic, at least for many years, and osmosis is not generally considered a factor in GRP boat hull durability, although GRP swimming pools have been severely affected, possibly because of higher temperatures. The much less common, deep seated and large blisters located within the structural laminate may weaken the hull and therefore threaten durability.

Osmosis has been defined [41] as 'the equalization of solution strengths' (i.e. concentrations) 'by passage of a liquid (usually water) through a semi-permeable membrane'. When a laminate is immersed in water, moisture is slowly absorbed through any gelcoat into the underlying structural laminate. For a hollow structure such as a boat hull, some moisture will pass harmlessly through the laminate and collect in the bilges or simply evaporate. However, some of the water may interact physically or chemically with materials within the laminate to produce breakdown products which are hygroscopic (water absorbing) or hydrophilic (water loving). These concentrated solutions will then act to dilute themselves by drawing further water into the laminate. The contents of blisters are usually acidic fluids and therefore capable of extracting metal cations from the glass fibres.

Figure 7.8 shows a typical process for the development of osmosis. Fresh water is drawn into the laminate more rapidly than sea water, because it starts at a lower solution concentration and hence the concentration gradient is higher. Osmosis is also accelerated by the GRP being in warm water. When the water is especially warm the resin is nearer to its T_g, and the processes of diffusion and absorption become very much quicker.

As the trapped solution becomes more dilute, it increases in volume and creates a hydraulic pressure within the laminate. This process can take many years, and in well manufactured structures it may never result in visible damage. But the internal pressure built up in the laminate may eventually manifest itself as blisters on the surface of the structure, usually echoing delaminations between the gelcoat and the structural laminate. This problem is of especial concern to the yachting fraternity, which has a significant remedial treatment industry. Repairs can be expensive and unless the root cause is eliminated, remedial work will not completely solve the problem.

Internal disk cracking (blistering) in resins was first reported by Steel [42] in 1967, and attributed to swelling, followed by rupture due to hydrolytic attack on the strained polymer chain. Ashbee et al. [43] found that fully cured polyester resins developed 'disk'-shaped internal cracks when exposed to hot water and proposed an osmosis hypothesis for blister formation.

Birley and co-workers [44,45] suggested that the stress at the gelcoat backing (structural laminate) resin interface was the most important factor in the initiation of blistering. Chen and Birley [46] classified blisters according to their origins:

- contaminants
- bubbles
- precracks and
- glass fibres.

They regarded the last two as the most significant contributors to the formation of blisters in chopped strand mat (both emulsion and powder bound CSM) in polyester resins.

The principal chemical factors implicated in the osmosis and blistering phenomenon have been claimed to include:

- residual glycol from the unsaturated polyester [4,41,47]
- laminating resins of high acid value [48]
- inadequate control of mixing [4]

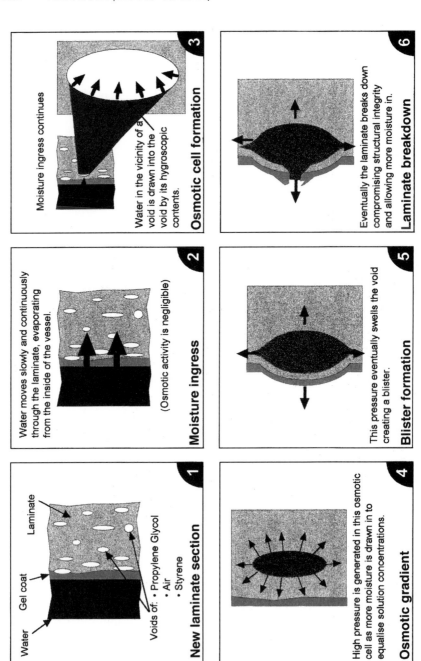

Figure 7.8 Onset of osmosis and consequential laminate damage

- the degree of resin cure achieved [4, 46] (blister resistance follows the same ranking as degree of cure [49])
- insufficient styrene, leading to undercure [49]
- excess styrene and its soluble oxidation products [48]
- excess/residual catalyst and/or accelerator [47,49]
- the catalyst carrier [50]
- soluble (polyvinyl acetate) binder on chopped strand mat [4,41,46–48,50,51,53]
- use of PVA release agent [48]
- pigments (especially dark ones, as they are often supplied in a hydrolysable diluent rather than an inert one [4,47,48,51,52])
- excessive time elapsing between successive layer applications [49,51].

Measures to reduce the problem, other than those implied by the factors above, include:

- avoidance of moisture on the mould and the reinforcement [4,50]
- complete wet-out of the reinforcement fibres [52]
- thorough consolidation to minimize voids [4,50]
- gel coat permeability should be lower than that of the structural laminate [4]
- the use of a light glass scrim in the gelcoat [4]
- use of a primer between the gelcoat and structural laminate [48]
- control of gelcoat thickness and quality [50].

7.3.1 Effect of using better quality resins on blistering

Unsaturated polyester resins can be adapted to improve the resistance to osmotic blistering [41]. One of the ways of improving a resin is by changing one or more of the starting materials, that is the glycol, the saturated acid, the unsaturated acid or more often anhydride, or the reactive diluent. The last-mentioned item is overwhelmingly styrene, and there is not usually any demand to change it. Considering the saturated acid component first, isophthalic polyester generally performs better than orthophthalic in water, but is more expensive. Chen and Birley [49] suggest that one of the reasons for the alleged superiority of isophthalic resins is the fact that they will normally be processed to higher molecular weights and will show lower acid numbers and hydroxyl numbers than equivalent (ortho)phthalic anhydride formulations. Furthermore, as is evident in the molecular structure, orthophthalate ester linkages are more accessible and hence more susceptible to attack than the ester linkages of isophthalate polyester. The tendency of the starting material, phthalic anhydride, to sublime during resin manufacture also provides a mechanism for undesirable low molecular weight materials to be incorporated in the formulation, leading to reduced corrosion resistance because of the relatively high water solubility.

The unsaturated acid used in the resin, that is, maleic or fumaric, also affects the blister performance of GRP laminates [49]. Laminates made from fumaric acid-based resins have better resistance than those from maleic resins, but it is well known that the great majority of maleic unsaturation normally changes anyway to the fumaric form during the resin synthesis, although the extent of the change depends on several factors, notably the nature of the glycols used. All-fumaric resins become more completely crosslinked and hence give higher heat distortion temperatures and higher chemical resistance than maleate resins, although a high level of crosslinking brings with it the disadvantages of greater brittleness.

Improvements can also result if the cheaper glycols are replaced by 2,2,4-trimethyl-1,3-pentanediol or 2,2-dimethyl-1,3-propanediol (the latter is usually referred to as NPG, because of the old name, neopentyl glycol). NPG has two pendant methyl groups, rather than just one in polypropylene glycol, and can offer better protection of the ester groups, and thus improves the hydrolysis resistance of the cured resin [49]. Isophthalic/NPG materials are strongly recommended for marine laminates, or for at least the first few millimetres of lay-up [52]. Changing to a marine isophthalic gelcoat and laminating resin would add ~0.5% to the cost of a typical 10 m sailing yacht [50].

7.3.2 Importance of fabrication in countering blistering

Skilled lamination and best industrial practice are very important in the achievement of optimum blistering resistance [46]. Guidance on manufacturing marine laminates is available from the British Plastics Federation [54]. Laminates using the same resin for both gelcoat and structural laminate are said to offer better blister resistance as there is no sudden change in the water penetration rate [49]. Another area of concern is the thickness of the gelcoat. When industrial laminates were made with uncontrolled gelcoat thickness, blisters formed first in those areas where the gelcoat was very thin [49].

Castaing et al. [55] have presented a mechanical model for evaluating the development of internal pressures and stresses in blisters with time. The model is based on the theory of isotropic plates, and represents a circular blister as a thin membrane, clamped elastically around its edge and uniformly loaded over the entire internal surface. That a blister can be simulated as a circular plate, imperfectly clamped on its circumference (the elastic edge) can be verified by measurement of the tangents of the blister deflection curve. The tangents are zero for perfect clamping, as no strain is allowed at the edge. The model has been validated [55] experimentally by ageing tests on a gel-coated polyester laminate (the structural laminate consisted of 10 layers of $290 \, \mathrm{g\,m^{-2}}$ woven E-glass in orthophthalic acid/polypropylene glycol polyester resin, with a $300 \, \mu\mathrm{m}$ thick isophthalic acid/NPG gelcoat cured under ambient conditions for two months) in distilled water. At 'quasiequilibrium', it appears that the final dimensions of the blister depend only on the mechanical characteristics and the thickness of the blister.

In a parallel study [56] with four resin systems in both distilled water and natural sea water ($35 \, \mathrm{g\,l^{-1}}$ sodium chloride) at 60°C, the osmotic blisters appeared first on (the gelcoat side of) the samples exposed to distilled water and (two to three times) later on those exposed to sea water. This confirms the observations of other workers that the rate of blistering is reduced as the solute gradient decreases. The gelcoat blistering appeared to have no effect on the structural integrity of the laminates at times up to 2500–3000 h at 60°C. Subsequent osmotic delamination of the first ply (5000 h at 60°C, which was taken to be equivalent to about 20 years in sea water at 20°C) resulted in a 25–30% decrease in the ultimate strength of the isophthalic laminate, and 40–65% decrease in the ultimate strength of the orthophthalic laminate after 5000 h. The mechanical model has recently been extended to elliptical/orthotropic blisters [57].

Dow and Bird [58] state that 'The bugbear of "osmosis", or gel coat blistering, . . . is now much better understood and should be regarded as entirely avoidable by careful choice of materials and fabrication procedure'.

The treatment of existing osmotic blisters requires careful planning and preparation. If the hygroscopic/hydrophilic materials responsible are not completely removed from the laminate, then a recurrence of the problem is highly probable. *The Osmosis Manual* [41] provides a clear explanation of the procedures for rehabilitation of a blistered laminate. The manual has also been published in summary form in the open literature [39].

Blistering has been associated with galvanic corrosion adjacent to the light glass-weft fibre in unidirectional carbon fibre reinforced plastics (see Section 7.9).

7.4   Mechanical deterioration of FRP laminates after sea water exposure

7.4.1 Glass fibre reinforced plastics

The availability of direct evidence about the durability of laminates in sea water was scanty until fairly recently, when anecdotal evidence from actual vessels which had been subject to destructive testing was supplemented by an increase in laboratory data. The following section exemplifies the kind of information obtained, mainly since the late 1970s. Short *et al.* [59] exposed GRP laminates to the sea at Bovisand, Plymouth Sound

Table 7.1 Percentage weight gain and percentage increase in shear modulus
after 100 days marine exposure in Plymouth Sound

Panel	Exposure	Weight gain (%)	Shear modulus (%)
A1/1	Untreated	0.34	90.3
A1/2	Sealed edges	0.26	95.4
A1/3	Sealed edges and faces	0.47	92.6
A1/4	Laboratory control	zero	97.1

(50°22′N 4°7′W) from July 1980 to February 1981. The panels consisted of
five plies of Marglas 280 (610 g m^{-2}) unidirectional glass rovings, woven with
a light weft thread at 5 mm intervals. The panels were hand laid unsaturated
polyester without a gelcoat. The exposed panels had a 16% reduction in
Rockwell hardness at resin-rich (inter-tow) positions relative to the control
panels. Exposed samples tested in four-point flexure surprisingly had a
higher Young's modulus and fracture stress than the control specimens.
However, it is possible that some postcuring took place during drying of the
exposed samples. Evaporation of residual styrene monomer would also
contribute.

Cargen *et al.* [60] exposed GRP laminates to the sea in the Hamoaze,
Plymouth Sound (50°22′N 4°9′W) from 15 December 1983 to 15 March
1984 (100 days). The panels were 12 plies of ECK25 (825 g m^{-2}) woven glass
rovings in Crystic 625TV unsaturated polyester, without a gelcoat. No
significant weight change and no deterioration in the panels assessed by
visual examination could be detected after exposure. The panels were
evaluated before and after exposure, using the anticlastic bending method
to determine the shear modulus. The results, referenced to the initial mean
values, are summarized in Table 7.1.

7.4.1.1 Effect of sea water immersion on impact resistance of GRP

Strait *et al.* [61] used instrumented drop weight impact tests to find the
effects of synthetic (ASTM D1141-52 Formula A) sea water immersion at
60°C on glass fibre/epoxy laminates.

Two materials were considered: 3M SP1002 (53 v/o continuous non-
woven E-glass fibres in DGEBA epoxy resin with both cross-ply and quasi-
isotropic lay-ups) and Cyanamid 5920 (two-layer 25 μm thick 8-harness
satin style 7781 E-glass faces over 64 μm thick nine-ply aligned woven
rovings in rubber toughened epoxy resin). Impact was achieved with a
101.6 mm square coupon clamped in a 76 mm diameter fixing ring and
impacted by a 12.7 mm diameter hemispherical tip with an impact velocity
of 3.57 m s^{-1} and impact energy of 208.8 J. All the dry laminates had equivalent values of energy required for incipient damage. Saturated SP1002

required 20% more energy than Cyanamid 5920 for onset of incipient damage, suggesting increased ductility resulting from water plasticization of the matrix without deterioration of the fibre/matrix interface. The peak load and energy absorbed were higher for Cyanamid 5920 than for SP1002 laminates, although all systems experienced a substantial reduction from their dry values. The total energy absorbed was reduced by between 14% (Cyanamid 5920) and 25% (cross-plied SP1002).

7.4.1.2   Changes in GRP caused by sea water in a hot climate (Bahrain)

Al-Bastaki and Al-Madani [62] exposed GRP (consisting of a single layer of chopped strand mat sandwiched between two layers of woven glass roving in a polyester matrix) to outdoor weathering (monthly average temperatures: 14°C minimum, 38°C maximum, and RH 37% minimum, 87% maximum). Further samples were put 'in a bucket filled with sea water and placed indoors' in Bahrain. After one year, the samples were tested in tension and three-point bending. The main change was a deterioration in flexural strength and an increase in brittleness (reduction in failure strain in tension and deflection flexure) and an increase in the flexural modulus. These effects were more pronounced after immersion in sea water and especially when the specimens were tested wet, rather than dried at room temperature for 5 days.

7.4.1.3   Temperate climate exposure

Adams and Singh [63] immersed carbon, glass and (thermoplastic textile type) polyester woven cloth laminates in natural sea water taken from Poole Harbour (Dorset, UK). The resin system was a flexibilized bisphenol-A epoxy resin, cured with polyoxypropylene diamines of two different molecular weights. The resin, designated L-1007, was developed at DRA Holton Heath. A glass woven roving in A2785 polyester resin was used as a reference material and designated 'Navy GRP'. The specimens were soaked at $10 \pm 0.5°C$ and $20 \pm 0.5°C$ with precautions against increasing solution concentration caused by evaporation. Three specimens of each material were soaked and tested at each temperature, being measured at weekly intervals for 3 months and less regularly for up to 15 months.

The flexibilized and unreinforced resin was also tested at 20°C. Moisture absorption in the unreinforced resin and in the Navy GRP was consistent with Fick's law. The Navy GRP showed the least total takeup of water. The epoxy laminates deviated from ideal Fickian behaviour. At 20°C, both the glass and polyester fibre laminates reached a peak mass and then decreased, suggesting that some material was being leached out into the water. The mechanical properties were determined by dynamic mechanical thermal analysis (DMTA). All the laminates experienced a reduction in the effective flexural modulus and a corresponding increase in their flexural loss factor. The flexural modulus fell by most (25–35%) in the polyester fibre laminates and the carbon fibre laminates changed very little in this respect at 10°C. The carbon fibre laminate modulus fell by about 20% at 20°C. The Navy GRP modulus was reduced by 10% at both temperatures and it performed better than the glass/epoxy. The latter showed behaviour intermediate between that of the carbon and polyester epoxy laminates. It was suggested that the behaviour of the epoxy laminates was caused by water diffusing into the matrix for carbon laminates, by water diffusing into both the matrix and the interface for the glass laminates, and by diffusion into the matrix and the fibres in the case of the laminates with thermoplastic polyester fabrics.

7.4.1.4 Shear

Davies *et al.* [64] studied the influence of water absorption on the interlaminar shear strength and end notch flexure (ENF: mode II shear) fracture toughness of quasi-unidirectional (88% 0°, 12% 90°) E-glass fibres in DGEBA epoxy resin cured with an amine hardener. Laminates were immersed in (a) distilled water and (b) the Atlantic Ocean (at Boca Raton, Florida) for up to 8 months at temperatures of 20°C, 50°C and 70°C. Sea water was more slowly absorbed than distilled water. This observation has frequently been made and it provides some reassurance that laboratory tests using distilled water are useful, if slightly cautious, estimates of behaviour in the ocean.

The shear strength was not affected by weight gains below 1%, but higher weight gains (up to 3%) reduced the shear strength *pro rata* by up to 25%. The mode II fracture toughness, G_{IIc}, also decreased with increasing immersion time (Fig 7.9) and clearly demonstrated the degrading influence of thermally assisted breakdown of the fibre/matrix interface, as indicated by SEM (scanning electron microscopy) fractographic examination.

7.4.1.5 Other mechanical property changes

Similar ENF tests, together with transverse tension tests [65], were conducted on unidirectional E-glass/polypropylene (ICI Plytron® ZM4350PA) after exposure to distilled water (ambient and 50°C) and Boca Raton sea water (ambient temperature). Maximum stable moisture content after 5 months exposure was 0.065% for sea water and 0.17/0.30% for ambient/ 50°C distilled water, respectively. The transverse tensile strength was virtually unaffected by water absorption. The mode II fracture toughness was far more sensitive to moisture absorption. G_{IINL} (onset of non-linearity) values for all water exposure conditions drop and can be fitted to a single curve.

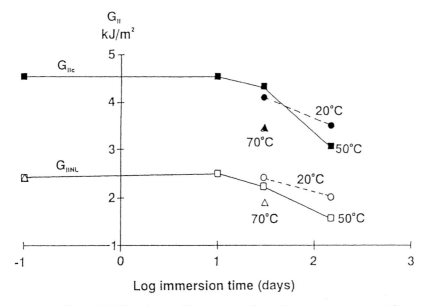

Figure 7.9 Toughness (G$_{IINL}$: onset of non-linear response and G$_{IIc}$: mode II delamination) plotted against log (immersion time). (Reproduced from 'Influence of water and accelerated aging on the shear fracture properties of glass epoxy composite', page 84 in *Applied Composite Materials* 1996 **3** (2) with the kind permission of Peter Davies and Kluwer Academic Publishers.)

Sea water appears to cause a significantly greater reduction in G$_{IIc}$ against moisture content but this is 'corrected' by referencing to the immersion time, suggesting that degradation is governed by local moisture contents adjacent to the notch tip and not by global moisture pickup levels.

7.4.2 Carbon fibre laminates

7.4.2.1 *Impact behaviour, debonding and fatigue in carbon fibre laminates*

Karasek *et al.* [66] conducted drop-weight impact tests on IM7 carbon fibre/ epoxy resin laminates after sea water exposure. Four resin systems (Table 7.2) were used, three of them being based on the same Shell Epikote 828 with different hardeners and/or impact modifiers. The volume fractions were in the range 60–69% fibre and the lay-up was [0/ ± 45/90°]$_{xs}$. Samples were about 6.35 mm thick for all materials (different cured ply thicknesses necessitated the use of differing numbers of plies) and cut to 100 × 100 mm squares for testing. The samples were saturated with ASTM D1141–52 Formula A synthetic sea water at 95°C, and then conditioned at ambient

Table 7.2 Matrix epoxy resin systems and layup repeat value in [0/ ± 45/90°]$_{xs}$ layup for Karasek *et al.* [66] laminates

Resin matrix system	Value for x
100 pbw Epon 828/5 pbw piperidine	4
100 pbw Epon 828/5 pbw piperidine/5 pbw CTBN (carboxy-terminated liquid copolymer of butadiene and acrylonitrile rubber modifier)	3
100 pbw DPL-1911 (85 pbw Epon 828/15 pbw Kraton rubber)/ 68 pbw nadic methyl anhydride hardener/0.85 pbw benzyldimethylamine accelerator	3
Fiberite 977	6

temperature in sea water until testing. Samples were clamped into a 76 mm diameter fixture and impacted, dry or wet, at −23°C, +21°C or 66°C with a 12.7 mm diameter hemispherical steel penetrator (energy about 275 Nm, velocity 3.57 m s^{-1}). The energy required to initiate damage decreased with the temperature, which is consistent with the reduction in matrix properties. Impact energy absorption was relatively insensitive to moisture at temperatures below T_g of the epoxy phase, but significant changes in behaviour occurred above T_g.

In a subsequent paper [67], subpenetration impact testing and ultrasonic C-scanning/optical reflection microscopy were used to characterize the damage in the three Epikote 828 resin-based laminates. At impact energies as low as 1.75 J mm^{-1}, the damage zone was identified as the classic 'cone-of-fracture', that is more extensive delamination was sustained towards the unimpacted surface.

Bradley and Grant [68] conducted initial screening and microindentation debonding of seven candidate 55–65 v/o unidirectional laminates for oil production risers. Degradation was measured through transverse tension after 3 months immersion in pure water at ambient pressure, simulated sea water (Instant Ocean) at ambient pressure and simulated sea water at 20.7 MPa (3000 psi). The materials are listed in Table 7.3. The moisture-induced degradation of transverse tensile strength was similar for the three immersed sets of conditions for each material. Both the graphite/vinyl ester laminates were eliminated through both low wet and low dry strengths. The degradation of interfacial shear strength (by microindentation) for F263 and 977-2 resin laminates supports the hypothesis that degradation is primarily interfacial rather than within the matrix. Failure in SP500-2 laminate was matrix dominated, in both wet and dry conditions. Tests were conducted in transverse three-point bend [69] and *in situ* fracture observations of these materials were made using a scanning electron microscope. The results are summarized in Table 7.3.

Table 7.3 Laminates tested by Bradley and Grant [68] with failure locations for selected materials

Laminate	Dry failure location	Wet failure location
E-glass/Dow Derakane 411 vinyl ester	—	—
E-glass/Dow Derakane 510 fire retardant vinyl ester	—	—
Carbon fibre/Dow Derakane 411 vinyl ester	—	—
Carbon fibre/Dow Derakane 510 fire retardant vinyl ester	—	—
Hexcel T2C145 carbon fibre/F263 TGDDM-DDS epoxy	Matrix	Interface
ICI IM7 carbon fibre/977-2 (TGDDM-aramid-DDS) epoxy	Interface	Interface
3M IM7 carbon fibre/SP 500-2 tricyclic hydrocarbon fluorene	Matrix	Matrix

The SP500–2 is the most brittle of the resin systems and it achieves a marginally higher tensile strength after immersion, which is attributed to plasticization or softening by the absorbed moisture.

Chiou and Bradley [70] conducted fatigue and static EDT (edge delamination tests) on dry CFRP (carbon fibre reinforced plastics) laminates and on similar materials that had been presoaked in sea water. The system studied was made from 46 v/o quasi-isotropic Hercules IM7 fibres in Dow Tactix 556 hydrocarbon epoxy novolac resin. Sea water saturation changed the dominant edge-cracking mode from the 45/90° interlaminar delamination in the unaged specimens to intralaminar cracking in the 90° plies. It did not significantly reduce the maximum available strain energy release rate, nor did it accelerate the growth of edge cracking in fatigue.

7.4.2.2 *Effect of immersion depth on uptake behaviour*

The moisture absorption of graphite fibre/vinyl ester laminates (described in Section 7.9) has been measured in Narragansett Bay sea water at surface atmospheric pressure and at pressures equivalent to a submerged depth of 610 m (2000 feet) [71]. The diffusion coefficients were the same in both cases, but the moisture content curve was higher for the deep submergence samples. Samples were pressurized for 'an interval' or for 40 h, 46 days and 133 days, then tested in bending according to ASTM D709. The shallow submergence samples showed an increase of 6–8% in modulus and no significant change in strength. After 4 months at 2000 feet, the modulus had decreased to 86% and the strength to 81% of the initial dry values. The changes were attributed to mechanical damage associated with the initial void content (this was 1%, with voids up to 3.2 mm diameter) induced by the increased pressure. No evidence was presented for such damage.

7.4.3 Sandwich structures

Very little has been published on the degradation of sandwich structures after wet ageing, although a very few sandwich hull GRP vessels are known to have been in use and to have performed satisfactorily for over 40 years. One concern is the durability of the adhesion between skin and core. A number of workers have developed simple tests to measure the fracture energy associated with skin–core debonding in sandwich materials: a climbing drum test, the shear of a cracked sandwich beam, a modified double cantilever beam specimen and top surface peeling on a sliding carriage [72]. Cantwell *et al.* [72] developed a new technique in which the lower surface skin is peeled from the core in a controlled fashion. Four glass fibre skin (consisting of two-ply $850 \, \mathrm{g \, m^{-2}}$ stitched quadriaxial fabric in an isophthalic polyester resin) with balsa core structures and various core/skin interlayers were tested after 45 days immersed in sea water.

The introduction of a layer of chopped fibres between the core and the skin did not improve the fracture toughness of the interface. The interface toughness was limited by the delamination resistance of the laminate skins. Water immersion increased the interfacial fracture energy, significantly in the case of three of the four interlayers. The absorbed water is believed to degrade the fibre/matrix interface, resulting in more fibre bridging of the crack and greater energy absorption after immersion.

7.5   Stress corrosion and fatigue in water

The stress corrosion, that is the corrosion as a result of the combined action of chemical and mechanical action, of glass fibre reinforced plastics in aqueous media has been reviewed by Roberts [73], Hogg and Hull [74] and Menges and Lutterbeck [75], although none of the work referenced is specific to sea water exposure. The subject of the corrosion of FRP under static loading is discussed in some detail in Chapter 3, and cyclic loading or fatigue is the subject of Chapters 5 and 11 in this book, but both these topics will be briefly mentioned here in the context of marine applications.

Some important experiments have highlighted the role of voids in laminates under cyclic loading when in water. To validate the design of GRP hulls for durability and performance in the tropical waters of the Persian Gulf, the Procurement Executive of the UK Ministry of Defence commissioned creep and fatigue tests at loads equivalent to one-eighth of the static ultimate strength of the laminate [76]. Creep tests were conducted for 7 days at 42 kN on $1000 \times 500 \times 25$ mm panels. Dynamic loading comprised two series of two million cycles at 50 load cycles per minute. Further dynamics tests on standard dumbbell specimens with deliberately high void contents at $31 \, kN \, m^{-2}$ at 200 load cycles per minute were conducted at ambient temperature in air, at 70°C in air and at 30°C in sea water. All the specimens survived two million cycles at one-sixth of the material static ultimate strength (much higher than the hull would experience in service) without visible damage. Those specimens with 5% deliberate voids tested in sea water at 30°C failed during the second series of two million cycles.

Sloan and Seymour [77] exposed unidirectional AS-4/3501 graphite/ epoxy samples to fresh flowing Pacific sea water under both static and fatigue loading conditions using precracked (parallel to the fibres) double cantilever beam specimens in the compliant load frame. Fatigue crack growth was accompanied by bridging of the crack tip by individual fibres. The bridging increases the fracture surface area and tends to hold the crack together, increasing the fracture resistance of the cracked specimen. The primary effect of sea water is to weaken the fibre/matrix interface bond ahead of the crack tip, causing an increase in crack bridging. Crack growth rates were five times slower in sea water than in air, and dropped immediately upon exposure to sea water. Exposure to sea water, without applied load, also increased the mode I fracture resistance of the material. The effect was completely reversible upon drying.

Hodgkiess et al. [78] exposed glass fibre reinforced laminates to (i) distilled water, (ii) simulated sea water ('Ocean Salt': 3.3% total dissolved solids) (iii) simulated spray/sunlight conditions in the laboratory and (iv) to tidal effects and permanent immersion in the lower Clyde estuary (55°45'N 4°55'W), where the sea water temperature was 6–12°C over the period of testing. The materials tested, which were exposed for up to 18 months with all sides and edges coated in the same resin, were:

- 9 mm thick chopped strand mat in Norpol 68–70 isophthalic polyester resin
- 8.5 mm thick chopped strand mat in Epikote 828/ancamine D aromatic amine epoxy resin
- 5.7 mm thick 263 mm internal diameter filament wound Epikote/ ancamine (as above) epoxy pipe

After 12–18 months, the three-point bend flexural strength (with a span-to-depth ratio of 10) of the polyester laminate was reduced by up to 20%, although this reduction occurred mainly in the first few (3–9) months of exposure. The polyester suffered large reductions in impact strength (50–70%, with a high level of scatter) after 6–9 months exposure, after which the values stabilized. The flexural strength of the epoxy was less severely degraded and impact strength was reduced by ~20% in the first 2–6 months before stabilizing. The relevance to the present discussion is that the continuous application of moderate stresses did not appear to cause enhanced deterioration. Nor did the absence of a gelcoat.

Immersion in water does not always have adverse effects. Sometimes the properties improve, or some properties may improve and others deteriorate. Kosuri and Weitsman [79] and Smith and Weitsman [80] conducted fatigue tests (5 Hz at 74–89% ultimate tensile strength) on cross-plied AS4/3501-6 prepreg carbon/epoxy laminates in three conditions:

- dry
- water-saturated but tested in air, and
- immersed in sea water till saturated and tested while still in sea water.

The saturated coupons had the longest fatigue life when tested in air, while the immersed coupons had the shortest life. Saturated coupons had significantly fewer transverse cracks than dry ones, possibly because of stress relief from sorption-induced swelling. The immersed fatigue coupons had more extensive delaminations than the dry ones. The reduced fatigue life of these materials is believed to be due to a synergistic effect whereby the adverse influence of capillary water movement is exacerbated by the mechanical cyclic loading. Comparative stress fields for the three test conditions were evaluated using a shear lag/finite element analysis, taking into account water trapped in the transverse cracks.

7.5.1 Rubber-lined glass-epoxy tubes

Chiou and Bradley [81] conducted hydraulic burst and stress rupture tests on 1.28 mm thick (58 v/o 87/± 35/87° hoop filament wound) tubes made from E-glass fibre/Brunswick LRF-571 DGEBA epoxy resin. There were 6% voids in the laminate. A co-cured nitrile rubber liner was employed, partly to keep the inner surface dry and partly to ensure that pressure could still be maintained if the GRP cracked during the tests. The tests followed 6 months immersion in static simulated sea water (Aquarium Systems Instant Ocean, $\rho = 1023 \, \text{kg m}^{-2}$, pH = 8.2). The tubes had a high (1.5%) moisture uptake, although some of this might have been free water in the voids, but saturation was not reached.

The exposed tubes exhibited a 20% reduction in burst strength, compared to dry tubes. The acoustic emission response suggested that damage started in the wet tubes at a much lower pressure. Stress rupture tests, that is, pressurize and hold, were conducted at three pressures (four were claimed in the text, but only three levels were stated) with the highest pressure being close to the rupture pressure of the wet tubes. There was no indication of moisture-induced degradation, even for the 'worst-case scenario', that is simultaneous application of a constant stress and immersion in sea water, which might be expected to cause eventual stress corrosion cracking. In fact, times to rupture were not much affected by water absorption.

7.6   Condition of US Navy FRP vessels and components after service

In the early 1960s, a search for the early US Navy 8 m (28 foot) fibreglass personnel boats that had been moulded in the early days of the reinforced plastics industry in 1946/47 succeeded in locating some craft that were still being used, but they 'were not readily available for inspection' [82]. Three 12 m (40 foot) 22 knot Coast Guard patrol boats with ten years service were located and identified to have 'incurred one-fifth of the maintenance costs of [comparable] steel craft of the same age and service duration'. The costs included some design modifications, notably to the transom and midship portions. The basic hull and deck laminates were in excellent condition and panels removed for testing had strength values consistent with data contemporary to the construction.

The fairwater of the submarine *USS Halfbeak* (SS-352) was replaced after 11 years of service by a more modern 'high-bridge' type of sail [2]. The laminate was nominally a 6 mm (1/4 inch) thick Style 181-Volan satin weave glass cloth in a general purpose room-temperature curing polyester resin, blended with 10% of a flexible resin for improved toughness. The fairwater was fabricated using a vacuum bag process. Two curved panels measuring approximately 1247 × 686 mm (49 × 27 inches), which were original sections kept in service for the full life of the sail, were evaluated by the Philadelphia Naval Shipyard. The overall condition was judged to be excellent, according to reports based on visual examination of the clear unpigmented laminate. Superficial burnt areas were found and presumed to be a result of inadvertent exposure to a welding or cutting torch. Distortion and out-of-plane roundness of the countersunk bolt clearance holes were noted and attributed to the supporting structure being modified during service, resulting in overstressing of the GRP connections.

The mechanical properties did not differ much after 11 years service from the original ones, and the average values were still within the specification requirements. Service reports indicated that very little maintenance had been required, except for the metal supporting structure.

Graner and della Rocca [83] conducted a condition survey of 74 US Navy boats of 21 different classes from 5–15 years old (the median hull, fabricated in 1958, was 10 years old). Defects were classified into 17 categories, most of which were the result of impact damage or abuse. The laminates were generally in excellent condition, regardless of the age of the boats, with material degradation limited to minor, surface skin delaminations, skin/core separation in sandwich panels and deterioration of the paper honeycomb cores. There was no indication of laminate degradation simply as a result of ageing in boats up to 16 years old. Even the earlier boats, while having poorer quality and strength characteristics, successfully withstood years of rigorous service.

7.7   Marine fouling and biodegradation

GRP hulls are subject to fouling by micro-organisms and this reduces the speed of the vessel or increases the power and fuel required to maintain a given speed, but plant growth is unlikely to cause structural deterioration of a laminate unless harmful degradation products are released. There may be a problem with hard shells penetrating the gelcoat or structural laminate after a long time. No reports addressing this potential problem have yet been identified.

Fouling is countered by using antifouling paints, which have traditionally contained biocides thought to be harmful to aquatic life. Other approaches have been tried, including protecting the hull with coatings that provide no foundation for the microorganisms.

7.7.1 Removal of fouling by brushing

Caron and Sieburth [84] studied the fouling of fibreglass reinforced 'cross-linked polystyrene resin' (*sic*) by bacteria and protists (unicellular organisms with both plant and animal characteristics). The experiments were conducted to simulate the planktonic ecosystem and associated flora and fauna of Narragansett Bay (41.5°N 71.7°W) in late summer/early autumn 1976 when fouling was at its worst. Periodic brushing at 3–4 day intervals drastically reduced protist species diversity and total protist cell density (Fig. 7.10). The fouled fibreglass surface was not completely smooth at a microscopic level and shallow microcrevices contained sedimented debris and microorganisms, even after brushing. Brushing effectively removes the secondary colonizers (fast-accumulating protists) and left a primary fouling film of more slowly accumulating bacteria. The fouling rate ($ng\,mm^{-2}day^{-1}$) with brushing was less than 4% of the rate for unrestricted fouling after 15 days. The mechanical disruption may substantially reduce the attachment of larval invertebrates, which require a substratum conditioned by microbial fouling.

Acorn barnacles at 1750 animals/m^2 were recorded as fouling species on the panels exposed by Short *et al.* [59] in Plymouth Sound (50°N 4°W), together with worms and algae. The full list of fouling organisms is given in Table 7.4.

Hodgkiess *et al.* [78] recorded an extremely thick fouling layer after 'several months submergence'. The layer included thick seaweed, other

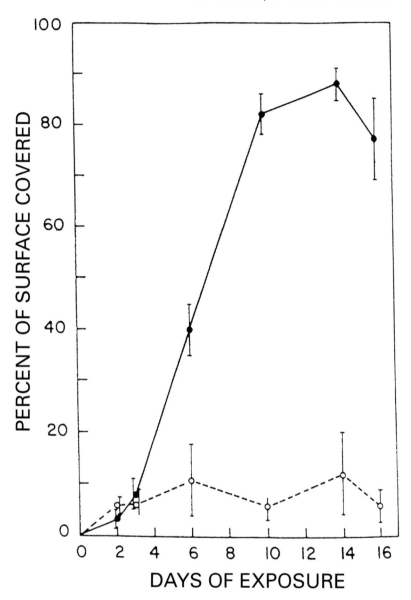

Figure 7.10 Estimates of percentage surface cover for brushed and unbrushed FRP fouling arrays, submerged in an estuary: ●, unbrushed fouling arrays; ○, brushed fouling arrays. (Reproduced from Caron and Sieburth [84], p 271, with kind permission of the authors and the American Society for Microbiology)

Table 7.4 Fouling organisms from panels exposed in
the intertidal region of Plymouth Sound

Algae: Myxophyceae
Cyanophyceae (blue/green)
Pleurocapsa spp. (Fuliginosa)
Symploca hynoides var.*fasciculatus*
Chlorophyceae (green)
Cladophora fracta (Siphonocladales)
Derbesia marina (Derbesiales)
Enteromorpha intestiralis (Ulotrichales)
Ulothrix speciosa (Ulotrichales)
Phaeophyceae (brown)
Ectocarpus confervoides (Ectocarpales)
Ectocarpus fasciculatus (Ectocarpales)
Rhodophyceae (red)
Porphyra linearis (Bangiaceae)
Porphyra umbilicalis (Bangiaceae)
Fauna *Phylum Annelida* (worms)
Pomotoceros triquetes
Phylum Arthropoda
Balanus crenatus (acorn barnacles)

---

algae and barnacles. The fouling layer was easily detached and there was no evidence of damage to the underlying laminate. Samples from the intertidal zone were lightly fouled.

Wagner *et al.* [85] examined reinforced plastics for their susceptibility to microbiologically influenced degradation. Laminates, resins and fibres were exposed to sulfur/iron-oxidizing, calcerous-depositing, ammonium-producing, hydrogen-producing and sulfate-reducing bacteria (SRB) in batch culture.

Surfaces were uniformly colonized by all physiological types of bacteria. However, the microbes preferentially colonized surface anomalies, including scratches and fibre disruptions. Epoxy and vinyl ester neat resins, carbon fibres and epoxy laminates were not adversely affected by the microbial species. SRB degraded the organic surface coating on the glass fibres. Hydrogen-producing bacteria appeared to disrupt the adhesive bonding between the fibres and vinyl ester resin and to penetrate the resin at the interface.

7.8 Cavitation erosion

Some materials can be damaged by collapsing air bubbles such as those in agitated water. It is not yet certain whether this affects FRP very much, but it is known to affect some metals.

Cavitation erosion damage can be defined as a loss of material from a solid surface, caused by the collapse of cavitation bubbles close to that surface. It is important that the possibility is considered in the design of any

Figure 7.11 Significant cavitation damage to a metallic marine propeller

Figure 7.12 Cavitation damage to a section of composite pipe flange

marine laminate structure, especially in hydrodynamic applications such as stern gear (propellers and rudders). Figure 7.11 shows significant cavitation damage to a metallic marine propeller blade and Fig. 7.12 shows the cavitation damage to the flange of a composite pipe section. There is very little published information in the literature.

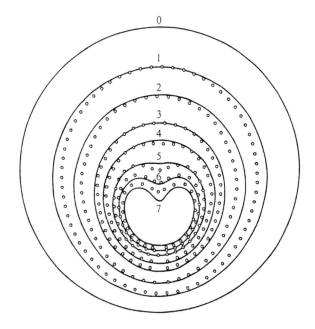

Figure 7.13 Collapse of a cavitation bubble close to a solid boundary, with theoretical shapes shown as solid lines (from Plesset and Chapman, [87]) compared with experimental observations as points (from Plesset and Prosperetti [88]) (reproduced from Lauterborn and Bolle [89], p 396 with kind permission of W L Lauterborn and Cambridge University Press)

Cavitation is normally due to the rapid formation, growth and subsequent collapse of bubbles or cavities in a liquid. These vapour bubbles are formed and grow when the local pressure falls below the vapour pressure in the liquid at the ambient temperature. When the liquid pressure exceeds this vapour pressure, the resulting bubbles collapse.

7.8.1 Mechanism of bubble collapse

Brennen [86] describes bubble collapse (Fig. 7.13) in a quiescent fluid as a developing spherical asymmetry, with a rapidly accelerating jet of fluid entering the bubble from the side furthest from the adjacent rigid wall. The re-entrant microjet achieves very high speeds such that its impact on the other side of the bubble generates a shock wave and thus a highly localized shock loading of the nearby wall. After the microjet disruption, the collapse of the remnant cloud of bubbles to its minimum gas/vapour volume generates a second shock wave which may impose two to three times the surface loading from the microjet. The microjet may be substantially modified and reduced by flow (e.g. rotation due to shear). Brennen notes 'that a bubble, collapsing close to a very flexible or free surface, develops a jet on the side closest to this boundary, and, therefore, travelling in the opposite direction!'

Cavitation causes high noise levels and severe vibration. It can be suppressed [90] by the following methods:

- Increasing the system pressure
- Reducing flow speeds
- Staged dissipation of the energy
- Lowering the temperature
- Introducing air into the flow
- Redesigning the system

Attrition of the surface by cavitation follows after an incubation period. The condition of the original surface influences the sites at which damage starts and grows. The process can be ductile (as in most metals) or brittle (as is the case of some body-centred cubic metals, ceramics, coatings and hard polymers). Mixed mode erosion has also been observed. Allen and Ball suggest that a compromise must be struck between hardness and microtoughness in order to control wear loss [91]. Materials with high ductility and low elastic modulus (such as rubbers and elastomers) possess good resistance to cavitation erosion, which can be attributed to their ability to store impact energy without exceeding the elastic limit [90]. However, these viscoelastic materials can theoretically behave in a brittle manner when stressed at very high frequencies.

The response of a specific material to cavitation attack is determined by the dynamics of the collapsing bubbles and their interaction with the solid material boundary. In a cavitating fluid, bubble dynamics is an extremely complex subject and the focus of considerable debate. Comprehensive reviews on the subject can be found in Knapp *et al.* [92], Morch [93], Lauterborn [94] and Prosperetti [95].

Cavitation damage has traditionally been considered to be dominated by either shock wave emission from off-boundary bubble collapse, or direct liquid jet impingement from the asymmetrical collapse of bubbles adjacent to the material surface. These phenomena are both observed during the collapse of a cavitation bubble. The majority of investigators now regard the liquid jet (water hammer) as the principal mechanism leading to damage. The global pressure magnification caused by the concerted collapse of a cluster of cavities has been suggested as an explanation for cavitation damage [96, 97].

The underlying premise of the cloud collapse theory is that the first bubbles to collapse are those at the periphery of a cavitation cloud, and they in turn transfer collapse energy to the core of the bubble cluster, thus intensifying water hammer pressures.

Numerical estimates of the pressures and jet impact times when considering the cavitation cloud model vary. Vyas and Preece [98] utilized a pressure transducer as a stationary wall adjacent to a cavitating ultrasonic horn and measured peak pressures of 900 MPa, that is one order of magnitude higher than the pressure values reported for single bubble collapse. The duration of the measured pulse peaks was greatly extended, of the order of 5 μs within the 50 μs oscillation period of the horn, implying the growth of bubbles over several cycles of horn oscillation.

Current understanding of the erosion of polymer laminates has come chiefly from rain erosion studies of composite aircraft materials. Analogies have been drawn between cavitation and liquid impingement processes on metallic materials; both involve extensive structural damage early in the incubation period as a result of short-term shock pulses. Furthermore, the deformed surfaces of materials which are subjected to both processes are very similar, and so are the methods used to improve their performance [99]. Attempts have been made to prioritize the factors affecting erosion of FRP by liquid impingement, with more or less predictable results [100] in respect of resin composition, fiber volume fraction and orientation, porosity and thermal conductivity. All these factors were found to affect performance.

Since it opened in 1981, the National Aquarium in Baltimore, Maryland has used 27 laminate pumps to circulate sea water. The pumps have operated continuously, except for three or four maintenance periods. The pumps were inspected by US Navy engineers after five years of continuous operation, and the pump-casings, impellers and other GRP components were found to be free of corrosion, erosion and cavitation damage [24].

A study of E-glass/PVC foam sandwich panels [101] ranked this kind of material above stainless steel in respect of cavitation resistance. On the other hand some investigations have reported damage, as the following examples illustrate.

Saetre [102] has reported cavitation erosion in GRP pipe bends guiding sea water at around $10 \, \text{m s}^{-1}$ and at pressures of less than 1 bar. Hammond et al. [103] examined the cavitation damage resistance of FRP before and after sea water immersion using a modified ASTM G32 method. A stationary specimen was placed beneath a Ti-6Al-4V tip oscillated at 20 kHz with an amplitude of 25 μm. The materials studied were:

- E-glass/5920 epoxy resin
- Scotch Ply 1002 glass/epoxy

- AS4 carbon/APC-2 polyetheretherketone thermoplastic matrix, and
- IM7 carbon in 977-2T thermoplastic toughened epoxy resin

For both dry and saturated conditions, the carbon/epoxy system performed best and the carbon/PEEK was the least resistant to damage. The damage resistance of the E-glass/epoxy and the carbon/PEEK immersed specimens decreased with immersion, whilst the Scotch Ply improved. The carbon/epoxy material was not affected by immersion. The topography of the damaged surfaces had a considerable influence on their subsequent erosion behaviour.

Lindheim [90] pointed out that the ASTM G32 standard has some disadvantages. In particular it is difficult to relate certain important test parameters (i.e. frequency and amplitude) to real life values such as static pressure and flow velocity. Also, the bubble sizes are smaller and the pressure gradients differ from those in hydrodynamic cavitation.

Lindheim used the rotating disk test method to measure the cavitation erosion performance of the following E-glass laminates:

- filament wound black pigmented 80 w/o roving in Epikote 827 epoxy
- filament wound 70 w/o 1200 tex roving in Epikote 828/MDA epoxy
- hand-laid/pressed 70 w/o 0/90° roving in Epikote 828/MDA epoxy
- filament wound 75 w/o 4800 tex roving in Derakane 411–45 vinyl ester
- hand-laid/pressed 69 w/o 0/90° woven roving in Derakane 411–45 vinyl ester
- resin transfer moulded 70 w/o ±45° unidirectional in Derakane 411–C50 vinyl ester
- resin transfer moulded 68 w/o ±45° unidirectional in Derakane 411–C50 vinyl ester
- continuous wound 43 w/o and 26 w/o chopped strand mat in BASF P69 polyester resin

The GRP generally showed higher cavitation erosion rates than 6 Mo-steel, titanium or high-density polyethylene. Damage incubation times for GRP ranged from a few minutes to almost 2 h. No detectable erosion loss was recorded for the other three materials after more than 20 h. However, the author cautions that there are no scaling methods available to relate the high intensity accelerated cavitation test results to in-service performance.

7.9   Galvanic corrosion

The corrosion of metals and semi-conductors involves the flow of an electric current within the material. Most of the constituent materials in fibrous laminates are insulators and in consequence, electrochemical corrosion is negligible. But graphite acts as a noble metal, lying between platinum and titanium in the galvanic series [104]. Carbon fibres have a structure which is essentially similar to graphite, being described as turbostratic (the layers of carbon atoms are essentially similar to those in true graphite, but have a greater spacing and greater disorder) and in consequence exposed carbon fibres adjacent to structural metals in the presence of sea water may behave like graphite and create a galvanic corrosion cell. In practice this can easily be prevented by insulating the carbon with a small quantity of glass cloth.

Hack and Macander [105] conducted a study to quantify the cathodic efficiency of a cross-plied low modulus carbon fibre/epoxy laminate. The laminate surfaces were made electrically conducting by belt-sanding to expose the fibres. The weight loss of sacrificial zinc anodes was monitored in static sea water and sea water flowing with a velocity of $10\,m\,s^{-1}$ for up to 270 days. The corrosion rate and current demand were similar for the carbon laminate and the nickel–aluminium bronze reference material. The graphite–epoxy specimens gained weight (maximum 0.4%) during testing because of water absorption and build up of a thin whitish deposit, pre-sumed to be calcerous material (see the reference to Tucker's work, below). Air-drying reduced the weight gain to 17–28 mg (from 15–70 mg). The authors did not consider the possibility of leaching damage resulting in weight loss, in spite of the higher weight of the dried material above!

Tucker [106] coupled CFRP laminates consisting of hand-laid 26% fibre unidirectional prepreg tape in vinyl ester resin and autoclave-cured 72% cross-plied Union Carbide T300 in Narmco 5208 epoxy, to mild steel im-mersed in sea water for intervals up to 3 months. Both laminates exhibited 'dramatic weight gains' when coupled to the steel under sea water. This was attributed to the growth of aragonite (a form of calcium carbonate) crystals on the surface of the laminate materials. Calcium carbonate occurs in natural sea water at 0.01 M concentration and is supersaturated by about 20% at pH 8, but requires a nucleation site. Blisters formed in the vinyl ester laminates after 3 weeks, developing in a regular pattern which coin-cided with the light glass weft in the unidirectional tape. All uncoupled strained samples gained approximately the same weight as the uncoupled, unstrained samples soaked in sea water. Epoxy and vinyl ester samples that were not coupled to the steel, but still strained and soaked under otherwise identical conditions, displayed neither crystals nor blisters.

Initial data [107] from coupling the vinyl ester laminates to aluminium indicated that rapid blistering of the laminate and rapid corrosion of the metal occurred, as might be expected, given their separation in the galvanic series. Blisters formed in the polymer where the undulating glass-fibre weft repeatedly approached the exposed surface and had a pH of 10 to 11. The degradation mechanism suggested was cathodic reduction of dissolved oxygen to form hydroxyl ions, which reacted with components of the lami-nate to form an osmotic cell.

Tucker [108] found that carbon fibre laminates acted as an extremely efficient cathode when coupled with metals in sea water. Deep submergence introduced the potential for increased moisture uptake through damage-dependent mechanisms.

Stafford *et al.* [109] studied the effect of electrolysis on graphite fibre/polymer matrix laminates used as electrodes in simulated sea water. The electrochemical behaviour was dependent on both the fibre orientation and the matrix polymer. Cathodic performance was stable. Anodic behaviour was affected by geometric considerations caused by fibre orientation, as well as by containment in a polymer matrix. The anode showed signs of interfacial attack after 30 min at an apparent current density of $25\,mA\,cm^{-2}$, with a reduction of about 15% in the average fibre diameter. This was attributed to a combination of chemical and electrochemical graphite oxidation, together with cavitation due to oxygen and chlorine evolution. In similar laminates with fibres parallel or perpendicular to the solution interface, the parallel orientation failed rapidly because it presented the greatest fibre–resin interfacial area to the solution. The laminate with perpendicular fibres decayed more slowly.

Comparisons were also conducted between five matrix systems: polyamide, which failed first; epoxybutadiene and unsaturated polyester, which both showed rather poor voltage stability; epoxy, and polyphenylenesulfide, the last-mentioned surviving longest. All were reinforced with random fibres.

Donnellan and Cochran [110] conducted controlled electrochemical experiments to elucidate the reaction mechanism in the galvanic degradation of carbon/bismaleimide laminates in electrical contact with aluminium alloy. The degradation progressed by surface cracking and removal of the gelcoat, exposing the fibres. Two values for each variable were used, that is salt concentration (0.1% and 3.5%), pH (7 and 11), temperature (25°C and 83°C), current density (40 and $160\,\mu A\,cm^{-2}$) and duration (1 and 4 days), as well as two environments (aerated or deaerated water). Significant degradation was associated with high temperature, high current density and long duration.

Oxygen concentrations were not a significant factor, whereas work by BASF, reported in the same paper, suggested that oxygen was required for degradation. Scrim cloths of glass/bismaleimide dramatically reduced corrosion currents. Polysulfide sealants and polyurethane coatings reduced degradation rates, especially when edges were covered.

Alias and Brown [111] identified two types of damage in carbon fibre/vinyl ester laminates (the same materials as described by Tucker and Brown) after long term galvanic coupling in sea water. After 90 h exposure in simulated sea water (3.5% NaCl) at −0.65 V (SCE: standard calomel electrode) regions of polymer surface dissolution were found above the carbon fibres. After 720 h, blistering was found again at high points in the glass weft accompanying the dissolution. A potential of −1.2 V (SCE) resulted in the carbon fibres becoming exposed after the covering polymer layer was rapidly removed, and carbon fibres were floating in the solution after 30 h exposure. The simulated testing accurately reflected the long term surface damage caused by galvanic exposure in sea water.

Alias and Brown [112] reported some further work on carbon/vinyl ester laminates in 3.5% sodium chloride and carbon/epoxy laminates in 0.5 N sodium chloride. Measurements were conducted for up to 90 h for open circuit and applied cathodic potentials of −100 mV (SCE) and up to 30 days at potentials of −1200 mV (SCE), with intermediate tests at −300 mV, −650 mV and −900 mV (SCE). Both types of laminate suffered severe damage at potentials more cathodic than −650 mV (SCE), suggesting hydrogen production (which occurs at −626 mV (SCE) at pH 6.5) acting in combination with the cathodic potential by direct reduction of the polymer. The vinyl ester laminates exhibited blistering, followed by regularly spaced local removal of the resin. The epoxy laminates exhibited a uniform pattern of porosity.

Sloan and Talbot [113] cathodically coupled 26-ply quasi-isotropic T300/934 carbon/epoxy laminate coupons to magnesium in natural Pacific sea water at 40°C for 140 days. In the uncoupled state in tapwater or natural sea water, the moisture uptake was ~0.85% and calcerous deposit was 0.14% and 0.36%, respectively. For the coupled materials, the moisture uptake was ~1.3% and the calcerous deposit was 16%. There was some leaching loss from the coupled material. The shear strengths in four point bend tests showed negligible degradation for the soaked samples, but a 30% reduction was measured for electrically coupled materials. This reduction was correlated to delaminations, reducing the effective specimen width by 20–40%.

Fourier transform infrared (FTIR) analysis of the damaged epoxy suggested that phthalate linkages in the 934 epoxy resin had been disrupted, probably by nucleophilic attack by hydroxide ions in the high pH environment. (Resin 934 includes 11.2% diglycidyl orthophthalate to lower the viscosity and improve fibre wet-out). Note that most marine grade polyesters are based on phthalic acid precursors as major constituents, and thus are likely to be even more prone to this form of degradation than epoxies when within cathodically protected systems.

Sloan and Talbot [113] reported a study of the anodic exposure of autoclave-cured 30-ply unidirectional AS-4/3501–5a carbon epoxy laminates in unaerated 0.5 M pH 7 sodium chloride at ambient temperature, against a platinum counter electrode. Crack formation was observed at potentials above 600 mV (SCE) at currents as low as 1 μA cm^{-2}. Discoloration (yellowish brown) of the electrolyte was observed at potentials above 900 mV with both the carbon/Pt and Pt/Pt electrode systems. The reinforcement fibres were attacked by atomic oxygen, which is an intermediate in the oxygen evolution reaction.

Chukalovskaya et al. [114] studied the corrosion behaviour of steel, stainless steel, brass, titanium and aluminium alloys in contact with 60 v/o carbon fibre reinforced phenolic resin in a variety of aqueous solutions, including simulated sea water (25 g l^{-1} sodium chloride, 3.4 g l^{-1} magnesium sulfate, 2.1 g l^{-1} magnesium chloride, 1.3 g l^{-1} calcium chloride and 0.2 g l^{-1} sodium bicarbonate) at 30°C and 50°C. The behaviour was predicted by combining the potential–current density curves for CFRP cathodes and the corresponding metal anode. Corrosion enhancement was noted below a critical value of S_c/S_m (the ratio of the surface areas of the composite, or laminate, to that of the metal) and passivation above the critical value. $[S_c/S_m]_{cr}$ is defined as the ratio of the critical metal passivation current density (anodic curve maximum) to the composite cathodic current density at this passivation potential. The brass/composite cells were unable to passivate anodically in sea water and corrosion increased with increasing S_c/S_m ratio. At 25°C, the stainless steel/composite and titanium/composite cells were steadily passive and increasing S_c/S_m ratios should enhance the passivity. Stainless steel/composite and aluminium/composite were susceptible to anodic activation and, in particular, to pitting. This is especially detrimental for aluminium. Aluminium in sea water is passive ($E_c = -0.62$ V and K ~60 mg m^{-2}h^{-1}). After making contact, the electrode potential of aluminium rises rapidly to a stationary value of -0.45 V and the pitting corrosion mass loss increases to 350 mg m^{-2}h^{-1}). The actual behaviour of this cell is in good agreement with the prediction from analysis of the diagrams (Fig. 7.14). At normal temperatures, the potential for local anodic activation of the stainless steel is higher than the free corrosion potential of the composite and their contact causes no damage. At elevated temperatures, stainless steel/composite contact becomes dangerous because of pitting.

Thermoplastic matrix composites have not yet been mentioned much in this review. The hydrophobic (low water absorption) materials such as the polyolefins may be more suitable than thermosets for use in marine applications and have the advantage of low cost. However, many of the engineering thermoplastics, notably polyamides and polyesters, have significant moisture absorption rates.

Pakalapati et al. [115] investigated some carbon/thermoplastic laminates. The materials were pultruded and they consisted of 50 v/o unidirectional continuous polyacrylonitrile-based carbon fibres in DuPont J-2 aromatic polyamide-based thermoplastic matrix. They were subjected to anodic and cathodic currents in sea water. Dynamic mechanical analysis was carried out in situ to measure the shear storage modulus (G') and shear loss modulus (G'') of 1.27 mm diameter rod shaped samples, subjected to small amplitude torsional oscillations. The moduli were constant with time in air,

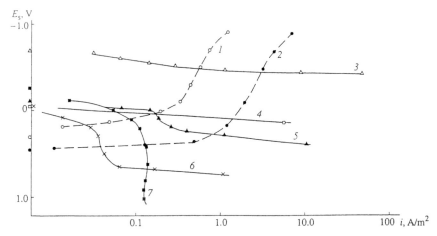

Figure 7.14 Polarization curves for the CFRP–metal cells in simulated sea water. 1 = Composite cathodic curve at 25°C, 2 = composite cathodic curve at 50°C, 3 = aluminium anodic curve at 25 °C, 4 = brass anodic curve at 25°C, 5 = stainless steel anodic curve at 50°C, 6 = stainless steel anodic curve at 25°C, 7 = titanium anodic curve at 25°C. (Reproduced from [114] by kind permission of TV Chukalovskaya and IAPC 'Nauka')

and submerged in sea water in the absence of applied currents. A constant anodic current (density $4140\,\mu A\,m^{-2}$) for 2 h caused both moduli to fall to about 45% of the original value (Fig. 7.15). A cathodic current of the same density and duration caused the moduli to fall to about 63% of the original value (Fig. 7.16). The moduli remained at these low levels after the current was turned off. Scanning electron microscopy showed matrix cracking after anodic current exposure and some fibres were separated from the matrix after cathodic current. Calcerous deposits were seen on the surface of the cathodically polarized composites (as observed earlier by Tucker [106]).

7.10 Concluding remarks

The overall impression given by the scientific and technical literature can be negative compared with the generally favourable experience of using FRP laminates in marine environments. This is simply because materials scientists are generally employed to investigate problem areas and to test new materials to their limits. In fact, fibre reinforced plastics have very appropriate properties for use in small and medium sized marine structures and the durability of FRP in marine environments to date has been excellent, considering that until fairly recently there was little experience on which to base materials optimization and structural design.

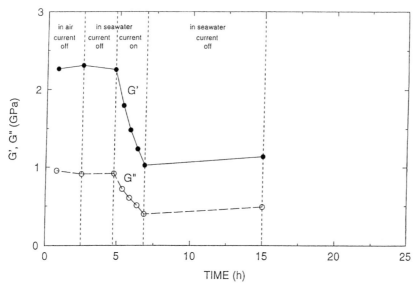

Figure 7.15 Variation in the storage modulus, *G'*, and the loss modulus, *G''*, with time, under a constant anodic current of density 4140 µA m^{-2} [115] (reproduced with kind permission of Raju Pakalapati and the Society of Plastics Engineers)

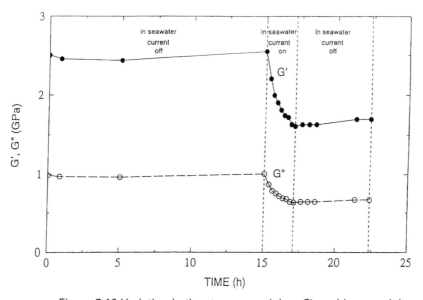

Figure 7.16 Variation in the storage modulus, *G'*, and loss modulus, *G''*, with time under a constant cathodic current of density 4140 µA m^{-2} [115] (reproduced with kind permission of Raju Pakalapati and the Society of Plastics Engineers)

Acknowledgement

The authors wish to acknowledge helpful discussions with the late Derek Sargent in respect of the text on galvanic corrosion.

References

1 K B Spaulding, 'Fiberglass boats in naval service', *Naval Engineers J* 1966 **78** 333–340.

2 N Fried and W R Graner, 'Durability of reinforced structural materials in marine service', *Marine Technol* 1966 **3**(3) 321–327.

3 Engineers of Gibbs & Cox Inc., *Marine Design Manual for Fiberglass Reinforced Plastics*, New York, McGraw-Hill Book Company, 1960.

4 C S Smith, *Design of Marine Structures in Composite Materials*, London, Elsevier Applied Science, 1990.

5 *NES 140: GRP Ship Design*, Naval Engineering Standards, Issue 2, undated.

6 *Rules for Building and Classing Reinforced Plastics Vessels*, American Bureau of Shipping, New York, 1978.

7 *Rules for Yachts and Small Craft*, Lloyd's Register of Shipping, London, 1983.

8 *Rules and Regulations for the Construction and Classification of Textile Glass Reinforced Polyester Vessels*, Nippon Kaiji Kyokai, 1984.

9 *Rules for Classification of High Speed Light Craft*, Det Norske Veritas, Hovik – Norway, 1985.

10 *Rules for the Construction of Reinforced Plastics Hulls*, Registro Italiano Navale, 1988 (in Italian).

11 *Provisional Rules for the Classification of High Speed Catamarans*, Lloyd's Register of Shipping, London, 1990.

12 *Tentative Rules for Classification of High Speed and Light Craft*, Det Norske Veritas, Hovik – Norway, 1991.

13 *Rules for Building Fibre-Reinforced Plastic Vessels*, China Bureau of Shipping, Beijing, 1991.

14 *Rules for the Classification and Construction of Sea-Going Ships, Part XVI: Hull Structure of Glass-Reinforced Plastic Ships and Boats*, USSR Register of Shipping, undated.

15 J Summerscales, 'Marine applications', *Engineered Materials Handbook 1 – Composites*, ed. T J Reinhart, ASM International, Metals Park OH, USA, November 1987, Chapter 12G, pp 837–844.

16 D C Hicks, C M Pleass, W A Fearn and D Staples, 'Development, testing and the economics of a composite/plastic seawater reverse osmosis pump', *Desalination* 1989 **73**(1–3) 95–109.

17 T Searle, J Chudley and D Short, 'The composite propeller', *Impact of New Technology on the Marine Industries*, Warsash, 13–15 September 1993, paper 59, pp 1–10.

18 T Searle, J Chudley and D Short, 'Composites offer advantages for propellers', *Reinforced Plastics* 1993 **37**(12) 24–26.

19 T Searle and D Short, 'Are composite propellers the way forward for small boats?', *Mater World* 1994 **2**(2) 69–70.

20 T Searle, J Chudley, D Short and C Hodge, 'The use of FRP composites for ships propellers', *International Conference Structural Materials in Marine Environments*, Institute of Materials, London, 11–12 May 1994.

21 T Searle, J Chudley, D Short and C Hodge, 'The composite advantage', *Propellers and Shafting*, SNAME, Virginia Beach VA, 20–21 September 1994.

22 T Searle, D Short, J Chudley, A Tate and S Bucknell, 'Resin transfer moulding of marine products', *19th International Composites Congress*, British Plastics Federation, NEC Birmingham, 22–23 November 1994, Session 2 paper 4. Publication 293/7, pp 27–35.

23 T J Searle, J Chudley, S M Grove and D Short, 'Manufacturing of marine propellers in composite materials', *4th International Symposium on Advanced Composites* and *5th Europe-Japan Bilateral Colloquium on Composite Materials 'High-Technology Composites in Modern Applications'*, Corfu – Greece, 18–22 September 1995, Session V.

24 A J Sedriks, 'Advanced materials in marine environments', *Mater Perform* 1994 **33**(2) 56–63.

25 L G P Dalmolen, J Willing, A Goksoyr and P Osen, 'Para-aramid fibre-reinforced composite bellmouths for a floating production platform', *Conference on Structural Materials in Marine Environments*, London, IoM/MTD/IMechE/IMarE, 11–12 May 1994, pp 109–118.

26 Composite holds back seawater at 13 fathoms *Civil Eng* 1996 **66**(6) 87.

27 K Hirose, T Yamamoto, S Kuroda and T Fujii, 'A floating pier system for marinas using glass fibre mat-reinforced stampable sheets', *J Adv Mater* 1996 **28**(1) 32–41.

28 A Weaver, 'Composites begin to bridge the gap', *Reinf Plastics* 1996 **40**(11) 22–26.

29 T Carlean, 'Composite grating is right for today's offshore market', *Reinf Plastics* 1996 **40**(12) 38–44.

30 F J Barnes, 'Composite materials in the UK offshore oil and gas industry', *SAMPE J* 1996 **32**(2) 12–17.

31 M Skrifvars, H Manninen and K Strang, 'Recycling of production waste from the fibre-reinforced plastics industry', *Rivestimenti & Materiali Compositi* 1997 (18) 36–40.

32 A R Mableson, R J Osborn and J A Nixon, 'Structural use of polymeric composites in ships and offshore', *Proceedings 2nd International Conference Polymers in a Marine Environment*, Institute of Marine Engineers, London, October 1987, paper 9, pp 79–86.

33 S Nagae and Y Otsuka, 'Effect of sizing agent on corrosion of glassfibre reinforced plastics (GFRP) in water', *J Mater Sci Lett* 1996 **15**(1) 83–85.

34 G S Springer, 'Moisture absorption in fibre-resin composites', *Developments in Reinforced Plastics – 2*, ed. G Pritchard, Applied Science, London, 1982, Chapter 3, pp 43–65.

35 A Apicella, L Nicolais, G Mensitieri and M del Nobile, 'Environmental degradation of polymeric matrices for high performance composites', *Proceedings International Conference Advanced Materials*, Strasbourg, May 1991, *Symposium A4: Composite Materials*, North-Holland, Amsterdam, 1992, pp 11–21.

36 E M Nechvolodnova, E V Prut, I M Belgovskiy, A T Ponomarenko and V G

Shevchenko, 'The effect of moisture absorption on mechanical and dielectric properties of epoxy resins and composites', *Acta Polymerica* 1992 **43**(3) 191–192.

37 A R Bunsell, 'The effects of moisture on the electrical properties of epoxy composites', *Developments in Reinforced Plastics – 3*, ed. G Pritchard, Elsevier Applied Science, London, 1984, Chapter 1, pp 1–24.

38 J Summerscales, 'Non-destructive measurement of the moisture content in fibre-reinforced plastics', *Br J Non-Destructive Testing* 1994 **36**(2) 64–72.

39 N Clegg, 'Osmosis – the truth!' *Yachts and Equipment*, Spring/Summer 1995, 40–44.

40 J R M D'Almeida, 'Effects of distilled water and saline solution on the interlaminar shear strength of an aramid/epoxy composite', *Composites* 1991 **22**(6) 448–450.

41 N Clegg, *The Osmosis Manual: Understanding and Treating Osmosis in GRP Yachts*, Manual OT1 V3.0 (06/06/96 13:09), Nigel Clegg, County Durham, 1996. Tel: +44.1740-620489.

42 D J Steel, 'The disk cracking behaviour of polyester resins', *Trans J Plastics Institute* April 1967 **35**(116) 429–433.

43 K H G Ashbee, F C Frank and R C Wyatt, 'Water damage in polyester resins', *Proc Roy Soc London A* 1967 **300** 415–417.

44 A W Birley, J V Dawkins and H E Strauss, 'Blistering in glass fibre reinforced polyester laminates', Preprints, *14th Reinforced Plastics Congress*, British Plastics Federation, Brighton, November 1984, paper 37, pp 167–172.

45 A W Birley and F Chen, 'Blistering in glass fibre reinforced polyester laminates', Preprints, *16th Reinforced Plastics Congress*, British Plastics Federation, Blackpool, November 1988, paper 12, pp 49–54.

46 F Chen and A W Birley, 'Blistering of GRP laminates in water, 1: Characterisation and classification of blistering', *Plastics, Rubbers and Compos Proc Applic* 1991 **15**(3) 161–168.

47 J S Ghotra and G Pritchard, 'Osmosis in resins and laminates', *Developments in Reinforced Plastics – 3*, ed. G Pritchard, Elsevier Applied Science, London, 1984, Chapter 3, pp 63–95.

48 J S Ghotra and G Pritchard, 'Blistering of fibreglass boat hulls', *Proceedings 1st International Conference, Polymers in a Marine Environment*, Institute of Marine Engineers, London, October/November 1984, paper 3, pp 15–22.

49 F Chen and A W Birley, 'Blistering of GRP laminates in water, 2: Further study of the factors affecting blister formation', *Plastics, Rubbers Compos Proc Applic* 1991 **15**(3) 169–176.

50 G Tipping, 'Osmosis: predicting blister resistance', *Reinf Plastics* 1987 **31**(10) 269–270; 1987 **31**(11) 298–302.

51 L S Norwood and E C Holton, 'Marine grade polyester resins for boat building in the 1990s', *Polym Polym Compos* 1993 **1**(5) 381A–390A.

52 L S Norwood, 'Blister formation in glass fibre-reinforced plastic: prevention rather than cure', *Proceedings 1st International Conference, Polymers in a Marine Environment*, Institute of Marine Engineers, London, October/November 1984, paper 4, pp 23–29.

53 T Abraham, 'Environmental effects on polyester laminates reinforced with

chopped strand glass fibre mats', *Kautschuk Gummi Kunststoffe* 5/1996 **49**(5) 350–353.

54 British Plastics Federation, *Guidance notes for the manufacturing of glass fibre polyester laminates to be used in marine environments*, BPF Publication No 220/3, London, August 1983.

55 P Castaing, L Lemoine and A Gourdenne, 'Mechanical modelling of blisters on coated laminates I – theoretical analysis', and 'Mechanical modelling of blisters on coated laminates II – experimental analysis', *Compos Struct* 1995 **30**(2) 217–228.

56 P Castaing and L Lemoine, 'Effects of water absorption and osmotic degradation of long-term behaviour of glass fibre reinforced polyester', *Polym Compos* 1995 **16**(5) 349–356.

57 P Castaing, L Lemoine and A Gourdenne, 'Mechanical behaviour of blisters on laminates', *Proceedings ECCM-7, IoM & EACM*, London, May 1996, volume 1, pp 153–158.

58 R S Dow and J Bird, 'The use of composites in marine environments', *Iron and Steelmaking* 1994 **21**(6) 431–444.

59 D Short, A W Stankus and J Summerscales, 'Woven glass-fibre reinforced polyester resin composites exposed to the marine environment', *Proceedings 1st International Conference Testing, Evaluation and Quality Control of Composites*, Guildford, September 1983. Butterworth Scientific, Guildford, 1983, pp 212–220.

60 M R Cargen, R T Hartshorn and J Summerscales, 'Effect of short-term continuous sea water exposure on the shear properties of a marine laminate', *Trans I Mar E (C)* 1985 **97** (Conf 2, Paper 21) 131–134.

61 L H Strait, M L Karasek and M F Amateau, 'Effects of seawater immersion on the impact resistance of glass fibre reinforced epoxy composites', *J Compos Mater* 1992 **26**(14) 2118–2133.

62 N M S Al-Bastaki and H M N Al-Madani, 'Effect of local atmospheric conditions in Bahrain on the mechanical properties of GRP', *Polym Testing* 1995 **14**(3) 263–272.

63 R D Adams and M M Singh, 'The effect of immersion in sea water on the dynamic properties of fibre-reinforced flexibilised epoxy composites', *Compos Struct* 1995 **31**(2) 119–127.

64 P Davies, F Pomies and L A Carlsson, 'Influence of water and accelerated aging on the shear fracture properties of glass/epoxy composite', *Appl Compos Mater* 1996 **3**(2) 71–87.

65 P Davies, F Pomies and L A Carlsson, 'Influence of water absorption on transverse tensile properties and shear fracture toughness of glass/polypropylene', *J Compos Mater* 1996 **30**(9) 1004–1019.

66 M L Karasek, L H Strait, M F Amateau and J P Runt, 'Effect of temperature and moisture on the impact behaviour of graphite/epoxy composites, Part I: impact energy absorption', *J Compos Technol Res* 1995 **17**(1) 3–10.

67 M L Karasek, L H Strait, M F Amateau and J P Runt, 'Effect of temperature and moisture on the impact behaviour of graphite/epoxy composites, Part II: impact damage', *J Compos Technol Res* 1995 **17**(1) 11–16.

68 W L Bradley and T S Grant, 'The effect of moisture absorption on the interfacial strength of polymeric matrix composites', *J Mater Sci* 1995 **30**(21) 5537–5542.

69 T S Grant and W L Bradley, 'In-situ observations in SEM of degradation of graphite/epoxy composite materials due to seawater immersion', *J Compos Mater* 1995 **29**(7) 852–867.

70 P Chiou and W L Bradley, 'Effects of seawater absorption on fatigue crack development in carbon/epoxy EDT specimens', *Composites* 1995 **26**(12) 869–876.

71 W C Tucker and R Brown, 'Moisture absorption of graphite/polymer composites under 2000 feet of seawater', *J Compos Mater* 1989 **23**(8) 787–797.

72 W J Cantwell, G Broster and P Davies, 'The influence of water immersion on skin-core debonding in GFRP-balsa sandwich structures', *J Reinf Plastics Compos* 1996 **15**(11) 1161–1172.

73 R C Roberts, 'Environmental stress cracking of GRP: implications for reinforced plastics process equipment', *Composites* 1982 **13**(4) 389–392.

74 P J Hogg and D Hull, 'Corrosion and environmental deterioration of GRP', *Developments in GRP Technology – 1*, ed. B Harris, Elsevier Applied Science, London, 1983, Chapter 2, pp 37–90.

75 G Menges and K Lutterbeck, 'Stress corrosion in fibre-reinforced plastics in aqueous media', *Developments in Reinforced Plastics – 3*, ed. G Pritchard, Elsevier Applied Science, London, 1984, Chapter 4, pp 97–122.

76 Endurance of composite hulls, *Marine Eng Rev* May 1988 16–17.

77 F E Sloan and R J Seymour, 'The effect of seawater exposure on mode I interlaminar fracture and crack growth in graphite/epoxy', *J Compos Mater* 1992 **26**(18) 2655–2673.

78 T Hodgkiess, M J Cowling and M Mulheron, Durability of glass reinforced polymer composites in marine environments, *Conference: Structural Materials in Marine Environments*, IoM/MTD/IMechE/IMarE, London, May 1994, pp 58–72.

79 R Kosuri and Y J Weitsman, 'Sorption process and immersed fatigue response of graphite/epoxy composites in seawater', *Proceedings ICCM-X*, Whistler, BC, Canada, August 1995. Cambridge, England, Woodhead, Volume VI, pp 177–184.

80 L V Smith and Y J Weitsman, 'The immersed fatigue response of polymer composites', *Internat J Fracture* 1996 **82**(1) 31–42.

81 P-L Chiou and W L Bradley, 'Moisture-induced degradation of glass/epoxy filament-wound composite tubes', *J Thermoplastic Compos Mater* 1996 **9**(2) 118–128.

82 B Cobb, 'A report on the long-term durability of resin glass boats', *Ship and Boat Builder*, February 1963, 41–42.

83 W R Graner and R J della Rocca, 'Evaluation of US Navy GRP boats for material durability', *26th Annual Technical Conference, SPI Reinforced Plastics/Composites Division*, Washington DC, February 1971, Session 7-F.

84 D A Caron and J M Sieburth, 'Disruption of the primary fouling sequence on fiberglass reinforced plastic submerged in the marine environment', *Appl Environ Microbiol* 1981 **41**(1) 268–273.

85 P A Wagner, B J Little, K R Hart and R I Ray, 'Biodegradation of composite materials', *Internat Biodeterioration Biodegradation* 1996 **38**(2) 125–132.

86 C E Brennen, *Hydrodynamics of Pumps*, Concepts ETI, Norwich VT, 1994. Also: Oxford University Press, Oxford, 1994.

87 M S Plesset and R B Chapman, 'Collapse of an initially spherical vapour cavity in the neighbourhood of a solid boundary', *J Fluid Mech* 1971 **47** 283–290.

88 M S Plesset and A Prosperetti, 'Bubble dynamics and cavitation', *Ann Rev Fluid Mech* 1977 **9** 145–185.

89 W Lauterborn and H Bolle, 'Experimental investigations of cavitation bubble collapse in the neighbourhood of a solid boundary', *J Fluid Mech* 1975 **72** 391–399.

90 T Lindheim, 'Erosion performance of glass fibre reinforced plastics (GRP)', *Revue de l'Institut Francais du Petrole* 1995 **50**(1) 83–95.

91 C Allen and A Ball, 'A review of the performance of engineering materials under prevalent tribological and wear situations in South African industries', *Tribology Internat* 1996 **29**(2) 105–116.

92 R T Knapp, J W Daily and F G Hammitt, *Cavitation*, McGraw-Hill, New York, 1970. LC 77–96428.

93 K A Morch, 'Dynamics of cavitation bubbles and cavitating liquids', *Treatise Mater Sci Technol* 1979 **16** 309–355.

94 W Lauterborn, 'Cavitation bubble dynamics – new tools for an intricate problem', *Appl Sci Res* 1982 **38** 165–178.

95 A Prosperetti, 'Bubbles dynamics: a review and some recent results', *Appl Sci Res* 1982 **38** 145–164.

96 K A Morch and I Hansson, 'The dynamics of cavity clusters in ultrasonic (vibratory) cavitation erosion', *J Appl Phys* 1980 **51**(9) 4651–4658.

97 K A Morch, 'Energy considerations on the collapse of cavity clusters', *Appl Sci Res* 1988 **38** 313.

98 B Vyas and C M Preece, *J Appl Phys* 1976 **47** 5133–5138.

99 C M Preece, 'Cavitation erosion', *Treatise Mater Sci Technol* 1979 **16** 296–297.

100 J H Brunton and M C Rochester, 'Erosion of solid surfaces by the impact of liquid drops', *Treatise Mater Sci Technol* 1979 **16** 185–248.

101 V Djordjevic, J Kreiner and S Zivojin, 'Cavitation erosion examination of composite materials', *33rd International SAMPE Symposium*, March 1988, pp 1561–1570.

102 O Saetre, 'Testing of composite pipes in high velocity seawater', *10th International OMAE Conference*, Stavanger, 1991, Volume IIIB, p 577.

103 D A Hammond, M F Amateau and R A Queeney, 'Cavitation erosion performance of fiber reinforced composites', *J Compos Mater* 1993 **27**(16) 1522–1544.

104 M Ohring, *Engineering Materials Science*, Academic Press, San Diego, 1995, p 510.

105 H P Hack and A B Macander, 'The effect of cathodic protection on graphite fiber-epoxy composite in natural seawater', *Mater Perform* 1983 **22**(3) 16–20.

106 W C Tucker, 'Crystal formation on graphite/polymer composites', *J Compos Mater* 1988 **22**(8) 742–748.

107 W C Tucker and R Brown, 'Blister formation on graphite/polymer composites galvanically coupled with steel in seawater', *J Compos Mater* 1989 **23**(4) 389–395.

108 W C Tucker, 'Degradation of graphite polymer composites in seawater', *Trans ASME: J Energy Resources Technol* 1991 **113**(4) 264–267.

109 G R Stafford, G L Cahen and G E Stoner, 'Graphite fiber-polymer matrix composites as electrolysis electrodes', *J Electrochem Soc* 1991 **138**(2) 425–430.

110 T M Donnellan and R C Cochran, 'Galvanic degradation of polyimide based composites', *3rd International Conference on Polymers in a Marine Environment*, Institute of Marine Engineers, London, 23–24 October 1991.

111 M N Alias and R Brown, 'Damage to composites from electrochemical processes', *Corrosion* 1992 **48**(5) 373–378.

112 M N Alias and R Brown, 'Corrosion behaviour of carbon fibre composites in the marine environment', *Corrosion Sci* 1993 **35**(1–4) 395–402.

113 F E Sloan and J B Talbot, 'Corrosion of graphite fibre reinforced composites 1: galvanic coupling damage', *Corrosion* 1992 **48**(10) 830–838. 'Corrosion of graphite fibre reinforced composites 2: anodic polarization damage', *Corrosion* 1992 **48**(12) 1020–1026.

114 T V Chukalovskaya, A I Shcherbakov, L A Chigirinskaya, V V Bandurkin, I L Medova and P A Chukalovskii, 'Electrochemical corrosion of metals in corrosion cells with carbon-fiber-reinforced plastic', *Protection of Metals* 1995 **31**(2) 136–141. Also: Zaschita Metallov, 1995 **31**(2) 149–154.

115 S N R Pakalapati, F Gadala-Maria and R E White, 'Dynamic mechanical analysis of a unidirectional carbon fibre-thermoplastic composite subjected to anodic and cathodic currents', *Polym Compos* 1996 **17**(4) 620–626.

8

Survey of long term durability of fiberglass reinforced plastics tanks and pipes

BEN BOGNER

8.1   Introduction

This chapter is a compilation of information on the long term durability of fiberglass reinforced plastic structures for civil engineering and infrastructure applications. Included in the survey are data from fluid containment tanks and pipes with up to 26 years of service. Correlations are obtained for the variations in static fatigue strength or creep, cyclic fatigue strength and burst strength of pressure piping. The relationship between static fatigue strength and residual burst strength is explored.

The effects of exposure to typical exposure environments on strength and modulus retention for both stressed and unstressed materials are examined and implications for design are discussed. Strength retention for gasoline (petrol) storage tanks and fiberglass piping are documented and analyzed after many years of service. The overall structural changes during the service life of a structure are examined and the basis for the designs discussed.

Analysis of the changes in strength and stiffness properties with time for large size composite structures indicates that structures exposed to a high moisture environment can sustain their strength for long periods of time. However, when exposure to weathering and ultraviolet radiation is present, appropriate surface protection appears to be required for the long term durability of the appearance. Several case histories have been documented which prove the durability of these structures and they provide a guide for the use of composites in new applications.

8.2   Composites

Composites, and more specifically, fiberglass reinforced plastics, have been used as structural materials for a wide range of applications in several fields of technology and construction such as water and sewage transport and treatment, storage, ships, pedestrian bridges, column overwraps and building facades. To achieve low overall cost performance and long term durability, these structures must have at least a 50 year service life. The question of the long term durability of fiberglass reinforced plastic structures, therefore, becomes an important consideration. Obviously, composites could be designed for an extremely large number of structures, but the economics significantly reduces this potential.

Long term durability (i.e. retention of strength and stiffness properties with time) is a concern because of the potential degenerative effects of such exposure factors as atmospheric moisture and chemicals, abrasion, ultraviolet radiation and weathering. These exposure factors are additional to, and can exist in conjunction with, the customary strength and modulus degradation tendencies due to continuous loading and cyclic loading or fatigue. Designers need to understand the use of fiberglass materials when subjected to exposure and fatigue effects for long periods of time and service conditions.

Fatigue strength of reinforced plastic materials is generally explored with laboratory tests of specimens in cycling machines. Test specimens are specially prepared or cut from the full structure. Such tests, however, cannot provide accurate information about the durability of a complete structure under actual operating conditions. Thus composite structures are sometimes built and subjected to full scale testing. The subject of fatigue is discussed fully in Chapter 5.

A useful general insight into the long term behavior of fiberglass reinforced plastic structures can be gained from examination of actual structures in real service over a long period of time. Such information is often difficult to obtain because of the requirement for accurate documentation of the material properties and service history of the components. Furthermore, the likelihood of finding structures that have been in use for the time period of interest (30 years) might seem rather small. Nevertheless, there are several specific documented instances where complete structures have been evaluated for real aging effects. Data are also available for complete structures tested under laboratory simulation of long term effects.

It is the purpose of this chapter to document the information available on the physical property degradation of these complete fiberglass reinforced structures. The scope includes identification of sources, tabulation and plotting of data and the correlation and analysis of results.

8.3   Fluid containment vessels – tanks and pipes

8.3.1   Underground gasoline (petrol) storage tanks

A large number of early fiberglass reinforced plastic (FRP) tanks were constructed in the 1960s for the underground storage of gasoline. The tanks

Table 8.1 Properties of FRP after use in a fiberglass underground storage tank in continuous service in Texas, USA for 6 years, 10 months

Property	Units	1964 Test Results[a]	1971 Test Results[b]
Flexural strength	MPa	245	230
Flexural modulus	GPa	10.34	12.60
Barcol hardness	—	40	48

[a] Average values.
[b] Single value.

were filament wound, with E-glass and high molecular weight isophthalic polyester resin and with a C-glass interior surface veil. The type of resin used provided a good barrier to both water and gasoline. Overall dimensions were of the order of 2.4 m in diameter by 6 m in length, with a 5 mm wall thickness. Two of the underground tanks were unearthed after different times of service and pieces were removed for determination of flexural strength properties.

One tank, installed in Houston, Texas, was removed in 1971 after 6 years and 10 months of service [1]. The tank was sectioned for observation of the interior and a section was cut from the tank wall on the bottom for measurement of material properties. Analysis of the bedding material under the tank indicated a pH of 9.75. Visual inspection revealed the tank to be in excellent condition. There was no evidence of crazing or cracking of the inner surface. Only a staining of the inner surface from gasoline colorants was found. The outer surface was similarly without any signs of attack. There was, however, a light chalking (due to long term exposure to moisture) which was easily scrubbed off.

A comparison of measured flexural properties of the sample cut from the tank with original values is shown in Table 8.1. Considering that the aged properties are single values compared to average values for the original properties, the measured changes are not statistically significant. Mechanical properties appear to be effectively unchanged after the time of exposure.

Another tank was installed in Stony Ridge, Ohio, USA and was removed in 1977 after 12 years and 9 months of continuous service. A photograph of the tank after removal is shown in Fig. 8.1. The exterior surface of the tank ends and the center joints had a white, somewhat chalky film, which easily scrubbed off with an abrasive cleaner. This surface effect was typical of air-cured fiberglass surfaces exposed to moisture over long periods of time. No areas of cracking or crazing were found on the inner surface. A section was cut from the tank wall on the side, and flexural properties and hardness

Figure 8.1 Photograph of underground tank removed after 12 years, 9 months of continuous service (see text)

Table 8.2 Fiberglass underground storage tank after continuous service in Ohio, USA for 12 years, 9 months [2]

Property	Units	1964 Test Results	1977 Test Results
Flexural strength	MPa	180	190
Flexural modulus	GPa	9.035	8.200
Barcol hardness	—	37	35

were measured (five samples for each measurement). Average results [2] are shown in Table 8.2. Effectively, there was no significant change in properties over the nearly 13 year time period. Small decreases were observed only for the flexural modulus and the Barcol hardness.

For more than 26 years, another buried composite tank, UL-1, kept gasoline and diesel fuel from leaking into surrounding ground water. The tank's alphanumerical designation means it was the first of its kind to achieve Underwriters Laboratories (UL) recognition [3] for corrosion resistance and structural integrity.

Excavated in the summer of 1990, the 23 000 l (6000 gallon) vessel looked so good that engineers decided to use it again. The primary reason for the tank's long term performance is the correct selection of materials of construction and a structural design that could handle the soil loads.

The UL-1 tank was buried in 1964 under a service station in Schaumburg, northwest of Chicago, Illinois, USA. When new state highway plans overlapped onto the service station property, the tank had to be excavated. The tank was as good as the day it was first buried, according to the site engineer.

The resin specification for UL-1 fell upon the tank fabricator, Owens-Corning Fiberglass Corporation of Toledo, Ohio. This was a new material for UL, so considerable testing was completed and documented [4]. Unsaturated polyesters based on isophthalic acid were chosen because they offered a broad range of chemical resistance. However, before finalizing a resin specification, the engineers tested three different resin systems: orthophthalic and isophthalic polyesters, and epoxies. The isophthalic resins proved to be the optimum choice based on outstanding performance, reasonable cost and good processability. The orthopolyester resins did not provide enough corrosion resistance for this application, especially in acidic soils. Epoxy resins were too costly, more difficult to process, and slower curing.

Processors fabricate underground fuel storage tank by combining isopolyester resin with fiberglass and surfacing veils on a cylindrical mandrel. Since 1964, almost 350 000 isophthalic polyester tanks have been buried under service stations around the United States, according to the Fiberglass Petroleum Tank and Pipe Institute.

Evaluation of the 26 year old tank was as follows. The results (essentially no change in flexural strength with time) can be seen. First, gasoline storage tanks are vented and, therefore, are not subject to internal fluid pressure, but they do have significant external pressure caused by the earth load. Also, an underground tank is not exposed to ultraviolet radiation or weather erosion. However, there is contact on the outside with moisture from the surrounding soil and on the inside with the fluid contents. Therefore, the tank wall can be regarded as an earth load stressed material, immersed in water and gasoline. The tank was reinstalled with an expected service life of more than 30 years.

8.3.2 Durability of fiberglass underground storage tanks demonstrated by a 25 year old tank

The corrosion resistance of unsaturated thermosetting polyesters based on isophthalic acid, or isopolyesters, has been well documented [5]. A gasoline storage tank was buried at a Chicago, Illinois service station on May 15, 1963. When dug up on May 11, 1988, the tank was in excellent condition. It is believed to be the oldest reinforced plastic, or composite, tank in the world. The tank still had gasoline in it, so it had been in constant service for 25 years. There were no signs of leakage into the soil, structural distress or corrosive or chemical attack. It was strong enough to be lifted from the hole by a crane, using the center outlet pipe as the only source of support. It is believed the tank could have been used for another 25 years. The installation date and location of the tank were recorded because it was only the second known reinforced plastic underground gasoline tank to go into service. It was one of about 60 tanks developed in the mid-1960s for the Amoco Oil Company (then known as the Standard Oil Division of the American Oil Company). The oil company was looking for an alternative to steel tanks, which were prone to leakage due to corrosion.

The unearthed tank had been molded in two sections by the former Fibertron Company, of Cedarburg, Wisconsin. They fabricated sections by laying up resin-impregnated fiberglass woven roving and chopped strand mat on a mandrel. The composite semi-cylinders, along with foam-filled end caps, were bonded together with composite lap joints. The design and construction of today's composite tanks are more sophisticated. For example, the cylindrical portion of today's isophthalic polyester tanks has a unitized construction.

Two steel tanks were removed from the same site and at the same time as the composite tank. There are no records to show how long these steel tanks were in service. One of the steel tanks was dusted with white metal oxide corrosion, while the other had signs of corrosion at the weld line. Rust had weakened the joint so much that the weld line could be scraped away with a pocket knife, revealing holes. Corrosive attack on an underground gasoline tank comes from both the inside and outside. An inside problem is moisture, which is often present in the fuel and settles at the bottom. Depending on various conditions, the moisture can create severe internal corrosion. But an inspection of the 25 year old composite tank showed no internal corrosion, even though there was moisture with the fuel on the bottom.

The exterior of a steel tank can be broken down by a galvanic, electrochemical reaction. However, the exterior of the 25 year old composite tank was in excellent condition. No corrosion was noted, even though the soil was saturated with moisture to about a third of the way up the tank wall.

Steel tanks are often protected by sacrificial anodes. These anodes need to be monitored and replaced regularly. Tanks made from reinforced isopolyesters are inherently corrosion resistant and non-conductive. They do not require cathodic protection.

The isopolyester tank was cut into sections and tested for mechanical properties. This was a unique opportunity to make time-based analyses of properties of reinforced isopolyester composites in a corrosive environment. This was because two other tanks of the original 60-tank Amoco project were dug up and tested in the early 1970s.

A comparison of data from all three tanks is given in Table 8.3. These data show that the 25 year old isopolyester tank had remained strong and showed no unusual deterioration. Taking standard deviations into account, it is doubtful any significant change had taken place in the last 17 years.

Table 8.3 Properties of unearthed FRP tanks (isophthalic polyester resin SG-10, described in bulletin IP-86[a]) with glass reinforcement. Values are averages of several samples

	Tank A	Tank B	Tank C
Age at testing (years)	5.5	7.5	25
Buried . . . excavated	1/7/65 . . . 8/21/70	4/4/64 . . . 10/24/71	5/15/63 . . . 5/11/88
Flexural strength, psi (MPa)	19 500 (134)	24 200 (167)	22 400 (154)
Flexural modulus, psi (MPa)	725 000 (4992)	795 000 (5482)	635 000 (4378)
Tensile strength, psi (MPa)	10 700 (74)	13 600 (94)	10 500 (72)
Tensile modulus, psi (MPa)	1 160 000 (7260)	1 053 000 (8000)	1 107 000 (7630)
Tensile elongation, %	1.11	1.25	1.13
Notched Izod impact strength, ft-lb in^{-1} (J m^{-1})	9.7 (518)	11.0 (587)	14.1 (753)

[a] 'Make corrosion resistant unsaturated polyesters with Amoco PIA', Amoco Chemicals, Chicago, Illinois, USA, March 1995.

8.3.3 Laboratory tests of corrosion resistance

The reliability of this real life experience is corroborated by the results of laboratory research. American Society for Testing and Materials (ASTM) C581 [6] is designed to reflect the resin deterioration when a laminate is totally immersed in a corrosive medium. Central to the procedure is the strict specification of a standard laminate and of standard conditions for exposing samples to corrosive liquids. The method specifies determining the hardness and flexural properties of the laminates at 1, 3, 6 and 12 months.

If it is assumed, as many experts believe, that resin deterioration occurs at a decreasing rate, approximated by a logarithmic function, then when the first year's data are plotted on a log–log grid, the best straight line through them can be extended to predict future retention of properties. Figure 8.2 (discussed later) shows a hypothetical example of this operation. This analytical technique discounts minor fluctuations in the data, yet provides a long term prediction from short term testing. Hardness, flexural strength and flexural modulus are measured because they show the resin's properties better than properties such as tensile strength, which are dominated by glassfiber properties.

Because of the unavoidable fluctuations of data that result from destructive testing after immersion, we frequently report a composite rating that represents the range in which retention of properties falls when the best straight line is extended to 10 years. If any of the three measured properties falls below 50% retention, that resin/medium combination is rated as unacceptable. Resins that are so badly deteriorated as to be unmeasurable are labeled complete failure. The rating is the average of the ratings for the three measured properties if no failure occurs.

8.3.4 Conclusions (tanks)

Underground tanks must contain petroleum products for 30 years without undue maintenance. They must resist soil corrosion and gasoline, including wet gasoline, while maintaining structural integrity. The 25 year old tank and the laboratory tests of corrosion resistance prove that reinforced isopolyester tanks are meeting these demanding requirements. An isopolyester tank buried today can be expected to perform without corrosion well into the 21st century.

8.4   Pipe studies

8.4.1 Cured-in-place pipe

Cured-in-place pipe relining is a proven method for repairing underground conduits such as water pipes and sewer pipes. The method involves the technique of inversion lining. An inversion lining is formed by inserting into the pipe a resin-impregnated felt tube, which is inverted against the inner wall of the pipe and allowed to cure. (Cure is the chemical reaction that permanently sets the polymer from a viscous liquid state to a rigid material) [7]. Most of the resins used are unsaturated polyesters, with a small portion of the relinings using epoxy resins. Cured-in-place pipe relining is successful in dealing with a number of problems, particularly in sewers needing structural reinforcement. The process is well suited to pipes under buildings, near large trees or under highways where traffic disruption must be minimized.

The process has been used for over 25 years and this section will document one of the older pipes that was evaluated after 20 years service. The service environment that the pipelines experience are sewer water and corrosive soils. The process is used to restore corroded steel, cast iron and concrete pipes. Steel and cast iron pipe are susceptible to corrosion by sewer water through normal acid and salt attack, and by the galvanic

Table 8.4 Average test results of cured-in-place pipe after 20 years in service

Sample	Mean width, (mm)	Mean thickness (mm)	Flexural stress (MPa)	Flexural modulus (MPa)
Left side	14.72	6.82	49.26	2990
Right side	14.92	5.61	43.15	2870
Current standard, UK-WIS 4-34-4	—	—	—	2200

corrosion caused by soil conductivity. Concrete shows the most corrosion from hydrogen sulfide/sulfuric acid attack on the inside of the pipe. The inherent advantages of the cured-in-place pipe process are in the use of resins that will not be corroded by the environment and process flexibility to fit a wide variety of existing pipes, even where they have been distorted by high soil loads over time.

Test laminates taken from a 20 year old pipe section in the UK showed that cured-in-place pipe had not undergone any significant change. The pipe had physical properties higher than the original requirements for the job, see Table 8.4.

In 1971, one of the first pipes was installed in the UK in London in the form of 1.17 × 0.61 m egg shaped profiles. A polyester felt mat tube impregnated with an isophthalic polyester resin was installed and cured in place. After 20 years, a section of the pipe was taken out of service and the strength properties were measured. The cured-in-place pipe retained its properties and still had a flexural strength that was 30% over the initial installation requirements. The parts showed no sign of wear or deformation. Several other pipe sections have been checked after years of service and found to be virtually unchanged.

8.4.2 Pressure pipe – effect of 25 years of oil field flow line service on epoxy fiberglass pipe

Glass fiber reinforced epoxy, polyester and vinyl ester piping systems have been used for over 40 years to control corrosion problems in oil fields, chemical plant, industrial plant and municipal water and wastewater transport [8, 9].

The following section documents 25 years service life of an epoxy pipeline used to transport crude oil from oil wells to a collection station. This work shows the predictability and the long term service life of the pipes under cyclic load.

Because of the excellent cooperation of a major oil company, buried oil field flow lines in Crane County, Texas are the most thoroughly documented application of fiberglass pipe. In 1961 and 1962 oil field flow lines in this major production area were installed using unlined aromatic amine cured epoxy, filament wound pipe. At that time, this type of pipe was a relatively new product, which had a good short term record as an economical solution to corrosion problems in oil field applications. Prior to the development of epoxy fiberglass pipe, cement lined steel with cathodic protection was the standard piping system for this type of oil field service.

In 1961, unlined epoxy fiberglass pipe had five years of successful trial installations in oil field flow lines in the southwest of USA and the economic advantages of this type of pipe were becoming well known in the petroleum industry.

Two lines, installed in 1962, were selected by the owner and the pipe manufacturer as a site for obtaining much needed data which could be used to determine the long term corrosion resistance of unlined aromatic amine cured epoxy fiberglass pipe in typical oil field flow line service. It was agreed that sections of pipe would be removed at regular intervals and tested so that the physical properties could be compared with the original properties of the pipe. This section documents the results of the 25 year history of these pipelines.

8.4.3 Operating conditions

The 50 mm (2 inch) diameter unlined amine cured epoxy fiberglass pipelines chosen for this study were installed in June 1962. These buried lines were in continuous service handling salt water, sour crude oil and natural gas at 275–410 kPa (40–60 psig) and 25°C (80°F). The lines ran from pumping wells to a unit which separates the oil, gas and salt water. The pipe was rated for 2070 kPa (300 psig) cyclic pressure service at 65°C (150°F). These lines were occasionally hot oiled under increased pressure at 105°C (220°F) for paraffin control.

8.4.4 Test results

Pipe specimens were removed after 4, 8, 15, 20 and 25 years of oil field flow line service. These specimens were subjected to short term burst tests, cyclic pressure tests and visual inspection. Test results are shown in Tables 8.5 and 8.6. After 8 years of service, the pipe had essentially the same strength and appearance as when manufactured.

After 15 years, there was an apparent loss of strength. However, the 15 year loss of strength was not severe; the burst strength still exceeded the

Table 8.5 Cyclic pressure test results

Corrosion exposure (Years)	Hoop stress (MPa)	Cycles to failure
4	141.934	1412
4	146.757	2390
4	141.934	2775
4	150.202	302
4	150.202	805
4	146.068	835
8	153.647	1845
8	148.135	892
8	153.647	1077
15	144.690	2259
15	147.446	262
15	147.446	956
15	150.202	580
15	155.025	303
15	142.623	316
20	147.446	44
20	149.513	192
20	151.580	287
20	150.202	192
20	147.446	160
20	149.513	120
25	155.025	2
25	157.092	5
25	150.202	3800
25	126.776	4008
25	126.087	3055
0		780

Test pressure: 8275 kPa (1200 psig).

minimum values for pipe produced in 1962, and the cyclic test data, while below the regression curve for 1962 pipe (ASTM D2143), was above the statistical lower confidence level for 1962 pipe. The appearance of the interior surface of the 15 year specimens was unchanged, but the exterior of the pipe had darkened. There were no signs of chemical attack. The exterior darkening was probably the result of hot oiling the lines. Aromatic amine cured epoxy resin darkens with age when exposed to air and this process is accelerated by an increase in temperature.

After 20 years of service, the burst strength appeared unchanged, but the cyclic pressure data continued to show a loss of strength. For the first time in this study, all specimens failed after less than 300 cycles at 8275 kPa (1200 psig) and for the first time one specimen failed after less than 100 cycles at 8275 kPa (1200 psig). These low values (typical 1962 pipe withstood

Table 8.6 Burst strength of pipe

Corrosion exposure (Years)	Burst hoop stress (MPa)	Type of failure
4	336	Wall weeping and end-cap failure
8	348	Wall weeping
15	345	Wall weeping and end-cap failure
20	371	Wall weeping
25	277	Wall weeping and major wall fracture
0	275	Wall weeping

1000 cycles at 8275 kPa (1200 psig) at the time of manufacture) did not give a true statistical indication of strength loss in the pipe.

Figure 8.2 shows that all but one of the 20 year specimens were above the 97.5% statistical lower confidence limit for the cyclic pressure regression curve produced from pipe manufactured during the time that the pipe was made. The appearance of the interior surface of the 20 year specimens was unchanged, but the darkening of the exterior surface was increased. There were no signs of chemical attack.

After 25 years of service, the burst strength appears unchanged. The average burst pressure for the 25 year exposure specimens is 19 MPa (2700 psig) and the average burst stress is 3200 MPa (43 000 psi), but the mode of failure has changed. In all previous burst tests, failure had been evidenced by pipe wall weeping. All of the 25 year burst test specimens failed by major pipe wall fracture. Because of the significant drop in 8275 kPa (1200 psig) cyclic performance at the 20 year exposure level, cyclic

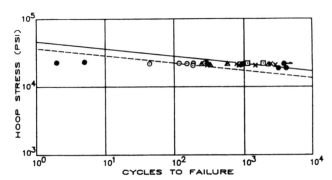

Figure 8.2 Cyclic pressure test data: unlined filament wound epoxy pipe in oil field service for 25 years. ——, Normalized regression curve of original pipe ASTM D-2143; ------, statistical confidence limit. ●, 25 year old pipe; ○, 20 year old pipe; △, 15 year old pipe; □, 8 year old pipe; ×, 4 year old pipe

pressure data at the 25 year interval were taken at two pressure levels. The 8275 kPa (1200 psig) data are erratic; two specimens failed at less than 10 cycles while the third specimen exceeded 3800 cycles. These are the low and high values for 8275 kPa (1200 psig) cyclic testing during the entire 25 year study! Cyclic pressure data taken at 6750 kPa (980 psig) produced data points which fall within the 97.5% statistical lower confidence limit for the cyclic pressure regression curve produced from pipe manufactured during the time when this pipe was made. All of the cyclic pressure test failures were by pipe wall weeping. The 25 year specimens still have good appearance. The inside surfaces are still smooth and glossy; the same as with all previous specimens. The exterior darkening does not appear to have increased.

Split disk hoop tensile strength retention testing (ASTM D2290) was performed on rings cut from the 15 year and 25 year specimens. Both of the values obtained, 85% at 15 years and 75% at 25 years, exceed the minimum strength retention requirement. The burst pressure data, 6750 kPa (980 psig), the cyclic pressure data and the split disk hoop strength retention data indicate that this pipe should still give many years of useful life in this service.

The 8275 kPa (1200 psig) cyclic pressure performance of this pipe has definately been affected by 25 years of oil field flow line service. The two specimens which failed after less than 10 pressure cycles follow the trend which started at the 15 year exposure level. The specimen which exceeded 3800 cycles at 8275 kPa (1200 psig) appears to be an anomaly. The 8275 kPa (1200 psig) cyclic test pressure is four times the 2070 kPa (300 psig) cyclic pressure rating of this pipe. Therefore, the drop in performance at this pressure level, while significant, is not an indication that the pipe is nearing the end of its useful life.

The change in mode of failure in the burst tests also indicates that the laminate has been affected by 25 years of service. The major pipe wall fractures are typical for pipe made from less flexible resins. It appears that significant changes in resin properties have begun to occur.

The use of logarithmic extrapolation methods offers an excellent method of predicting the long term service length of composite pipe. This is similar to the extrapolation shown in the underground fuel tanks testing.

8.5   Reinforced thermoplastics pipe

Reinforced thermoplastics pipe can currently be made using E-glass fiber or carbon, aramid or polyethylene fiber reinforcement. The development of these new materials for high performance applications is still relatively recent compared with thermosetting resins and there are no long term case histories of product performance available yet. The pipes are made by winding fiber reinforced thermoplastic tapes over an extruded thermoplastic liner pipe and, optionally, protecting the tapes with a co-extruded thermoplastic outer layer. The tapes can be bonded to the rest of the structure using infrared heating.

Frost [10] has carried out some development work at Shell, in cooperation with several other companies and universities. He has also outlined a design procedure for predicting engineering performance, stress–strain response and strength under specified conditions. The pipes investigated included aramid/polyethylene, carbon/PVDF and glass/polypropylene. All these thermoplastic types failed in the burst mode, although not in identical ways, when subjected to short term burst tests. (Glass-epoxy pipes fail by weeping, with matrix cracking, before bursting). Longer term burst tests with slow increases in pressure were then performed on the aramid/polyethylene pipes. The effect of temperature was considerable, and burst pressure fell from about 210 bar at 23°C to 125 bar at 67°C. (It was therefore recommended that the material should not be used above 60°C).

By performing idealised low speed loading rate tests at two different rates, a regression curve was generated from the two failure points, with the form

$$\sigma = A - B \ln (t)$$

where A and B are constants.

The long term reduction in tape strength with increasing load was taken to be the same as the reduction in aramid fibre strength, given by the logarithmic decay expression

$$\sigma = \sigma_0 (0.915 - 0.0175 \ln(t))$$

where t is in minutes and σ_0 is the short term burst strength.

Miner's Law is given by the equation:

$$\int dt / \{t(\sigma(t))\} = 1$$

where the integral is from time $t = 0$ to $t = t_f$, the failure time, and $\sigma(t)$ is the applied stress or pressure, as a function of time.

Miner's expression can be integrated to give the time to failure and the failure stress as a function of the regression curve constants, A and B, which are identified from the two tests. The paper cited gives considerable detail and proposes a qualification procedure for reinforced thermoplastics pipes and joints, in which manufacturers would be required to provide either a static or a cyclic regression curve, as appropriate, with data up to 10000 h or 1 million cycles. A qualification test would then be required to demonstrate experimentally the regression curve to 1000 h. The verified regression curve would be extrapolated to the design lifetime, typically 20 years.

8.6   Concluding comments

Fiberglass composite materials have a long history of proven performance. Several of the long term, documented case histories have been reviewed. The initial faith placed in these materials was well founded and this has resulted in a large and still growing industry with activity throughout the world.

References

1 L E Pearson, *FRP Gasoline Tank – Six and a half year Case History*, Owens Corning Fiberglass, TC/TOS/RPD/71–21, May 7, 1971.

2 Owens-Corning Fiberglass, *Original Fiberglass Underground Tank Stands Test of Time and Use*, Toledo, Ohio, Publication 15-PE-8465, 1978.

3 Underwriters Laboratories Inc, UL 1316 *Glass-fiber-reinforced plastic underground storage tanks for petroleum products*, IL, USA.

4 L E Pearson, *FRP Underground Storage Tanks – Long Term Corrosion Resistance for Petroleum Fuels*, Owens-Corning Fiberglass, TC/TOS/RPD/71–17, June 30, 1971.

5 Amoco Chemicals, *25 year Old Tank Showcases Corrosion Resistance of Isopolyesters*, Amoco Chemical, Bulletin IP-88e, 1993.

6 *Standard Practice for Determining Resistance of Thermosetting Resins Used in Glass-Fiber Reinforced Structures intended for Liquid Service*, Annual Book of ASTM Standards, 1988.

7 K D James, Technical Service Report SR 24914/1/A, MTS Pendar Ltd, London, 1991.

8 K J Oswald, Evaluation of epoxy and vinyl ester filament wound pipe removed from chemical and petroleum service, *Managing Corrosion with Plastics*, NACE, 1977, Volumes 1, 2, 3, pp 240–244.

9 K J Oswald, 'The effects of chemical and petroleum service on epoxy and vinyl ester filament wound pipe', Paper 18, 18/1–18/8, *NACE Corrosion '84*, NACE, Houston, USA, 1984.

10 S R Frost, Shell International Oil Products BV., Amsterdam.

9
Epoxy vinyl ester and other resins in chemical process equipment

PAUL KELLY

9.1   Introduction

The use of composite materials in the chemical process industry began in earnest during the mid to late 1960s. During this decade the major economies grew at an unprecedented pace and a willingness to experiment, allied to the need to do something radical to limit the ravages of corrosion in the chemical industry, led engineers to take a serious look at the widespread use of glass reinforced plastics (GRP) corrosion resistant materials.

Within the Dow Chemical Company, one area stood out, the chlor-alkali production and utilizing plants. The plants were expanding at such a pace that the repair and replacement programmes needed to run the plants at full capacity could no longer be supported. It was also found that as the need for more and more product arose, we were obliged to increase the reaction temperatures and thus speed up the reactions. This also accelerated the rates of corrosion.

In these plants high pH caustic streams are typically found running beside low pH acid streams. There are incoming brine and outgoing hypochlorite lines, not forgetting the wet and dry chlorine and hydrogen gas streams. These chemicals are then used to manufacture a wide range of chemical intermediates which require chlorine or caustic alkali in their processing. Chlorinated solvents are one such product, as are the glycol ethers. Within these secondary processing plants there are frequently mixed streams containing both a caustic alkali, or a mineral acid, and an organic material. It is these mixed streams which give the most problems with corrosion.

A need therefore existed for a material with good resistance to a broad spectrum of corrosive chemicals. Dow's development laboratories identified composite materials based on the use of a corrosion-resistant matrix such as epoxy vinyl esters. We believe that such composite materials have filled that gap admirably for us.

The use of GRP has progressed since those early days and the objective of this chapter is to highlight some of the pitfalls, to offer some advice on how to get the best out of composite materials and to document some of the outstanding performances that have been encountered over the years.

9.2   Raw materials

The first step in the selection and design of a successful GRP structure is the correct choice and specification of the raw materials suited to the application in mind. Therefore, the major raw materials will be considered and the areas of successful application will be pointed out in each case.

The fundamentals of reinforced plastics have already been outlined in Chapters 1 and 2. They are basically a combination of two materials. There is a fibre reinforcement (normally glass) which is embedded in a resin. The applications discussed here have historically employed thermosetting resins. The fibre bears most of the mechanical load placed on the structure, while the resin is there essentially to protect the fibre from chemical attack.

9.3   Resins

There are five different families of resins typically used to manufacture pipes and tanks for use in the chemical process industry.

9.3.1   Polyesters

Unsaturated polyester resins are in general polycondensation products of dicarboxylic acids or anhydrides and dihydroxy alcohols (together with diols or glycols) the product being dissolved in styrene monomer. The anhydrides most frequently used today are maleic and phthalic. They hydrolyse to form acids during the reaction. The maleic acid also changes to fumaric acid. The diols are selected from ethylene, propylene, diethylene, dipropylene and neopentyl glycols, along with bisphenol A.

The uses of polyesters are many and varied and they have traditionally held a major part of the anticorrosion market in the broad sense. For corrosion-resistant chemical process equipment, more expensive and specialized polyesters are used. The acids are then generally isophthalic or fumaric acids, with the hydroxyl functionality being provided by neopentyl glycol or bisphenol A.

Polyester resins are easy to work with. They have viscosities at room temperature of between 300–700 mPa s, which is sufficiently low to allow fast and efficient wet out of the glass fibres. They cure well at room temperature to afford acceptable mechanical and thermal properties and resist a wide range of dilute chemicals at moderate temperatures. They are cheap, varying in price from approximately 2–3 DM kg^{-1}.

The so-called isopolyesters are based on the use of isophthalic acid rather than phthalic anhydride and are widely used for their ease of handling and their resistance to many chemicals at low concentrations and to moderate temperatures (<40°C). Although isopolyesters give good dependable service in these areas, the vessels manufactured from them will not last as long as those made from epoxy vinyl esters.

When bisphenol A is used in conjunction with fumaric acid, a bis-A fumarate resin is produced. The best known trade marks for this type of resin are the Crystic® 600 series supplied by Scott Bader, and the Atlac® series, commercialized by DSM-BASF. This type of resin is noted for its excellent chemical and thermal resistance. It performs particularly well in contact with high pH solutions, and to a lesser extent with acids. It is, however, rather brittle, with only 2–3% tensile elongation at break, making it difficult to use for the construction of pressure vessels. Bis-A fumarates retail at around $4 DM kg^{-1}$.

9.3.2 Epoxy resins

The most widely used epoxy resins are reaction products of either bisphenol A or a novolac phenolic resin with epichlorhydrin. When used to manufacture corrosion-resistant structures for use in the chemical process industry, epoxy resins are generally hardened with either aromatic or cycloaliphatic amines. The hardeners for epoxy resins are, with few exceptions, added at levels varying from 20 phr (parts per hundred resin) to 100 phr. This means that the hardener is actually quite a high proportion of the matrix resin and has quite a profound effect on the mechanical and corrosion properties of the cured resin. Thus the selection of the most suitable hardener is critical to the eventual success of the application. Epoxy resins have viscosities of several thousand mPas at room temperature, which makes it much more difficult to wet out glass fibre efficiently with them than with polyesters. Wet-out therefore involves heating the resin formulation to between 40°C and 60°C to reduce the viscosity to less than 1000 mPas.

The hardeners offering the best corrosion resistance with epoxies, combined with excellent mechanical properties, are the aromatic amines. The most frequently used one is MDA (methylene dianiline). However, MDA is classified in the European Union as a Category 2 carcinogen, that is a substance that should be treated as if it is carcinogenic to man. Because of this, there is now a trend to replace MDA whenever possible with IPD (isophoronediamine), a liquid cycloaliphatic amine with a better toxicity profile. IPD provides good corrosion and mechanical properties, although not achieving as good a combination of mechanical and corrosion properties as MDA. Both MDA and IPD are used at around the 25 phr level with typical low molecular weight liquid epoxy resins.

Epoxy composites cured with IPD offer resistance to a wide range of chemicals up to about 100–120°C. They offer better corrosion resistance in general than polyesters, but in many respects they are not quite as good as epoxy vinyl esters. This is due to the presence in the cured resins of the more sensitive amine hardeners. They nevertheless outperform other resins in their resistance to water at temperatures above about 80°C. Because of the excellent mechanical properties associated with epoxy resins, they are the system of choice when high pressure pipes are required.

There are three large suppliers of epoxy resins leading the global market both technically and commercially. They are Dow Chemical, Ciba Speciality Chemicals and Shell Chemical. For composite applications the most commonly used grades are the low molecular weight, low viscosity liquids such as DER® 330 or DER® 331 and Shell Epikote® 828. They sell in the marketplace at prices in the range 4–6 DM kg^{-1}. The hardeners are manufactured by other companies, for example Air Products, BASF or Huls. Depending on which grade is selected, the price is expected to be in the 5–15 DM kg^{-1} range.

9.3.3 Epoxy vinyl ester resins

When epoxy resins are used to make composite structures, three major drawbacks are encountered:

1 Because of their two-step hardening process, they are slow to cure, and they require a minimum postcure of 2–4 h at 120°C to achieve 70–80% of optimal properties.
2 Their viscosity makes it difficult to wet the glass fibres efficiently.
3 The use of amine hardeners renders the cured resins susceptible to acid attack.

With these issues in mind, the so-called epoxy vinyl ester range of resins was developed in the 1960s. These resins are made by reacting a liquid epoxy resin with methacrylic acid to terminate their molecules with reactive unsaturated ester groups and then dissolving them in styrene. By doing this the curing process is changed from the two-step epoxy process (including obligatory postcure) to a one-step free radical polymerization, similar to that used with polyesters. This means that the resins can be cured at room temperature to achieve 80–90% of their optimal mechanical properties. The addition of styrene allows the viscosity to be dropped to 100–500 mPa s at room temperature, thus permitting a rapid and efficient wet-out of the glass fibre. As the resin does not require significant quantities of amine hardeners, its chemical resistance is significantly improved.

Unsaturated resins such as polyesters and epoxy vinyl esters have ester groups in their structures, as their names suggest. Esters are susceptible to hydrolysis and this process is accelerated and catalysed by the presence of acids or bases. But epoxy vinyl esters contain substantially fewer ester linkages than do polyesters. They contain only one at each end of the resin molecule, whereas polyesters contain several at various points along the molecular chains. As already noted, methacrylic acid is used to manufacture the epoxy vinyl esters. This means that next to each ester linkage is a large methyl group. This group occupies a lot of space and sterically hinders any molecules approaching the ester group by impeding their access. These two aspects of the design of the epoxy vinyl ester molecule combine to make them more chemically resistant than polyesters.

Epoxy vinyl ester resins exist as two distinct families, depending on which type of epoxy resin is used in their manufacture. The first and most commonly used family is manufactured from standard bisphenol A-based liquid epoxy resins. This group of vinyl ester resins is typified by a tensile elongation at break of 5–6%, combined with a HDT (heat deflection temperature) of 100–120°C. From the corrosion standpoint, they resist a wide range of aggressive chemicals well. In particular, they outperform other resins of the family in their resistance to high pH caustic solutions. Bisphenol A-based epoxy vinyl ester resins retail in the 6–8 DM kg^{-1} price range.

The second vinyl ester family uses a novolac epoxy resin as its starting point. The resulting epoxy novolac vinyl ester resins have a higher crosslink density than the bisphenol A epoxy vinyl ester resins. This means that it is more difficult for chemicals to penetrate the matrix, and they have improved resistance to organic solvents and mineral acids. The heat resistance is also higher, with the commercial grades having a HDT of 140–150°C. Some special grades have been developed for use in contact with gas at temperatures up to 220°C. Depending on the cure cycle and measurement technique, they have glass transition temperatures (T_g) between 200°C and 220°C. They are also a little more brittle than the previous family, with a tensile elongation at break of 3–4%. Because of the increased raw material costs, epoxy novolac based vinyl ester resins sell in the 8–10 DM kg^{-1} range.

A third category of vinyl ester resin is formed when tetrabromo bisphenol-A (TBBA) is used in the manufacture of the base epoxy starting resin. This affords a resin with up to about 20 wt% bromine bound into its structure and is designed to achieve good fire retardancy. An added advantage of the bromine is that because of the large size of bromine atoms, it improves the chemical resistance, particularly towards caustic and hypochlorite solutions [1].

Epoxy vinyl ester resins are the resin of choice for industrial corrosion-resistant GRP equipment and the leading supplier is the Dow Chemical Company, with its brand name of Derakane®. Some properties of various resin families are given in Table 9.1.

Table 9.1 Resin mechanical and thermal properties

Property	Bis-A EpVE	Nov EpVE	IsoPoly	OrthoPoly	EpMDA
HDT (°C)	100–120	140–160	70–80	70–75	155–165
Tensile					
Strength, MPa	85–90	85–90	60–70	50–60	70–75
Elongation (%)	5–6	3–4	2–3	1.5–2.5	4–5
Modulus (GPa)	3.4–3.5	3.5–3.9	3.3–3.5	3.8–4.0	—
Flexural					
Strength (MPa)	135–145	140–150	110–130	80–90	90–100
Modulus (GPa)	3.3–3.5	3.6–3.9	3.5–3.7	3.8–4.0	3.8–4.0
Compressive					
Modulus (GPa)	2.0–2.2	2.0–2.2	—	—	3.5–3.8
Strength (MPa)	120–130	145–155	—	—	115–120
Deformation (%)	7.5–8.5	10–11	—	—	—

Bis-A EpVE = Bisphenol-A epoxy vinyl ester (Derakane® 411), Nov EpVE = Novolac epoxy vinyl ester (Derakane® 470), Isopoly = Iso Polyester, Orthopoly = Orthopolyester, Ep MDA = Bisphenol-A epoxy resin cured with methylene dianiline.

9.4   Glass

Attention also needs to be paid to the correct selection of the reinforcing fibre if the maximum longevity of the vessel is to be achieved. When reinforcing fibres are mentioned in the chemical process industry, three kinds are concerned: glass, aramid and carbon or graphite. For the manufacture of cost competitive corrosion-resistant vessels for use in industry then the selection may be reduced to just one, glass.

There are several different types of glass (E, C, S, A, Ar, etc.) but again, in the context of the subject of this chapter, the choice is immediately reduced to just E and C glasses.

E-glass was originally developed for the construction of electrical laminates for use in the printed circuit board industry. Hence its name E-glass. It has however adapted itself well to the needs of the composite industry and today is the most commonly used reinforcing fibre. It comes in two distinct varieties:

- E-glass itself, which is used in the structural wall of most vessels and in the inner layers when the corrosion requirements are not too strenuous.
- A subcategory of E-glass is ECR-glass. This is a trade name of Owens Corning Fiberglass (OCF). ECR-glass is more resistant to mineral acids

than the standard E-glass and is used when contact with these acids is expected.

C-glass stands for chemically resistant glass and resists a wider range of chemicals than does E-glass and in particular mineral acids.

The glass fibre reinforcement comes in several different forms. The largest proportion is supplied as continuous rovings wound onto bobbins. This glass is used in the filament winding industry to construct the structural wall of the vessel being made. It is also much appreciated by the pultrusion industry, as this process is continuous and the continuous nature of roving allows uninterrupted production. People using a chopping gun also appreciate continuous roving and chop it into lengths of 10–50 mm before spraying it onto the mould with the resin. It is the simplest and cheapest form of glass reinforcement.

Chopped strand mat (CSM) is manufactured from roving by chopping the fibres into strands of 20–30 mm length and binding them into a mat using a binder resin. The fibres are placed so that equal weights of them are to be found aligned in all directions. For the purposes of this chapter we need only consider two weights, $300 \, \text{g m}^{-2}$ and $450 \, \text{g m}^{-2}$. It comes with either emulsion resin binder or powder binder resin. For optimal corrosion resistance, only the powder bound varieties should be considered as the emulsion applied binder resins may absorb, or even be dissolved by water. CSM is used to build up the inner corrosion barrier and when hand laid up pieces are made it is interspersed between the woven rovings to construct the structural laminate. When laminates are made with CSM alone, a fibre content of 25–35 wt% is typically achieved.

Woven roving (WR) is made by weaving strands of continuous roving together to form a bidirectional fabric. Typically this affords a mat of higher density than a CSM, but with the fibre oriented in one or more directions. Weights typically used for the manufacture of industrial equipment vary between $450–600 \, \text{g m}^{-2}$. In cured laminates, a fibre content of 40–50 wt% is typically achieved. It is used essentially to build up the structural wall of the vessel, where the increased fibre content is appreciated and it is usually applied in conjunction with alternating CSM layers.

Surfacing veil is a light tissue of surface density $20–50 \, \text{g m}^{-2}$ and is normally used in the build up of the inner corrosion barrier or to give a smooth resin-rich outer surface. For certain corrosion conditions, where glass may be attacked, surfacing veils made from synthetic fibres are used. Both polyacrylonitrile and polyester fibres give good results, with a slight advantage for polyester where corrosion resistance is concerned, and where workability is an issue, then polyacrylonitrile may outperform. When surfacing veils are used, a resin-rich laminate is formed with a resin content in the order of 90 wt%.

9.5   Laminate buildup

When building a laminate for industrial use, it is generally made in at least two stages. First the so-called resin-rich corrosion barrier is laid down. This consists of two veil layers followed by two chopped strand mat layers. The resin content in the veil layers is usually around 90 wt% while in the chopped strand mat layers it is between 70–80 wt%. The thickness of this layer is normally between 2–3 mm. The purpose of this first section is to protect the subsequent layers from any chemical attack and it is not expected to contribute to the structural integrity of the vessel.

The function of the second layer is primarily to bear the mechanical load required of the vessel. It is therefore constructed with a higher glass content. If it is made by hand lay-up then WR is interspersed with CSM. This will normally result in a glass content between 45–55 wt% depending on the skill of the laminator. If the layer is made by filament winding, then the glass content will be between 65–75 wt%.

To maintain the resin content gradient through the laminate thickness and to avoid unwanted resin flow between the layers, the corrosion barrier is allowed to gel before the structural laminate is applied. After the gelling of the corrosion barrier another CSM is applied before either filament winding commences or WR is applied. The resin in the corrosion barrier is allowed to gel but not to cure completely. If it is completely cured it will be difficult to obtain good adhesion between the two layers and grinding of the outer surface of the corrosion barrier will be required prior to commencing the buildup of the structural layer.

A final layer consisting of one or two light surface veils is normally applied. The resin used for this outer layer may typically contain up to 0.5 wt% of paraffin wax and possibly a UV stabilizer. It main function is aesthetic but it also fulfils other purposes. The paraffin wax will 'bloom' to the surface of the uncured resin. This will seal the surface and prevent air contact, thus avoiding any difficulties which may arise due to oxygen inhibition of the resin curing process. The wax also acts as a UV stabilizer and protects the vessel from attack by the sun in climates where this is an issue.

9.6   Postcure

9.6.1   Why postcure?

Unsaturated resins such as polyesters and epoxy vinyl esters cure well at room temperature and reach 80–90% of their optimal mechanical properties when cured in this manner, although they still require a postcure to approach their published optimum corrosion or thermal properties. Epoxies have a completely different cure mechanism, and for it to function properly, in the context of the applications considered in this chapter, a postcure is essential.

9.6.2 For how long and at what temperature?

In the case of epoxies using MDA or IPD hardeners, a minimum of 2–3 h at 120°C is required to achieve good corrosion, mechanical and heat resistant properties. For polyesters and bisphenol A epoxy vinyl esters, a minimum cure of 2–4 h at 70–80°C should be sufficient to ensure optimal corrosion properties. For novolac-based vinyl esters, 4 h at 100°C will optimize the corrosion properties but to achieve a HDT of 150°C then a cure closer to this temperature is required.

9.6.3 When to postcure?

If the vessel being constructed is specified for service close to the corrosion or temperature limitations of the resin as described in the resin manufacturers' literature, then a postcure is necessary. If any doubt exists, then the resin manufacturer should be consulted.

9.6.4 Where to postcure?

The most appropriate place to postcure is at the vessel manufacturers' site, using hot dry air under well-controlled conditions. This is not always practicable, as some GRP fabricators do not have an oven or the vessel is too big to place in their oven. In such cases, the next best alternative is to postcure at the end-users site after the vessel has been installed. This should again preferably be done with hot dry air. If not, a steam lance may be used. Failing all else, a vessel will postcure satisfactorily in service, if the service temperature is high enough. If in-service postcure is selected as the most appropriate method, then care should be taken to raise the temperature of the vessel slowly, say from ambient to the service temperature over a 24 h period.

9.7   Temperature limitations

Unlike metals, composites do not have a sharp melting point. They lose their mechanical properties progressively over a range of temperatures. To determine the key temperature range, two properties are used.

In the composites industry the most frequently used property is probably the HDT. It is measured by resting the sample on two fixed points and applying a specified load, or hanging a given weight from the centre. The whole assembly is placed in a thermostatically controlled oil bath heated at a controlled rate. The temperature reached when the centre of the sample has deflected by a given distance is reported as the HDT [2]. The HDT of a resin describes the temperature where the resin has lost a small part of its rigidity and at which the rate of loss is beginning to accelerate.

The second property used is the better known glass transition temperature (T_g). It is a more theoretical measurement defining the temperature at which the material passes from the glassy state to the rubbery state. It is measured by determining the point where the modulus of the material is most sensitive to temperature change [3]. In general it is some 10–30°C higher than the HDT, and so it is further along the temperature/mechanical property curve, and the material being measured has lost more of its rigidity at this temperature.

Most design norms or standards advise that composites should not be used at temperatures higher than 20°C below the HDT of the resin used. This means that polyesters can be used up to about 70°C, bisphenol A epoxy vinyl esters to 80–90°C and novolac epoxy vinyl esters to 130°C. This principle works quite well for polyesters and bisphenol A epoxy vinyl esters. It is rare for applications in the chemical process industry to require temperatures above 80–90°C. While theoretically it is possible to use novolac epoxy vinyl esters up to 130°C, there are actually very few applications requiring stability above 110°C.

At this point we have to distinguish between gases and liquids. Gases have a much lower heat capacity than liquids and they do not transfer their heat as efficiently and quickly to a containing laminate as do liquids. This means that in low pressure gas contact, the inner laminate wall is more often than not some 20–30°C colder than the gas. It is the temperature of the composite wall itself which is important and not the temperature of the fluid it contains. For flue gas absorber towers, it has become common to use GRP gas inlet sections with a gas temperature as high as 220°C [4]. So while the guide given in the standards is quite good for constant liquid contact, GRP can be used quite safely in contact with gas at temperatures well in excess of the resin HDT.

Composites, as distinct from metals, are heat insulators. This means that if the vessel has no outer wall insulation, then despite the inner surface of the composite laminate being at a higher temperature than the resin HDT, the full thickness of the structural wall is often at a substantially lower temperature. When analysing the structural integrity of a vessel, it is the temperature that the structural laminate is likely to experience which is most important.

Again, because of the heat insulating properties of GRP, if the vessel overheats for a limited period of time, it may still not be damaged because the structural wall has not had time to attain a temperature sufficiently high to cause problems. If a high temperature excursion does occur, a thermosetting GRP vessel will not simply melt like a metal or a thermoplastic construction. What will happen is that as the resin gradually loses mechanical properties it will slowly transfer the mechanical load to the glass fibres. A novolac epoxy vinyl ester resin based composite at say 40°C above the HDT of the resin will still have about 25–35% of its room temperature mechanical properties, and as GRP vessels are usually designed with safety factors of 8 to 10, there is more than sufficient strength left in the laminate to maintain the vessel's structural integrity.

9.8   Manufacturing processes

There are many methods used to convert a liquid resin and glass fibres into useful solid structures. Fabrication processes for reinforced plastics as a whole, including thermoplastics, have already been described in Chapter 2. This section focusses on the more common methods used in the chemical process industry and indicates their place in that market.

9.8.1   Hand lay-up

Hand lay-up is still the most commonly practised manufacturing method used to make corrosion resistant GRP for use in the chemical process industry. To begin, a mould is made. It can be made from a variety of materials, including GRP itself, and more frequently wood. Having made the mould, it is ground and polished to a smooth surface. This is important for ease of demoulding and also it helps to achieve a smooth high gloss inner surface on the vessel being manufactured. The mould is then covered with a release film, usually a saturated polyester. The best known trade name is probably Mylar® from Du Pont. A layer of resin is placed onto the film followed by the fibre. The fibrous reinforcement is worked manually into the resin by a person (laminator) with a bristle roller. One layer is laid after another, until the required thickness is attained. The vessel is then put aside for the resin to cure and the mould extracted. The fibres used in the process are normally discontinuous and in the form of a mat. CSM and WR are interspersed to form the structural layer, achieving a glass content of 45–55 wt%.

9.8.1.1   Advantages

- The cost of the mould is relatively low and practically any shape can be made. This means that hand lay-up is a cost competitive method to make individual articles and short production runs of complicated shapes.

- Because of the high resin-to-glass ratio and the attention to detail paid by the laminator, a high quality corrosion-resistant vessel is produced. It is the manufacturing method of choice when corrosion resistance is the prime concern.
- The process is adaptable using very little machinery and so can easily be used for on-site fabrication.

9.8.1.2   Disadvantages

- It requires quite a lot of skilled manpower and so where labour is expensive, hand lay-up suffers.
- As the products are hand made, each piece is different and variation in a series is high.
- The relatively low glass contents achieved means that high wall thicknesses are required if a given mechanical strength is to be attained.
- The quality of the vessel made depends to a large extent on the quality of the lamination and hence the degree of skill and experience of the laminator.

9.8.2 Filament winding

Filament winding is the most common mechanized process used in the composites manufacturing industry. It encompasses taking a polished and tapered steel mould known as a mandrel, onto which is laid a demoulding film such as Mylar®, mentioned above. The mould is placed horizontally in the machine and rotated. Onto this is placed the resin and the veils and mats necessary for the buildup of the corrosion barrier. This layer is carefully rolled by the operator to remove any unwanted air. The barrier is allowed to gel and then a CSM is rolled onto it. The winding of the structural wall begins directly on the CSM. Continuous glass rovings are pulled through a resin bath and then wound onto the mandrel. The machine turns the mandrel at a constant speed and the turning mandrel pulls the fibres through the bath and onto itself. The bath and winding head are moved up and down the mandrel to ensure complete coverage. Each layer is placed at an angle to the other. This angle can be altered to allow more or less hoop or axial strength in the vessel. When the required thickness of vessel wall has been achieved, the resin is allowed to harden and the mandrel withdrawn. The glass content achieved by this process in generally 65–75 wt%.

Because of the continuous nature of the fibres and the mechanization of the process, filament winding lends itself particularly well to the production of pressure pipe. It is also used quite frequently for the production of larger diameter structures such as tanks and storage vessels.

9.8.2.1   Advantages

- It is a mechanized process and so each production piece is the same and consequently series approval can be obtained.
- The fibres are continuous, offering advantages when pressure is a premium.
- A high fibre-to-resin content is achieved, leading to optimal mechanical properties and reduced raw material cost.
- The continuous roving fibre used is cheaper than mat.

9.8.2.2   Disadvantages

- It is difficult to make complicated shapes.
- At the end of each piece, where the resin and fibre application head changes direction, the fibre orientation changes and becomes unusable; it has to be cut off and thrown away.
- Quite a lot of capital investment is required in machinery and mandrills.
- The corrosion resistance, while excellent, may not be quite as good as that obtained by hand lay-up

9.8.3   Spray-up

In the spray-up process, a mould is made, just as in the hand-lay-up process. Continuous roving is fed into a chopper gun, which as its name suggests chops the fibres and sprays them onto the mould; resin is also sprayed at the same time. The fibre and resin streams meet each other on the mould surface. The laminate is then rolled to remove as much air as possible. The fibre content achieved with this process is typically between 25–35 wt%. This method is rarely used by itself for corrosion-resistant vessels but is used in combination with filament winding for larger vessels.

9.8.3.1   Advantages

- It permits the use of cheap continuous roving.
- The labour content is lower than for hand lay-up.

9.8.3.2   Disadvantages

- It is difficult to get trapped air out of the laminate.
- The fibre content is low.

9.8.4 Drostholm process

This is similar to the filament winding process, but the mandrel is made of a continuous steel band. The vessel is wound on the mandrel, at the end of which the steel band is fed up the centre of the vessel to the beginning, in order to recommence. Frequently, sand and chopped fibres are also dropped on to the mandrel to increase the thickness and thus stiffen the vessel.

It is usually used in order to produce large numbers of medium diameter tanks, or long runs of medium pressure (up to 15 bar) specification pipe. An example of a company currently associated with this process is Owens Corning Pipe.

9.8.4.1 Advantages

- Cheap continuous fibre rovings are used.
- The process is continuous.

9.8.4.2 Disadvantages

- It is difficult to use for diameters smaller than about 1 m.
- Air entrapment may be a problem.

9.8.5 Centrifugal casting

In centrifugal casting, the mould is rotated and the resin and chopped fibres are introduced via the centre of the mould. The resin and fibres are pressed against the inner wall of the mould by centrifugal force, such that the outer surface of the vessel being made is in contact with the mould surface. When the resin has cured, the mould is opened and the vessel withdrawn. A fibre content of 30 wt% is typically achieved. The method is most frequently used to manufacture large silos for use in agriculture. With this method it is possible to incorporate in excess of 50 wt% of sand. This produces a pipe of exceptional stiffness, which is very useful when the pipe is buried. The process is therefore used to produce low pressure underground pipe for sewage or industrial effluent. An example of a company currently associated with this type of pipe is Hobas.

9.8.5.1 Advantages

- Cheap continuous rovings are used.
- A well compacted laminate is produced.
- The labour content is low.

9.8.5.2 Disadvantages

- It is difficult to make a corrosion barrier.
- The fibre-to-resin ratio is low.
- It is a challenge to achieve full cure of the inner surface as it is in contact with the air during curing and subject to oxygen inhibition.

9.8.6 Pultrusion

During the pultrusion process, continuous fibre rovings are pulled from the bobbins through a resin bath and into a heated die. The resin, picked up by the fibres in the bath, cures in the die and the profile made is pulled out at the other end and cut to the required length. A fibre content of 65–75 wt% is achieved. Production rates vary, depending on the thickness and the complexity of the piece being manufactured, but are usually in the 0.5–$2.5\,\mathrm{m\,min^{-1}}$ range. The process is typically used to make profiles for ladders, stairways, grating and so on.

9.8.6.1 Advantages

- A high glass-to-resin ratio is achieved, resulting in good mechanical properties.
- The fibres are continuous.
- It is a relatively fast production process.
- It is a machine process, so consistent production is attained and type approval may be achieved.
- Cheap glass roving is used.

9.8.6.2 Disadvantages

- The process can only be used for straight objects.
- It is difficult to produce a good corrosion barrier.

9.9   Longevity and maintenance of GRP [5]

Near the town of Stade in northern Germany lies a major Dow Chemical production site. The plant lies on the southern bank of the Elbe river and occupies $5.2\,\mathrm{km^2}$ of land. Almost 2 billion DM worth of chemicals are produced each year on the site and 1600 people are in full time employment. The facility specializes in the production of chlorine and products requiring chlorine in their manufacture. The salt is mined locally and split electrochemically to chlorine and caustic soda.

Glass reinforced plastics materials have been used on this site since the late 1970s, and a considerable amount of experience in their use has been accumulated. They are most frequently used in chlorine production plants, where brine, hydrochloric acid, caustic alkali and sodium hypochlorite are handled at temperatures up to 100°C. They are also used in a wide range of other applications across the site.

During the summer of 1995, Dow undertook an extensive review of the usefulness and performance of these materials. The survey covered some 190 vessels, 10% of which were more than 18 years old, with the oldest having seen over 25 years' continuous service in contact with 15% HCl at 40°C.

9.10   Use of GRP at Dow Stade

9.10.1 Types of vessels

By the summer of 1995, when the survey of GRP began (Fig. 9.1), the number of GRP vessels in service had risen to 190. The bulk of the utilization of the GRP vessels was either in tanks or in gas scrubbers. Scrubbers formed 30% of the GRP, with 40% in the form of tanks. Chimneys contributed 3% of the vessels, filters 6%, distillation columns 5%, gas ducting 6%, and condensers 6%. There was also another 5% of a mixed bag including demistors, mixing vessels and sound absorbers. Piping was excluded from the survey.

9.10.2 Chemical service conditions (Fig. 9.2)

As already mentioned, most of the vessels concerned were in the chlorine production areas, but there are also some in the HCl recycling plant, the chlorinated polyethylene (Tyrin*) plant and also the glycerine and methylene diisocyanate plants. The most common application area was in acidic low pH media (37% of all the vessels); whereas 23% were in high pH caustic service, 27% were in neutral pH applications and the final 13% have been classified as mixed organic and inorganic streams.

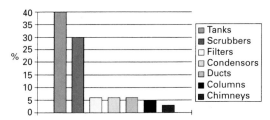

Figure 9.1 GRP vessels in use at Dow Chemical's site in Stade, north Germany

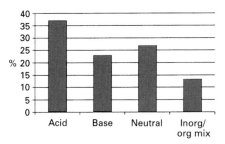

Figure 9.2 Chemical environment in which GRP vessels are used at the Dow site at Stade

9.10.3 Temperature (Fig. 9.3)

One gas quench vessel was operating at 200°C, but most of the vessels were being used at or below 100°C. The largest proportion of the vessels (40%) were in service at temperatures just above ambient, that is between 20–40°C. There were 26% in contact with fluids between 60–80°C, 14% between 40–60°C and 11% at ambient temperature, with the remainder (8%) at between 80–100°C.

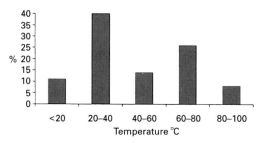

Figure 9.3 Temperatures of operation of reinforced plastics vessels in Dow survey

9.10.4 Resins used

In all cases, the resins used were Derakane® epoxy vinyl esters (Fig. 9.4). In 64% of the cases the resin was one of the higher temperature type, that is an epoxy novolac vinyl ester resin (Derakane® 470). This resin is particularly effective in containing acids and organic solvents. Its HDT of 150°C renders it the resin of choice for the higher temperature applications. Bisphenol A epoxy vinyl ester resins (Derakane® 411) were used for 25% of the vessels, in particular for the high pH caustic and hypochlorite streams. The brominated epoxy vinyl esters (Derakane® 510) were used for the remainder (11%) of the GRP vessels. These resins are important when both fire

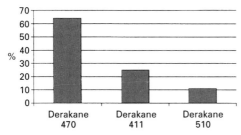

Figure 9.4 Types of vinyl ester resin used in the survey of Dow vessels at Stade, Germany

retardancy and corrosion resistance are sought in the one resin. They have been used mainly for gas (hydrogen) ducting and where low molecular weight organic solvents are present.

9.10.5 Life expectancy (Fig. 9.5)

The lifetime of GRP vessels is difficult to predict and the information presented here is based on a relatively young plant, with very few vessels having reached the end of their life. The time in service is also related to the investment programs put in place at various times in Stade. Nevertheless there is quite an impressive range. Of the vessels surveyed, 18% have seen less than 5 years service, 57% have been in service for between 5–10 years, 7% for between 10–15 years, 5% for 15–20 years, 10% for 20–25 years and impressively, there are 3% of the vessels which have had more than 25 years of continuous service.

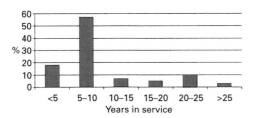

Figure 9.5 Service life of GRP vessels in Dow survey

9.10.6 Maintenance record (Fig. 9.6)

Of the 190 vessels surveyed, two thirds have required no maintenance whatsoever. The most interesting subgroup to look at are those vessels that have been in service for 20 years or more. In this group, 44% have never had any maintenance work performed. The most frequently performed maintenance operations are to replace mechanically damaged nozzles and

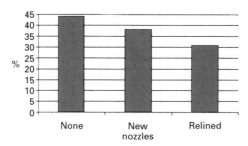

Figure 9.6 Maintenance history of vessels over 20 years old in Dow survey

connecting flanges. The next most frequent repair operation is the replacement of the inner resin-rich corrosion liner. For vessels with over 20 years in service, 38% have had one or more nozzles replaced, while 31% have been relined. Where relining has been necessary, it has been made on average between 15 and 16 years after the vessel was originally put into service.

The following pages discuss the main categories of GRP application found in chemical plant.

9.11 Ducts

The containment and transportation of low pressure gas streams by way of gas ducting is one of the most frequent industrial uses of GRP structures today. The reasons why GRP has become so popular for this type of application are many and varied, but will include some of the following.

9.11.1 Weight

GRP has a weight advantage over steel of about 4 to 1 or thereabouts, depending on the glass to resin ratio. As a lot of ducting is installed over ground, the weight reduction leads to savings in the dimensioning of the supporting structures. It also means that the installation is easier and faster to complete.

9.11.2 Corrosion

The corrosion resistance offered by GRP to such a wide range of chemicals means that it is the material of choice, particularly where a mixture of gases is being transported. In any gas stream, condensation resulting in accumulation and concentration will occur at some point in the line. This will lead to pitting corrosion in traditional materials and will result in failure of the

Figure 9.7 Plastilon built duct at an Enso Gutzerite pulp plant in Finland installed in 1997. It conducts gases from the bleach plant at 60°C

duct. The inherent corrosion resistance of GRP ducting also means that the exterior of the duct does not require corrosion protection, again offering an immediate cost saving for the construction project manager, and later on in the life of the duct, for the maintenance manager.

9.11.3 Insulation

GRP is a heat insulator. This means that when transporting hot gases, a reduction in the added outer insulation can be considered, and in certain conditions, no insulation at all may be required. One important area for this application is in flue gas desulfurization for coal fired electrical power generation plant. In this process, sulfur dioxide-containing gas is fed into an absorber tower, where it is washed with a limestone slurry. The sulfur dioxide in the gas stream forms calcium sulfite, which is subsequently oxidized to calcium sulfate (gypsum).

The raw gas contains humid mixtures of SO_2 and SO_3 and condensing sulfuric acid at temperatures of 80–100°C. The 'cleaned' gas contains the same contaminants, but at much lower concentrations [6]. Both duct systems have been successfully manufactured from GRP using an acid-resistant novolac epoxy vinyl ester resin. Large diameters are often used for high flow rates at low pressures. The photograph in Fig. 9.7 is of a Plastilon built duct at an Enzo Gutzerite pulp plant in Finland. It conducts gases coming from the bleach plant at 60°C. The diameter varies from 2000–2500 mm and the duct was installed in 1977.

Another area of growing importance for the use of GRP ducting is in the microelectronics industries. Here the need to maintain 'clean rooms', allied to the use of some quite aggressive chemicals for the etching process, has meant that corrosion-resistant GRP becomes the material of choice.

No article on the subject of ducts would be complete without some reference being made to duct fires. Fires do sometimes occur in duct systems containing flammable materials. The most frequent cause of ignition is a spark generated by static electricity. When this happens, the priority has to be to avoid the propagation of the fire and to limit the damage caused to both personnel and property. Sprinkler systems can be built into ducts or nitrogen can be bled into the gas streams. This can be done with any material. With GRP, consideration may be given to using corrosion-resistant brominated resin to contain any incident and to limit the damage it may cause to equipment.

9.12   Storage tanks and process vessels

Probably the most popular use of GRP in industrial circumstances is in the form of corrosion-resistant storage and process vessels. The reason for the use of GRP for this application is its inherent corrosion resistance. This means trouble free storage and processing of corrosive and possibly dangerous chemicals. It also means a material which does not need to be repainted every couple of years.

In the chemical process industry, GRP vessels are frequently used to store mineral acids and strong caustic solutions. It is of particular interest when chloride ion is present. There are very few metals or alloys which can offer the long term resistance to pitting, caused by the presence of chloride ion, which GRP can offer. The difficulty in the chemical process industry is that very few 'pure' chemicals are used. Most of the time we are dealing with mixtures of a wide range of chemicals. A considerable body of knowledge on the subject of the corrosion resistance of metals to pure chemicals is available and a metal can always be found to resist a particular chemical, but very few can offer the broad range of resistance which GRP can offer.

In the pulp and paper industry, in the specific area of pulp bleaching, there is widespread utilization of GRP materials. They provide corrosion-resistant containment for the pulp and for the bleaching chemicals they contain. When pulp is bleached chlorine dioxide, sodium hypochlorite, sodium hydroxide and chlorine are typically being used, and it is for the

Figure 9.8 Photograph showing an epoxy novolak vinyl ester resin tank which has required no maintenance while holding 15–30% HCl at 40°C since 1971. It is now being replaced (1998)

storage and processing of these reactive chemicals that the GRP is essentially used.

In the metals-refining industry, crude metals are typically dissolved in strong mineral acids. The solutions are then placed in large open vessels known as cells and an electric field placed across the vessel. The metal then concentrates on one pole and the impurities fall to the bottom or precipitate out. The temperatures are in the region of 50–90°C and GRP is used throughout the process, not only for its corrosion resistant properties, but also for its electrical resistance.

Filament winding or hand lay-up are the manufacturing methods most frequently used for tanks and process vessels. Where pressure resistance is critical, filament winding is the method of choice, and where corrosion resistance is the overriding factor, then hand-lay up is the most suitable method.

Figure 9.9 Glass reinforced vinyl ester resin tanks used in the pulp and paper industry (see text for details)

Tanks are regularly made up to 10m diameter. Above this size is exceptional, but by no means unknown. Vessels up to 20m diameter have been successfully made and used.

Manufacturing takes place at the fabricator's plant for the smaller vessels, but the larger vessels (where transportation is a problem) can be made on site.

Electrolytic cells for the metals refining market can be made from GRP but are also made from polymer concrete. For this material, sand is mixed with the resin and the vessel is cast with or without glass reinforcement.

Where fire retardancy is an issue, vessels can be made entirely from brominated resins, or more commonly (particularly in Europe), just the outer millimetre or two can be made from the fire retardant grade. If a

Figure 9.10 Polymer concrete electrolytic cells used in copper purification. The walls of the vessels are 80 mm thick (see text for details)

buildup of static electricity needs to be avoided and the vessel grounded, for example because it is to be used for hydrocarbons, then some conductive fibres such as carbon fibres, with or without some graphite filler, can be incorporated into the structure.

Figure 9.8 shows a 44 m³ tank installed at the Dow Stade plant in northern Germany. It was installed in 1971 and the photograph was taken in 1995. The tank is scheduled for replacement in 1998 after 27 years service. It was built by Hawodur in Holland using Derakane® 470 epoxy novolac vinyl ester resin. It had been used throughout its life to contain 15–30% hydrochloric acid at 40°C. At the time of writing (early 1998) it had not required any maintenance.

Figure 9.9 shows a typical pulp and paper industry application, where GRP is successfully used to make large storage tanks for corrosive bleaching chemicals that are difficult to maintain in more traditional materials. It consists of a series of 400 m³ tanks, measuring 7.5 m diameter and 9 m high, mounted on a flat concrete base. They are used for the storage of chlorine dioxide solution at 10°C. The solution is needed to bleach ECF (elemental chlorine free) paper grades. They were designed by Enzo Gutzerite for the Schauman Ab. pulp plant in Pietarsaari in Finland, and built by Muotekno (now known as Plastilon) in Imatra by hand lay-up, using Derakane® 411-45 epoxy vinyl ester resin. They have been in service since 1976.

Figure 9.10 shows some polymer concrete electrolytic cells used in the electrowinning process to purify copper. The cells are designed to contain a solution of 15% sulfuric acid and 3% copper sulfate at 65°C. They were manufactured by Wallcrete Ltd in Zambia by a casting method and have been in use since 1975. They are 3.85 m long by 1.03 m wide by 1.9 m high. The walls of the vessels are 80 mm thick. They are usually manufactured from a special low viscosity epoxy vinyl ester resin (Derakane® 411-C50) intended to facilitate the incorporation of the sand.

9.13   Pipes

Pipes are another frequent application of GRP in an industrial context. GRP pipes offer superior corrosion resistance in comparison to metals and they are lighter in weight, thus making installation easier and less costly. Because of the weight advantage, they can be supplied in 13 m lengths, so fewer joints are required for long pipelines. The inner surface of a GRP pipe is smoother than either metal or concrete. This means that for sewer pipe in particular, a 30% higher flow rate can be achieved through a GRP pipe than through a concrete one of similar diameter.

Flow in a pipe is predicted by Hazen Williams [7] with the following formula:

$$\text{flow} = 1.318 \, ACR^{0.63} \times S^{0.54}$$

where R is the hydraulic radius, S is the hydraulic slope and A is the cross-sectional area. C is a flow factor related to the roughness of the inner surface. It is 160 for new GRP pipe, and degrades to about 150 after about 6 months service. This compares with 115 for concrete and 120 for new carbon steel, degrading to 65 with the buildup of oxide layers. R, S and A are constants when comparing pipes designed with similar dimensions.

GRP pipes are manufactured by a variety of methods, including hand lay-up, filament winding, the Drostholm process and centrifugal casting. Hand-lay up is used, as with tanks, when a higher resin content is required to optimize the corrosion resistance. Filament winding is the process of choice when pressure resistance is required and a high reproducible quality is important. The Drostholm process is used generally to manufacture large quantities of <15 bar pipe to be used at lower pressures, usually with a high wall thickness and producing a stiffer pipe. The centrifugal casting process can be used for atmospheric pressure waste pipe.

Piping made from GRP is used in many parts of the chemical process, pulp and paper and metals refining industries. The areas of use mirror those of GRP for vessels. Where there is a GRP vessel there is generally a GRP pipe leading into it or taking product from it. An additional area gaining popularity today is in water lines for safety showers and fire mains. With the traditional carbon steel, rust particles are produced in the pipe and when the water flows through it, they are transported with the water and cause problems such as clogging of the sprinkler heads. The problem can be solved because no rust can be generated inside a GRP pipe.

GRP pipe is also popular for the production of sewer pipe in areas where the sewage is corrosive. In the Middle East for example the ground is very flat. This results in a long residence time for the sewage in the pipe. Because of the lack of water in these areas, the sewage is also more concentrated than in western Europe. The temperature is also quite high and these three circumstances combine to cause fermentation of the sewage, resulting in sulfuric acid being formed inside the sewage pipe and hence the requirement for corrosion-resistant GRP. Because of the large amounts of pipe required and the need to be competitive with more traditional materials, the structures typically used incorporate an epoxy vinyl ester resin inner liner (approx. 2–3 mm thick) to contain the corrosion, with an outer structural wall using polyester resin.

Another area of growing interest is in the off-shore oil and shipping industries. Here, corrosive oxygen-containing salt sea water is used for fire water and ballast water lines. This causes corrosion in the carbon steel lines, leading the end-users to consider GRP. In this area a large proportion of the pipe uses an epoxy resin matrix. Epoxy resin offers some advantage over epoxy vinyl ester resin in mechanical properties, resulting in a better pressure resistance. Epoxy was also on the market 10–15 years before epoxy vinyl ester and had already become the resin of choice for the oil industry in particular.

One of the most important selection criteria for the oil industry is the performance of the material in a hydrocarbon fire. Here again the GRP industry can contribute, and over the last few years we have seen an increasing use of fire retardant phenolic resin applications.

Figure 9.11 is a photograph of the ring main on the Norske Shell Draugen off-shore oil production platform. The pipe was manufactured by Vetroresina in Italy and installed by Norwegian contractors. The system is designed to withstand both sea water and crude oil at temperatures up to 40°C, with excursions allowed up to 70°C. The pressure rating of the pipe is for 25 bar. The manufacturing process was filament winding and the resin used was a bisphenol A epoxy vinyl ester (Derakane® 411-45). GRP was chosen in competition with 6% molybdenum steel and titanium. The GRP solution was found to be 50% cheaper, and easier to install and assemble.

Figure 9.12 shows a GRP pipe system designed by John Taylor & Sons to contain sewage. The pipe is for installation in the United Arab Emirates where the particular local conditions make the use of GRP invaluable, and GRP is used with a vinyl ester inner liner and ECR glass, associated with an outer structural wall made from polyester resin.

Figure 9.11 The ring main on the Norske Shell Draugen offshore oil platform. Pressure rating, 25 bar. The design is intended to withstand sea water and crude oil up to 40°C and occasionally to 70°C

Figure 9.12 Glass reinforced polyester sewage pipe in the United Arab Emirates. The glass is ECR because of acid conditions and the liner is vinyl ester

(Pipe is discussed more fully in Chapter 8, which also contains case histories).

9.14 Gas absorbers (scrubbers)

9.14.1 Why they are used

In recent years, people and their governments have become more and more concerned with the quality of the environment we all live in. This has resulted in an increased awareness of the importance of good and responsible pollution control practices. This in turn has meant a large increase in the number of exhaust gas cleaning systems being installed. The centre of any such system is frequently the absorber tower or scrubber itself.

In an absorber, gas is injected usually at pressures not too distant from atmospheric and at temperatures usually below 100°C. A liquid cleaning medium, designed to react with/neutralize or absorb the contaminants in the gas, is sprayed into the gas stream. The tower in many cases is packed with a lightweight material designed to increase the surface area of the scrubbing liquid and thus to increase the scrubber efficiency.

In many cases, the raw gas contains acids, bases or other corrosive contaminants. The scrubbing liquors consist typically of a dilute acid, base or water. A solution of the contaminants and the scrubbing liquor collects in the bottom of the scrubber. A typical system operates under alternating acid/base conditions, with high humidity, with oxygen present and at temperatures between 50–80°C. The combined set of conditions is extremely difficult for any metal to handle, so the inherent corrosion resistance of GRP comes into its own.

The largest concentration of paper pulp industries in Europe is located in Norway, Sweden and Finland. These countries have precarious ecosystems, caused by their northerly latitudes, which dictates that they must pay particular attention to how they treat their environment. Chlorine gas, chlorine dioxide and sulfates are frequently used in the pulp bleaching process. These chemicals were released to the atmosphere in the past, but today are treated and absorbed in scrubbing towers.

The chemical industry itself is also a large user of absorber towers. They are used to clean a wide range of waste gas streams before releasing them to the atmosphere. These streams again contain acids, halogens, hydrocarbons and oxides of sulfur and nitrogen, all of which need to be removed before release.

The metals refining industries are also large manufacturers and users of strong mineral acids and they also clean their gas streams.

A more recent industry to begin to use chemical gas cleaning systems is the power generation industry. Fossil fuels burned to produce the power contain sulfur. On burning, this is converted to sulfur oxides which, if released untreated to the atmosphere, are suspected of being converted to sulphuric acid and falling to earth as 'acid rain'. The fuels are burned in air, and air contains nitrogen which is also partially converted to nitrogen oxides and nitric acid. These are also thought to contribute to the acid rain phenomenon. The volumes of gases produced by the power industry are such that in many cases quite large scrubbing vessels are required. They have been sucessfully made in GRP with diameters between 10–20 m and 50–100 m high.

Another of the newer industries to install gas cleaning has been the waste incineration industry, whether it be municipal or industrial waste. The challenge in this sector is that waste is waste, and as such when burned will contain a vast array of contaminants and their composition will vary daily. The resistance of GRP to a wide range of corrosive conditions helps to contain this difficulty. Waste incinerators also have the hottest incoming gases, with temperatures over 200°C being common. The special heat requirements of waste incinerator gas scrubbers can only be overcome by using newer high temperature-resistant resins and specialist laminating techniques.

9.14.2 Advantages

There are several advantages in using GRP for the construction of absorber towers, some of which are described below.

9.14.2.1 Corrosion

First and foremost among the advantages has undoubtedly to be the corrosion resistance of GRP over metals in the conditions generated inside a typical scrubber. The incoming gases almost always contain one or more strong mineral acids, such as sulfuric, nitric, hydrochloric and hydrofluoric acids. They are neutralized, using water or a weak solution of a base such as sodium hydroxide or lime/limestone. In the pulp industry, streams containing thiosulfates are also frequently encountered. All of this is in an atmosphere close to 100% humidity at a temperature normally between 50–80°C. As gas streams are being cleaned essentially before release to the atmosphere they also contain oxygen and because of their design, the liquid streams will also be saturated with oxygen. Alternative materials include high nickel alloys or rubber lined steel. The alloys are more expensive and their welds are particularly susceptible to chloride or fluoride pitting attack. Rubber lining is cost competitive, but it is limited to use below about 70°C. This means that it is operating in a scrubber at or just below its temperature limitation, whereas GRP can safely be used 20–30°C higher in contact with the liquid, or even 100°C higher when in contact with the gas.

The outer wall of the vessel will also be corrosion resistant and so will not require painting either at the beginning of its life or subsequently. This is a major maintenance saving compared with steel vessels.

9.14.2.2 Weight

The density of GRP is generally around two (depending on the glass to resin ratio) compared to eight for steel. This means that a large GRP absorber vessel will be substantially lighter than its metal counterpart. This weight advantage is all the more important, as the volume of absorber vessels usually contains quite a lot of gas. So the weight of the vessel determines the weight of the required supporting structures. A cost saving may thus be achieved. The reduced weight of the vessel also makes installation easier and cheaper.

9.14.2.3 Heat insulation

As GRP is a heat insulator, many such vessels have been made and installed successfully without any further insulation being added.

9.14.2.4 Fabrication

Hand lay-up can of course be used to fabricate scrubbers. While the method is competitive for smaller diameter vessels, the high labour content of this method renders it uneconomic for the larger units. As this particular application frequently requires large vessels, filament winding or centrifugal casting are the more usual fabrication methods employed. Vertical winding techniques are used for the larger diameter vessels, say over 7 m. Here the mould is fixed vertically and the glass and resin application unit is moved around it. For long high vessels in particular they are manufactured in 3–4 m high sections and dropped onto each other, connecting with a bell and spigot joint. The thicker joint area then doubles as a stiffening ring. The largest vessels are manufactured on site to avoid the difficulties and cost incurred in transporting these large, cumbersome structures over long distances.

9.14.3 Resins

Recent research by Bergman [8] at the Swedish Corrosion Institute has demonstrated that in the gas/acidic sections of absorbers, novolac epoxy vinyl ester resins associated with acid-resistant glass types perform best, while in the high pH liquid-containing section, bisphenol-A epoxy vinyl ester resin systems are the resins of choice. Large absorber towers take a relatively long time to construct, because of their size. The fabricators need to use resin systems with long pot lives and gel times. This fact, together with the large surface areas, results in the evaporation of more styrene monomer than in other applications. Not only does this affect the quality of the surrounding air, but it is also a material and financial loss. Modern resins have now been adapted to cope with this difficulty and today the so-called LSE (low styrene emission) grades are used. These resins contain a small amount (hundreds of ppm) of a film-forming additive which floats to the surface of the resin, impeding the evaporation of the styrene monomer (much as cream is placed on the top of coffee/whisky mixtures to keep the alcohol in the liquid).

Figure 9.13 PCK refinery in Schwedt, Germany. The flue gas desulfurization scrubbers are said to be the largest GRP ones in the world (see text for further details)

Figure 9.13 is a photograph of the PCK refinery in Schwedt, Germany. It is claimed to contain the world's two largest GRP flue gas desulfurization scrubbers, which were fabricated and installed by the Reinforced Plastic Systems (RPS) company, using low styrene emission resins. On-site production was necessary to handle the extremely large dimensions of 9.00 m diameter and 49 m height. Almost 200 tonnes of Derakane® 411-45 LSE and 470-36 LSE resins were used during the time necessary (only 7 months) to fabricate the two scrubbers. Five cylinder pieces per scrubber, with a diameter of 9 m, up to 11 m high and with a wall thickness of 30–60 mm, were fabricated with the vertical filament winding technique from RPS. One cylinder piece weighs approximately 50 tonnes and the total weight of one scrubber, including its inside installations, is 300 tonnes. The area where the electrostatic filters are installed was protected by a resin filled with aluminium powder to avoid electrostatic charging (see light coloured area on photograph).

A GRP fume scrubber and outlet ducts, shown in Fig. 9.14, have now been dismantled after 8 years of trouble free service at the tin smelting plant of Capper Pass Ltd., Humberside, UK. The scrubber, installed in 1982, was

Figure 9.14 GRP fume scrubber, dismantled after 8 years of trouble free service. The resins used were vinyl esters, including a brominated fire retardant grade. Dimensions: 11 m high, 5 m diameter

Table 9.2 Exhaust stream composition from
scrubber in gas cleaning system

Water vapour	12–15%	v/v
SO$_2$	1–3%	v/v
SO$_3$	0.1–0.15%	v/v
CO$_2$	1–3%	v/v
Fluorides	<0.1%	v/v
Chlorides	<0.05%	
Arsenic	<0.5 mg N^{-1}m^{-3}	

11 m high and 5 m in diameter, and was designed to British Standard 4994,
being hand laminated by local fabricator Garlway Ltd. Derakane® 470-36
novolac epoxy vinyl ester resin was used, with chopped strand mat and
woven roving reinforcements, and an internal corrosion resistance barrier
containing two layers of Nexus™ polyester fibre synthetic veil. A Class I fire
retardant rating (to BS 476) was obtained on all external surfaces, by using
Derakane® 510A-40 brominated epoxy vinyl ester fire retardant resin, with
5% antimony trioxide.

Blisters and cracks were found on the internal surface during a routine
inspection in 1984. Engineers from both Garlway and Dow believed that
this did not pose a serious threat to the performance of the GRP. The
blisters did not get any worse and they did not appear to reduce the
performance during the scrubber's 8 years of service life. The only maintenance carried out consisted of duct repairs following fire damage. The
scrubber was part of a gas cleaning system handling the exhaust gases from
the plant's sinter machine, which prepared the incoming raw materials
ready for the blast furnace. The plant, which had been in operation for 40
years, had become specialized in processing low grade ores and tin, containing scrap from various sources. Much of the 'contaminant' was burnt off in
the sinter machine, resulting in an exhaust stream of the composition shown
in Table 9.2.

These gases entered the scrubber at 100–110°C and at flow rates of up to
100000 m^{3}h^{-1} where they were cleaned by spraying with recirculated water,
thereby reducing the gas temperature in the scrubber and outlet ducts to
between 70–80°C.

9.14.4 Visual inspection/assessment of the internal surface of a scrubber

An internal 'corrosion-resistant layer' (approximately 2 mm thick, containing two layers of synthetic tissue) showed an overall deterioration of the
surface, which had darkened to a red/brown colour; the surface had

Figure 9.15 Ian Reid, an independent consultant, takes a Barcol hardness reading from the inside surface of the scrubber

blistered (typically the blisters were of 10–20 mm diameter) and cracked, particularly in the highest temperature zones, (e.g. opposite the gas inlet).

The structural GRP laminate behind this corrosion-resistant layer was still sound. All the internal structural joints, brackets and internal overlays were in sound condition, with no signs of delamination.

Analysis and testing of cut-outs

The darkened surface extended only through the outer 1–2 mm of the surface (containing two layers of synthetic tissue). There were visible signs of corrosive attack on the glass fibres in the corrosion-resistant layer immediately behind the synthetic tissue reinforcement; this attack was limited to a thickness of approximately 0.5–1 mm and was located mainly in the highest temperature zone. The underlying corrosion-resistant barrier and structural glass reinforced laminate were apparently still in excellent condition. There was no evidence of corrosion of the glass fibre reinforcement. Independent laboratory analysis and testing of the structural laminates showed 75% retention of design strength and around 150% of design modulus. Figure 9.15 shows an independent consultant taking a Barcol hardness reading from the scrubber's inside surface. Values varied from 15 to 30. After 8 years of continuous operation at temperatures in the range from 70–110°C, the blisters, cracks and corrosion were restricted to the 2 mm corrosion-resistant surface layer. The structural laminate was still in excellent condition. Despite inner surface 'deterioration', the scrubber would have continued to give reliable service for many more years.

Although the surface deterioration was predictable, other plant operators may wish (say after 10 years of arduous duty) to replace the internal

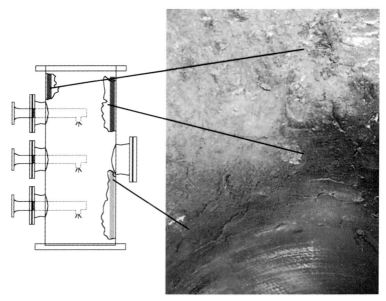

Figure 9.16 High temperature quench section to lower the scrubber inlet gas temperatures from 260°C to 60°C

corrosion-resistant layers by on-site relining to the original design specification. This action should realistically add a further 10 years of trouble free operating life.

Figure 9.16 shows a high temperature quench section designed by Dow, using three spray levels to lower the scrubber inlet gas temperatures from 260°C to 60°C. Brick-lined GRP was the only viable alternative to tantalum, a metal 100 times more expensive. The quench was installed in 1992. About a year later, the eroding mortar began loosening bricks near the top of the 2.4 m diameter quench. As bricks fell to the bottom of the duct, hot fluegas containing hydrochloric acid, nitrogen oxide and water vapour began affecting the GRP structure underneath. The problem continued undetected for as long as a year. Then, after Dow Plastics first noticed scorching in May 1994, a closer inspection in August showed just how deep the attack had been.

Significant cracking to a depth of 3–5 mm was found in the total circumferential area at the first and second spray nozzle levels. It would have been only a matter of time, possibly another one to two more years, before the part of the structure exposed to high temperatures had failed. But the Derakane® 470 novolac epoxy vinyl ester resin had held at temperatures far beyond its design limit.

The next step was to repair the quench, so Dow Plastics began working with Kurotec Kunststofrohrtechnik GmbH, the Stade-based GRP producer that had fabricated the quench. Up to the first spray nozzle, as many as five layers of scorched laminate were removed. New laminate layers were added, alternating a layer of $450\,g\,m^{-2}$ chopped strand mat with two C-glass veils made from Derakane® 470 novolac epoxy vinyl ester resin, filled with graphite. A support ring for the spray nozzle's ceramic sleeve was then installed with a 10 mm structural laminate, fabricated by hand lay-up.

From the support ring to the lower brick lining, the quench was relined with four layers of chopped strand mat and two C-veil layers. Finally, a ceramic sleeve was fitted within the quench's 950 mm inner diameter, completing repairs.

Today the quench is performing optimally, cooling fluegas as part of Dow Stade's clean air programme. Availability since the repair has been 100% and Dow Plastics engineers know from first-hand experience that GRP will continue to stand up to the rigours of their application.

9.15   Pultruded profiles

9.15.1 Applications

The last area of application for industrial composite to be discussed is the growing area of utilization in the manufacture of structural profiles. Today these profiles are finding more and more favour with engineers for the construction of grating-based flooring systems, stairways and hand rails [9].

Gratings can be made in one of two ways. They can be moulded by placing the reinforcement in a mould, filling it with resin and allowing the resin to cure. The advantage of this method is that a quite high resin-to-glass ratio is achieved, resulting in good corrosion resistance. There are equal amounts of fibre and therefore equal strength attained in each direction. The second method of manufacture is to construct them using purpose designed pultruded profiles. Using this method, a more adapted design is achieved with a higher fibre content and associated higher stiffness.

For this application, the corrosion requirement of the profiles is usually only for short term contact at ambient temperatures. Polyester resins dominate the market and have proved their value over many years of service.

For more corrosive environments, epoxy vinyl esters and phenolic laminating resins [10] are used quite successfully as well, particularly where fire retardancy is an issue, such as in the oil industry. The glass is normally E-glass roving, but some mats are used for more complicated shapes and an external surface veil is incorporated to produce an acceptable degree of outer smoothness.

Figure 9.17 FRP gratings installed in a chlor-alkali plant, exposed to sodium chloride and hydrochloric acid at temperatures from 80–100°C

Associated with any composite grating system is a hand rail and stairwell to go with it. These are made from suitably designed pultruded profiles. Such systems find favour where sound floorings and escape ways are essential. The electrochemical industry has long been a user of the systems and now we see more and more of their use on off-shore oil platforms.

9.15.2 Advantages of composite profiles

9.15.2.1 Corrosion resistance

As always with any use of composites in the chemical process and associated industries, the main driving force is corrosion resistance. Stairways in heavy industrial installations are used regularly and must maintain optimum functionality for the duration of the plant life. In case of accident they will in many cases constitute the main escape way. Metal profiles need to be continually painted to protect them from the corrosive environments found

Figure 9.18 Gratings made by a contact moulding method using a bisphenol A epoxy vinyl ester resin and its flame retardant, brominated equivalent. It was installed at the Enichem plant at Gela

where corrosive chemicals are used or where plants are located near the sea coast.

9.15.2.2 Weight saving

Composite profiles offer a considerable weight advantage compared to their metal counterparts. This translates into a cost saving when the structural design of the plant and its foundations is made. In the context of an offshore oil platform, where labour is expensive, one man can handle and install a much higher volume of profile than can be achieved with metals.

9.15.2.3 Insulating properties

Of particular interest to the electrochemical industry is the non-conductive nature of composite profile. Where electrochemistry is practised there is always the risk of escaping electrical current. The objective is to limit the damage that any such escape may cause. Metal profiles will conduct current, and by so doing they will speed up the rate of their own corrosive degrada-

Figure 9.19 Walkway, handrails, stairwells, cable trays and safety cages were made at the Dow Stade membrane chlorine plant using epoxy vinyl ester resin. Chemical environments include sodium hydroxide, sodium hypochlorite and brine

tion. Composite profiles, being electrically insulating, add to the safety package in operation and do not corrode.

9.15.2.4 Aesthetics

The pultrusion process by which the profiles are made lends itself exceptionally well to the inclusion of some pigment paste in the resin formulation. This allows brightly coloured profiles to be made, affording visually pleasant structures.

9.15.2.5 Cost

The pultrusion process has a low labour requirement and it can manufacture large amounts of profile with high glass-to-resin ratio, using continuous roving at a competitive cost.

9.15.2.6 Examples

Figure 9.17 shows gratings installed in a chlor-alkali plant exposed to various chemicals, including sodium chloride and hydrochloric acid, at temperatures between 80–100°C. They demonstrate clearly the reasons for the choice of such materials in an electrochemical environment.

The gratings in Fig. 9.17 and (same location) 9.18 were made by a contact moulding method using glass roving and both Derakane® 411 bisphenol-A epoxy vinyl ester resin and its brominated homologue Derakane® 510 for corrosion resistance and flame retardancy. They were fabricated by CO.MA.CO in Gorgonzola in Italy and are installed at the EniChem plant in Gela.

Figure 9.19 shows an installation at the Dow Stade membrane chlorine plant. It was built in 1996 by Techno Composites Domine Gmbh, using Bekaert profiles made with Derakane® 440-40 epoxy vinyl ester resin. The installation includes stairwells, handrails, walkways cable trays and safety cages. GRP and pultruded profiles in particular were used for their resistance to the chemicals being handled – notably sodium hydroxide, sodium hypochlorite and brine. Their electrical resistance was also a factor in this electrochemical plant [11].

References

1 T N Bishop and T W Cowley, 'Corrosion resistant fiberglass reinforced plastic composites for chlorine dioxide environments – Third report', *TAPPI Engineering Conference*, TAPPI, 1994.
2 ISO 75 Method A.
3 ASTM 3418–82.
4 A van Buren and P Kelly, 'The design and use of GRP for quench sections of gas scrubbers operating up to 220°C', *20th International Composites Conference*, British Plastics Federation, London, 1996.
5 P Kelly, M Jaeger and B Schwanewilms, 'The use of glass reinforced materials to solve corrosion problems in the chemical process industry', *EUROCORR '97, Proceedings of the European Corrosion Congress*, 1997, Vol II, pp 235–245.
6 S Youd and P Harrison, 'Epoxy vinyl ester resins – From the first 25 years to the 21st Century', *18th International Composites Conference 1992*, British Plastics Federation, London, 1992.
7 *Fiberglass Pipe Handbook*, New York, SPI Composites Institute, 1992.
8 G Bergman, *Corrosion of Plastic and Rubber in Process Equipment – Experiences from the Pulp and Paper Industry*, TAPPI Press, 1995.
9 Anon, 'Composite grating is right for today's offshore market', *Reinforced Plastics* 1996, **40**(12) 38–44.
10 A Weaver, 'Stronger phenolics find offshore uses', *Reinforced Plastics* 1996, **40**(12) 46–48.
11 Anon., 'Composites oust steel at Stade', *Reinforced Plastics* 1997 **41**(8) 34–36.

Repairs using fibre reinforced plastics

WING KONG CHIU AND RHYS JONES

10.1 Introduction

In today's competitive economic environment, one of the major challenges facing the transport industry is to produce vehicles (whether for land, sea or air) with reduced fuel consumption, thereby optimizing the world's scarce oil and gas reserves as well as reducing overall pollution. There are a variety of potential solutions for this problem; viz. improved aerodynamic behaviour and propulsion systems. Another solution is to reduce the overall weight of the vehicle. Here, reinforced plastics composites can play a significant role, especially if they can be introduced to primary structural components, that is wing and fuselage skins for aircraft structures. However, before this can occur it must be shown that they can adequately substitute for existing thin-skinned metallic components. The enhancement of durability by suitable repair techniques is a necessary requirement if fibre reinforced plastics (FRP) are to be cost effective.

The aeronautical industry pioneered the use of composites. In 1940 Dr N A de Bruyne and his team constructed an experimental Spitfire fuselage entirely out of unidirectional flax reinforced phenolic prepreg. In the marine industry composites have been used as early as the 1950s. Other successful applications have been found in the chemical, electrical, land transport and offshore industries.

This chapter will describe appropriate repair schemes for restoring the strength of structures made from FRPs. To demonstrate the versatility of FRPs, the use of this material for the repair of composite aircraft structures, composite boat structures, metallic boat structures and concrete civil structures will also be discussed.

Fibre reinforced plastics are commonly fabricated by embedding small diameter fibres with relatively high strength and stiffness into a relatively ductile polymeric binder or matrix. The main functions of this matrix material are to maintain the overall shape and geometry, to transfer the in-plane shear and transverse loads from one fibre to another, and to transfer load from one layer of fibres, which is frequently termed a lamina, to another by shear and peel. The properties of composite materials are generally highly directional. This enables the engineer to choose an appropriate lay-up and thereby 'design' a structure to achieve specified mechanical properties (i.e. stiffness, displacement, strength, frequency, damping, etc). Currently the most commonly used fibres are carbon, aramid, glass and ceramic. These fibres are generally embedded into matrix materials such as polyester resins, vinyl ester resins, epoxy resins, polyimides and various thermoplastics. However, within the aerospace industry there has been a recent move towards the use of both metal matrix and carbon–carbon, that is graphite fibres in a graphite matrix, composites.

Fibre reinforced plastics have many advantages for use as structural materials in aircraft, marine and civil structures, including their formability, high specific strength and stiffness, resistance to cracking by fatigue loading and their immunity to corrosion. However, whilst having these advantages, they are prone to a wide range of defects and damage which may significantly reduce residual strength [1,2]. The problem of low energy impact damage is of particular concern and can cause significant reductions in compressive strength (up to 65% of undamaged strength) [1]. Whilst impact can cause a significant amount of delamination, often the only external indication is very slight surface indentation. This type of damage is frequently referred to as 'barely visible impact damage' (BVID). An appropriate repair methodology has to be developed for the reinforcement of impact and delamination damage in composite structural components.

Composite materials have also been used to maintain and extend the safe life of bridges and reinforced concrete structures. Indeed, Australian, Japanese and US companies are already focusing on this technique. There are three principal ways in which composites can be used to rehabilitate ageing infrastructure:

1   By using a modified cement-based composite as a filler material.
2   As in (1), but with an external composite or metallic doubler also bonded over the damaged region.
3   Using a composite structural member, either to replace the deficient component or to provide an alternative load path. In the latter case, the repair patch, typically made from fibre reinforced plastics, would be directly attached to the structure, either mechanically or via an adhesive bond. These topics will be discussed further, below.

10.2   Repair of composite aircraft structures

Modern aircraft are constructed from a variety of materials, with aluminium and graphite epoxies now being dominant. The F/A 18 incorporates significant quantities of graphite/epoxy composite materials which, whilst forming 34% of the external surface area of the aircraft, only represent 9% of the weight. The components on the F/A 18 which are made from graphite/epoxy include the main wing skins, trailing edge flaps, vertical tailskins, horizontal stabilators, speed brake and access doors.

Graphite/epoxy offers numerous advantages over conventional metallic structures. This includes increased stiffness, reduced weight, resistance to corrosion and resistance to tension-dominated fatigue loading. However, fibre reinforced plastics do have their disadvantages. They are susceptible to damage by low energy impact caused by dropped tools, large hailstones, runway stones and bird strikes. Low energy impact damage can cause a significant amount of internal delamination. However, often the only external indication is a very slight surface indentation (Adsit and Waszczak [3], Starnes et al. [4]). Although BVID does not significantly reduce the tensile strength, the reduction in the residual compressive strength can be as much as 30%. To this end, a significant amount of work has been conducted to develop techniques to restore the residual strength of the impact damaged components. The standard repair methodology involves removal of the damaged region and its replacement with a new laminate joined to the structure via a scarf joint. An external doubler with a ply configuration involving ±45° is also commonly used with such scarf repairs. Scarf repairs are commonly used for repairing monolithic laminates up to 10 mm thick and the damage size is generally limited to 150 mm in diameter. When access is available to both sides of the structure, such a repair can achieve almost 100% of the original strength. At the moment scarfed repairs to composite structures are confined to components with a limit strain of under 4000$\mu\epsilon$. If the repair extends into a zone with a higher limit strain then the item is classed as irreparable.

The problem with scarf repairs is that they need extensive facilities and often require the structure to be sent to a major repair depot. To overcome this requirement, a new simplified repair methodology is currently under development. This methodology mimics that used in the composite repair of cracked metallic structures in that it only uses an external patch, or doubler, and is aptly illustrated in the work of Paul and Jones [1], Jones et al. [2] and Chalkley and Chester [5].

Paul and Jones [1] illustrated the potential for repairing a graphite/epoxy (AS4/3501) laminate using an externally bonded patch covering the impact damaged area. The dimensions of this test specimen were 300×900 mm. Specimens were fabricated with a ply configuration of $[+45_2/-45_2/0_4)_3/90]_s$. They were subjected to impact damage with impact energy in the order of approximately 7.5 J, and the damage area was subsequently measured by C-scanning. The damaged specimens were then repaired with a 16 ply patch $(0_2/+45-45/-45+45/)_2)_s$. The ends of the repair patch were scarfed as shown in Fig. 10.1. These patches were bonded to the damaged specimens

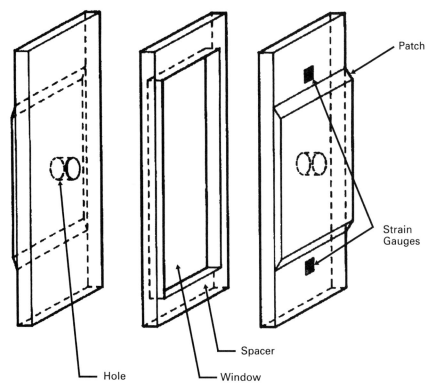

Patch

Strain
Gauges

Spacer

Hole   Window

Figure 10.1 Repair of damaged graphite–epoxy specimen

using cold-setting acrylic adhesive FLEXON 241. This adhesive was chosen for its shear strength and its ease of application. Since the tests were carried out at room temperature and ambient conditions, environmental effects on the adhesive were not an issue.

The results are given in Table 10.1. They show that the residual strength of the damage specimen can be improved significantly using the repair method described by Paul and Jones [1]. The distinct advantages of this simple repair technique are that it can be applied easily and rapidly and that there is no requirement for the existing internal delamination damage to be removed, unlike the scarf repair.

This repair technique was further illustrated using a damaged F/A-18 horizontal stabilator (Jones *et al.* [2]). The work was performed as part of the Composite Repair Engineering Development Program (CREDP) which is a joint program between the Canadian Forces, the Royal Australian Air Force and the United States Navy. The stabilator described in the report was initially classed as unserviceable and irreparable due to two fragment strikes from a tracer rocket.

Table 10.1 Results of static compression tests: single impact case [1]

Specimen number	Impactor diameter (mm)	Absorbed energy (J)	Damaged area (mm²)	Unrepaired (U) Repaired (R)	Failure load (kN)	Failure strain (με)	Predicted strain (με)
1	19.8	7.59	453	U	−191	−4503	—
2	19.8	7.55	453	U	−214	−4680	—
3	19.8	7.96	453	U	−213	−4826	—
4	19.8	5.6	479	R	−238	−4993*	−6164
5	19.8	8.21	428	R	−238	−4993*	−6164
6	19.8	7.84	448	R	−289	−6699	−6164
7	19.8	7.88	458	R	−289	−6699	−6164
8	30	7.46	733	U	−168	−4097	—
9	30	6.93	718	U	−192	−4375	—
10	30	7.88	665	U	−173	−4305	—
11	30	7.54	761	U	−196	−4350	—
12	30	7.74	761	U	−197	−4274	—
13	30	7.67	761	U	−178	−4025	—
14	30	7.66	800	R	−233	−5293	−5594
15	30	7.35	800	R	−233	−5293	−5594
16	30	7.45	739	R	−227	−5554	−5594
17	30	7.64	704	R	−227	−5554	−5594
18	39.7	6.25	1252	U	−178	−4061	—
19	39.7	5.76	1252	U	−204	−4445	—
20	39.7	5.87	1252	U	−181	−4090	—
21	39.7	7.83	1385	R	−238	−5337	−5542
22	39.7	7.58	1212	R	−238	−5337	−5542
23	39.7	7.7	1290	R	−222	−5099	−5542
24	39.7	7.59	1120	R	−333	−5099	−5542

Specimens exhibited extensive bending prior to failure (antibuckling rig was distorted).

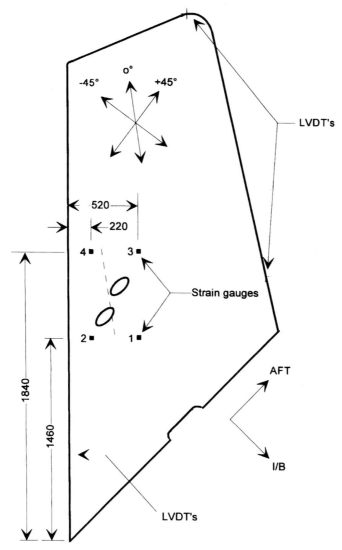

Figure 10.2 Location of impact damage, strain gauges and
displacement transducers on the CREDP F/A-18 horizontal stabilator

The F/A-18 horizontal stabilator comprises an aluminium honeycomb
sandwich structure with graphite/epoxy skins, with fibres oriented in the 0
and ±45 directions, and the cross-section of the stabilator is fully symmet-
ric. In the region of the damage the skin is 29 plies thick (3.68 mm) but
tapers off to 7 plies (0.89 mm) at the leading edge. This stabilator has
incurred extensive local damage to the composite skin and underlying

honeycomb. Figure 10.2 shows the locations of the two damaged areas, which are approximately the same size, and the ply orientation used in the stabilator.

This stabilator was repaired using an external doubler designed to match the stiffness of the stabilator skin. The ply configuration used for the repair patch was $(+45/-45/0_2/90/-45/+45/0)_s$. The loose fibres around the impact zone were removed and the edges of the damage zone were scarfed. Since the honeycomb was also damaged, the entire region was filled with packing adhesive and the repair laminate was subsequently bonded onto the surface using FM73 adhesive. It was reported that the repair successfully withstood both tension and compression loading. When the repaired section was loaded in tension, the majority of the region could be loaded to exceed the design limit bending moment. These results demonstrate that a simple repair design methodology can be used for interim field repairs to damaged composite structures. They also demonstrate that externally bonded doublers can be used to repair impact damage without the need to resort to extensive, resource intensive and costly scarf repairs, even for thick structural components.

10.3   Repair of ship structures with fibre reinforced plastics

10.3.1 Metallic ship structures

Bonded repairs using reinforced plastics as the repair material have also been used to repair metallic ship structures. Allan *et al.* [6] reported on the repair of an aluminium ship structure by adhesively bonding a composite doubler over the fatigue cracked area. The repair method used for this ship structure is similar to that used in the repair of aircraft structures (Jones *et al.* [7]). It was emphasized that adhesive bonding is a preferred method for the repair of aluminium ship structures because welding will affect the fatigue performance of the welded portion of the aluminium structure. The advantages of using materials like carbon fibre or boron fibre composite is that they are corrosion resistant. In the report, Allan *et al.* commented that the structure repaired using carbon fibre reinforced plastic with a simple surface treatment is 3–10 times more durable than one obtained by welding. The same authors also reported that they have been successful in using carbon fibre patches for repairing deck structures to accommodate both an undulating surface and changes in deck height.

The Royal Australian Navy has also conducted an extensive investigation into the use of bonded composite reinforcement for the repair of an undulated aluminium deck of their FFG-7 class frigate (Grabovac *et al.* [8]). This reinforcement, which is 5 m long and 1 m wide, was designed to reduce the cyclic stresses that had caused a cracking problem in the region of a weld between two aluminium alloy 5456 plates 12.75 mm and 6.375 mm thick. A wet lay-up resin application process was preferred for this application. Carbon fibres were chosen because of their excellent mechanical properties, and vinyl ester resin was chosen as the matrix material for the carbon fibre system and the adhesive because: (1) it was available and well-established commercially, (2) it performs very well as a structural adhesive and (3) it has a good storage life. Fatigue testing of representative test articles showed that the doubler system was highly resistant to fatigue damage.

10.3.2 Fibre reinforced boat structures

Glass fibre reinforced plastics have been used to make boats since the mid-1940s. Routine service reports of boats have indicated that fibreglass boats have stood up well during the early trials with the United States Coast Guard [9]. It was also reported that the cost of maintaining these fibre glass boats was approximately one-fifth that of the steel ones. Toghill and Flett [10] found that glass fibre structures in boats were easy to repair, for several reasons. From a structural point of view, one of the more important reasons is that the damaged area is usually fairly localized (as was the case for the F/A 18 stabilator). To this end, the original strength of the damaged laminate can be fully regained by applying a patch with generous overlap over the damaged region. This was aptly shown in the previous section. The repair philosophy outlined in the book by Baker and Jones [11] can also be used for the repair of fibreglass boats.

Willis [9] provided some detailed repair configurations that can be used for repairing damaged fibre glass boat structures both in the critical and non-critical areas. As an example, the suggested method of repairing a non-critical area accessible from both sides presented by Willis is shown in Fig. 10.3. Similar repair procedures are also outlined by Streiffert [12]. This figure shows that the repair schemes are essentially similar to those used in the repair of aircraft structures. In both instances it is important to match the stiffness of the parent structure to the repair patch,

$$E_s t_s = E_r t_r$$

where E_s, t_s, E_r, t_r are the moduli and the thicknesses of the structure and the repair, respectively. Failure to do this may affect the integrity of the repair and the parent structure.

10.4 Repair of concrete structures

The tendency of concrete bridges to corrode has led to considerable interest in the possibilities for using FRP as an alternative or as a repair material,

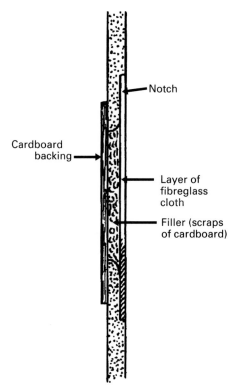

Notch

Cardboard backing

Layer of fibreglass cloth

Filler (scraps of cardboard)

Figure 10.3 Typical repair of fibreglass boat structure

whether the fibres are glass, carbon or aramid. The concrete itself corrodes as a consequence of, for example, chloride attack from salt, exposing the steel reinforcement which then has a limited life because of its vulnerability to atmospheric conditions.

Karbhari [13] reported that the deterioration of transportation related civil infrastructure represents one of the most significant challenges facing the world as we move into the twenty first century. This civil infrastructure includes bridges, motorways and buildings, mainly built with conventional building materials such as steel, concrete and wood. Although the materials have been in use for a long period of time, they are also susceptible to rapid deterioration resulting from the changing environment.

According to Marsh [14] the average life of a bridge in the USA is only 68 years, or as little as 35 years for the bridge decks, and there are always many structures requiring replacement or renovation. He describes several experiments with FRP, some involving building completely new composite bridges, and some concerned with the replacement and strengthening of old structures. In one example, glass/carbon/resin plates were bonded to the soffits of a bridge on a UK flyover, using a polymeric adhesive. The steel tendons inside grouting ducts had previously been deliberately cut to simulate advanced deterioration, and it was confirmed that the new composite strengthening plates were able to restore the previous strength. Obviously, material costs are higher than repairing with steel, but this is offset by savings in labour and equipment costs.

FRP offers many advantages such as the capacity to sheath and protect conventional materials against weathering and atmospheric pollution. Seismic protection is a major incentive in certain parts of the world [15]. But there remain widespread anxieties [14] about the very long term durability of FRP, especially in the strongly alkaline environment (pH 13) of concrete structures. Reductions in the strength of glass fibre reinforced resins in concrete beams have sometimes reached worrying levels in accelerated test programmes, especially under high stresses. The likelihood of corrosion is reduced when carbon rather than glass fibres are used, because carbon fibres are resistant to alkaline attack. Several companies have opted for vinyl ester resins as the matrix because of their strong record as anticorrosion materials. It must be said that experiments described by Sheard [16] and at the time of writing, still being carried out by a consortium of major international companies, suggest that properly selected composites may be able to withstand simulated concrete alkaline environments.

Redston [17] has described the development by ISIS Canada of smart sensing devices, based on optical fibres, which can automatically report changes in stresses and strains, using e-mail to alert highway engineers at a central location, eliminating the need for expensive site visits. Possible causes of changes would be corrosion, or big temperature changes, or exceptional loads. One example of an ISIS-related project was the strengthening of a 27 year old Manitoba bridge using carbon fibre reinforced plastics girders.

The attractive properties of fibre reinforced composite materials that include high strength-to-weight ratio, corrosion resistance and fatigue resistance made these types of materials suitable for use in reinforcing/rehabilitating civil infrastructure. For example, these high strength fibre reinforced plastics can be wrapped with relative ease around a bridge pier and a concrete column to reinforce the existing structure.

Kumar et al. [18] have reported on the fatigue performance of concrete slabs containing reinforced plastic (GRP) rebars. They concluded that cast in situ slabs performed better than precast slabs with respect to fatigue cracking and deflections. The bending response of similar beams has been studied by Faza [19].

Hanna and Jones [20] reported on the use of fibre reinforced plastics to reinforce concrete columns. Here it was demonstrated that by reinforcing a concrete column with a thin layer of glass fibre reinforced plastics, the load carrying capacity of the column can be increased by a factor of 7. It was found that the glass fibre reinforced plastic altered the mechanical behaviour of the concrete column and the combined system showed a significant amount of inelastic behaviour. In a separate experiment, a series of glass fibre reinforced columns were soaked in sea water for 140 days and then tested. It was found that although there was a reduction of approximately 20% in the load carrying capacity of an untreated specimen, the load carrying capacity of these conditioned specimens was still five times that of the unreinforced concrete column, thus showing that the reinforcement is indeed very resistant to the environment. These findings partially explain the current push for bridges in high seismic activity areas like California and Japan to be retrofitted with these types of reinforcement systems. A demonstration test article, the I-10 Santa Monica Viaduct in Los Angeles, was successfully retrofitted in this fashion with carbon fibre reinforced plastics [13].

Meier *et al.* [21] presented several examples of the use of carbon fibre reinforced plastics for reinforcing bridges and buildings. One of the examples given was the repair of the Ibach bridge. This bridge is located in the County of Lucerne and was completed in 1969. During core borings for the mountings of the traffic lights, the prestressing tendon in the outer web of the bridge was damaged. This resulted in the suspension of heavy convoys crossing the bridge until it was subsequently repaired with three carbon fibre reinforced plastic sheets (two with dimensions $150 \times 5000 \times 1.75$ mm and one with dimensions $150 \times 5000 \times 2.00$ mm). Subsequent loading tests proved that this simple rehabilitation work was conducted very satisfactorily.

Another example given by Meier *et al.* [21] was related to the reinforcement of the City Hall of Gossau. Here a hole had to be cut in a concrete slab to accommodate an elevator. It was reported that the edges of the hole were strengthened with carbon fibre reinforced plastics sheets instead of thick steel plates. This helps retain the aesthetics of the building and after painting, the presence of carbon fibre reinforced plastic sheets was completely disguised.

The Japanese are also active in the application of fibre reinforced plastics to reinforced concrete structures to improve their earthquake resistance. Kabatake *et al.* [22] reported on 14 cases in which chimneys were retrofitted with fibre reinforced plastics and were judged by the Building Disaster Prevention Association of Japan to be one of the most effective retrofitting methods. In this retrofitting programme, carbon fibre tapes impregnated with an epoxy resin were wound and adhered to the exterior of the chimney. Sumida *et al.* [23] have described experiences with aramid reinforced rods used as prestressing tendons in a floating bridge and in a concrete housing development in Japan.

The climatic environment in the Gulf States has been reported to have accentuated the deterioration of concrete buildings. Al-Gahtani *et al.* [24] conducted a series of experiments to determine the applicability of FRP for repairing these concrete structures in the Gulf States. Their studies showed that severe temperature fluctuations can reduce the strength of the resinous materials used in such repairs. This is despite the fact that resin-based repairs are widely publicized and frequently used in the Gulf region. This should serve as a reminder that repair with FRP materials should only be used after careful consideration has been given to the prevailing environmental conditions and their effects on the materials used in the repair.

10.5 Conclusions

Fibre reinforced materials are indeed very versatile. In addition to their use in primary load carrying structures, they have been used as repair materials for structural components made from composites and metals. This chapter highlights several examples which demonstrate the relative ease with which repair to structural components can be achieved with fibre reinforced plastics. It is also pointed out that, as with other materials, the designer of any repair scheme with reinforced plastics must be aware of the effects of the environment on the performance of the materials used. Although fibre reinforced plastics are used in corrosive environments, studies in the Gulf States have shown that severe temperature fluctuations can affect the performance of fibre reinforced plastics in a repair environment.

References

1 J Paul and R Jones, *Analysis and Repair of Impact Damaged Composites*, Defence Science and Technology Organisation, ARL-STRUC-R-435, Australia 1989.
2 R Jones, S C Galea and J Paul, *Assessment of Impact Damage in Composite Structures*, Defence Science and Technology Organisation, ARL-TR-23, Australia, 1993.
3 N R Adsit and J P Waszczak, 'Effect of near-visual damage on the properties of graphite/epoxy', *Composite Materials: Testing and Design*, ASTM STP 674, 1979, pp 101–117.
4 J H Starnes, M D Rhodes and J G Williams, 'Effect of impact damage and holes on compressive strength of a graphite/epoxy laminate', *Nondestructive Evaluation and Flaw Criticality for Composite Materials*, ASTM STP 696, 1979, pp 145–171.
5 P D Chalkley and R J Chester, '*Interim Report on Environmental Program – Durability of Graphite/epoxy Honeycomb Specimens with Representative Damage*

and Repairs' Defence Science and Technology Organisation, ARL-MAT-TM-397, Australia, 1998.

6 R C Allan, J Bird and J D Clarke, 'Use of adhesives in repair of cracks in ship structures', *Mater Sci Technol* 1988 **4** 853–859.

7 R Jones, L Molent, J Paul, T Saunders and W K Chiu, 'Development of a composite repair and the associated inspection intervals for the F111C stiffener runout region', *FAA/NASA International Symposium on Advanced Structural Integrity Methods for Airframes Durability and Damage Tolerance*, May 1994, pp 339–350.

8 I Grabovac, R A Bartholomeusz and A A Baker, 'Composite reinforcement of a ship superstructure-project overview' *Composites* 1993 **24**(6) 501–509.

9 M D C Willis, *Boatbuilding and Repairing with Fibreglass*, International Marine Publishing, Camden, ME, USA, 1972.

10 J Toghill and J Flett, *'Building Small Boats, Surf Craft and Canoes in Fibreglass'*, Reed Books, Terry Hills, NSW, Australia, 1990.

11 A A Baker and R Jones, *'Bonded Repair of Aircraft Structure'*, Martinus Nijhoff, The Hague, 1988.

12 B Streiffert, *'Glassfibre Boat Manual – Practical Repairs, Maintenance and Improvements'* London, MacDonald Queen Anne Press, 1989.

13 V M Karbhari, 'Application of composite materials to the renewal of twenty-first century infrastructure' *11th International Conference on Composite Materials, (ICCM-11)*, Gold Coast, Australia. Cambridge, UK, Woodhead, 1997.

14 G Marsh, 'Durability – a key issue for composites in infrastructure', *Reinforced Plastics* 1997 **41**(7) 26–31.

15 K Kageyama, I Kimpara and K Esaki, 'Fracture mechanics study on rehabilitation of damaged infrastructures by using composite wraps', *Proceedings ICCM-10*, Whistler, BC, Canada. Cambridge, UK, Woodhead, 1995, Volume 3, pp 597–604.

16 P A Sheard, 'Eurocrete uses composites to build durable structures', *Reinforced Plastics* 1997 **41**(7) 36–40.

17 J Redston, 'Smart technology for innovative infrastructure', *Reinforced Plastics* 1997 **41**(7) 32–34.

18 S V Kumar, S S Faza, H V S GangaRao and M Al Megdad, 'Fatigue performance of concrete slabs with glass fiber reinforced plastic (RP) rebars', *Proceedings 50th Annual Conference, Composites Institute*, SPI, New York, 1995, Paper 21-C.

19 S S Faza, *Bending Response of Beams Reinforced with Fiber Reinforced Plastic Rebars*, PhD Dissertation, West Virginia University, Morgantown, WV, USA, 1991.

20 S Hanna and R Jones, 'Composite wraps for aging infra-structure: concrete columns', *J Theor Appl Fracture Mech, Composite Structures*, 1997, **38**(1–4) 57–65.

21 U Meier, M Deuring, H Meier and G Schwegler, 'CFRP bonded sheets', *Fiber-reinforced Plastics Reinforcement for Concrete Structures: Properties and Applications,* Amsterdam, Elsevier Science, 1993.

22 Y Kabatake, K Kimura and H Katsumata 'A retrofitting method for reinforced concrete structures using carbon fibre', *Fiber-reinforced Plastics Reinforcement for Concrete Structures: Properties and Applications*, Amsterdam, Elsevier Science, 1993.

23 A Sumida, T Okamoto and M Tanigaki, 'Experiences with aramid FRP in concrete structures', *Proceedings 50th Annual Conference, Composites Institute*, SPI, New York, 1995, Paper 21-D.

24 A S Al-Gahtani Rasheeduzzafar and A A Al-Mussallam, 'Performance of repair materials exposed to fluctuation of temperature', *J Mater Civil Eng* 1995 **7**(1) 9–18.

Fatigue performance: the role of the interphase

NIKHIL E VERGHESE AND JOHN J LESKO

11.1 Introduction

The previous chapters have mostly surveyed broad aspects of reinforced plastics durability and have highlighted general principles. This chapter is rather different, being focussed more narrowly on recent research into a specific topic. It seeks to demonstrate that durability (or more precisely, fatigue life) can be affected, not just by the choice of resin or fibers, but by the nature of the boundary region between the two.

It should already be clear from earlier chapters that the generalized microscale details of continuous fiber composite materials exert a profound influence on a laminate's performance. For instance, simply altering the matrix toughness can significantly affect damage initiation and development during static and fatigue loading. The microstructure of the tough matrix maximizes the work of fracture and dissipates the local energy so as to evade premature failure. Enhancing fiber properties by extreme orientation and alteration of the graphitic crystallite size leads to increased tensile strength and stiffness. However, these gains in tensile performance are offset by significant reductions in compression strength and, once again, careful attention to the microstructure is required in order to control properties.

Interest has grown in trying to understand a region/constituent that can have exceptional influence on composite performance, despite making up less than 1% of the composite. This small but important region is called the interphase [1,2]. Sharpe here defines the interphase as a three-dimensional region which surrounds the fiber and extends into the adjacent matrix. This region possesses its own material properties and microstructure, distinct from those representative of either the fiber, matrix or couplant (if present). Its unique properties are influenced by the microstructure formed when the fiber and matrix are joined during manufacture.

It is the properties of this interphase that affect the mechanics of fiber–matrix interaction. In turn, the global performance of the composite material/structure is directly related to the interaction of fiber and matrix. Thus, it is important to develop an understanding of how to join fiber and matrix so as to produce advantageous interactions which benefit composite performance. More fundamentally, understanding how particular interphase properties influence fiber–matrix interaction, and thus composite performance, is of great interest as well. These ideas require some quantitative knowledge about the dimensions and properties of this transition zone. This is a more difficult problem because of the small scale of the interphase region (approximately 200–500 nm in thickness).

Here the interface is defined as a distinct two-dimensional boundary which defines, for instance, where a fiber ends and the binder begins. Most, but not all, investigators of composite performance believe that an interphase region and not just an interface is developed and exists in fiber composites. One needs only to look at a few of the most noted journals on composites to find extensive discussion of the interface and interphase. Yet, after approximately 20 years what is really understood about the interphase and its effects on composite performance? The authors claim that our knowledge is, at best, cursory! The authors believe that this is due to a lack of fundamental knowledge about how fiber and matrix are joined.

In addition, the thermomechanical properties of this region are not known, nor are they easily measured. The following questions still remain to be answered:

- How is an interphase formed when joining a solid surface possessing surface treatments and sizings with a matrix material? (Chemistry/ Materials)
- What type of microstructures are created at the interphase region? (Materials/Chemistry)
- What are the elastic and inelastic properties (E_{ij}, X^t_{ij}, X^c_{ij}, e^t_{ij}, e^c_{ij}, G_{ic}), where i and j = 1, 2, 3, within this interphase region, and how do they vary with radial distance from the fiber? Also, are they isotropic and uniform along the length of the fiber? How are the interphase properties related to the microstructure? (Materials/Mechanics/Chemistry)
- How do the elastic and inelastic properties and their distribution affect the interphase failure mechanisms/modes? What are the typical failure modes? How do we define interphase strength (the ultimate property which we use to define at what general loading state the fiber and matrix will separate parallel and perpendicular to the fiber direction)? (Mechanics/Materials)
- Finally, how do the mechanical properties and typical failure modes of a particular interphase microstructure influence stiffness, strength and durability under various conditions? (Mechanics)

Although the above mentioned points will improve our understanding of the interphase, there is already considerable experimental evidence in the literature [3–12] to support the idea that proper arrangement of material and microstructure within an interphase zone will, through a unique synergism, result in enhanced composite properties.

Attention is now given to the fatigue performance of these systems as influenced by the local property changes/gradients. As these materials will eventually make their way into structures that experience time-varying loads, questions must also be addressed about how a material's durability is affected by the character of the interphase.

Subramanian [9] investigated the effects of interphases on the tensile static and dynamic fatigue performance of $[0/90_3]_s$ cross-ply laminates. Subramanian observed differences in performance of the organic sized polyvinylpyrrolidone (PVP) fiber laminates compared with those composites where the fibers were sized with an unreacted epoxy oligomer. In tension ($R = 0.1$) fatigue of cross-ply laminates, the PVP interphase materials performed better in low cycle fatigue, while the epoxy interphase material showed greater durability in high cycle fatigue. Subramanian concluded that the observed phenomena were the result of one fundamental characteristic: the efficiency of stress transfer between the fiber and matrix as influenced by the interphase. By including the efficiency parameter, h, in the analysis of tensile strength and likewise in the analysis of tensile fatigue, Subramanian was able to predict the tensile fatigue performance with good agreement.

11.2 Recent work

This chapter will now outline two recent pieces of work that have been performed by the authors in the area of fatigue performance of composites with modified interphases.

11.2.1 Compression fatigue ($R = 10$) on 3 and 8 series composites

The study considered here employed materials obtained from McDonnell Aircraft Co., through Dexter Hysol, under the Air Force program 'Development of Ultra Lightweight Materials'. They contained the Courtaulds Research Apollo 45-850, consisting of $5\,\mu m$ diameter carbon fibers in Dexter Hysol's HC 9106-3 di-, tri- and tetrafunctional epoxy resin, toughened by addition of the thermoplastic polyether sulfone (PES). The thermoplastic PES toughening agent, present in this epoxy matrix, is presumed to phase separate out once it has cured. These panels were fabricated in an autoclave by Dexter Hysol.

For convenience, the two systems of interest are named 810-A, and 810-O taken from the nomenclature of Swain [7]. The leading number 8 represents the matrix system used (HC 9106-3) and the materials in this category are referred to as the 8 series composites. The trailing 10 denotes the level of surface treatment, 100%. The suffix A or O designates the type of sizing. Details of the oxidative surface-treatment levels are not known, but they are related to a standardized degree of application, designated 100% and deemed optimum by the manufacturer. The A sizing was an unreacted bisphenol-A based epoxy and the O sizing was the linear and amorphous thermoplastic PVP [10]. Again, these were the sizings, incorporated in Swain's and Subramanian's work, in which the A sizing is the 'standard epoxy sizing' and the O sizing is the 'organic size'. These sizings were reported to be present at approximately 0.7 wt% of the fiber.

The microstructures of the 810-A and 810-O interphases were revealed by an etching technique [10]. Under identical etching conditions the epoxy sized fibers did not reveal as distinct an interphase region as that observed in the PVP sized composite. The epoxy sized composite showed an irregular interface region between fiber and matrix. Particular regions around the epoxy sized fiber showed irregular gaps between fiber and matrix. On other portions of the A sized fiber perimeter, matrix material was observed in direct apposition to the fiber. In direct contrast, the PVP interphase extended into the matrix approximately 10% of the fiber diameter, or ~0.5 μm. Although this interphase was readily observed around many of the fibers, it was not always uniform in thickness for each fiber. In some cases, the interphase region was somewhat smaller or not present at all. The formation of a varied morphology in the PVP interphase might be explained through the simple process of diffusion. The maximum processing temperature of these composites was 177°C and, as previously stated, the T_g of the PVP size was approximately 109°C. Hence, the PVP was well into its rubbery regime and potentially able to disperse/dissolve into the epoxy [13]. This is not to neglect any other driving forces producing interaction of the two polymers. It has been established that the PVP is generally miscible with the matrix [14].

The 3 series composites, as they are now named, contained essentially the same Courtaulds fibers noted above except in Hercules' untoughened aerospace epoxy 3501-6 matrix. The 3501-6 epoxy possessed a tensile modulus of 0.64 Msi, (4.4 GPa) tensile strength of 10 ksi (69 MPa) and an elongation at break of 1.7%. The G_{Ic} of the neat material was 130 J m^{-2} and it showed a T_g of 185°C. The nomenclature for the two systems is similar to that of the 8 series composites where the 8 is replaced by the 3, now representing the 3501-6 epoxy matrix. The quasistatic properties of the 8 and 3 series composites are summarized in Table 11.1.

Comparing the strength and toughness of the A and O interphases for the

Table 11.1 Summary of the quasistatic properties of the 8 and 3 series composites

	Toughened matrix HC 9106-3		Untoughened matrix 3501-6	
	PVP Size 810-O	Epoxy Size 810-A	PVP Size 310-O	Epoxy Size 310-A
Unidirectional tension				
Strength, ksi (MPa)	444 (3061)	407 (2806)	416 (2868)	410 (2827)
Failure strain (%)	1.66	1.31	1.57	1.37
Unidirectional compression				
Strength, ksi (MPa)	160 (1103)	106 (731)	180 (1241)	186 (1282)
Failure strain (%)	0.74	0.45	0.93	0.79
Transverse flexure				
Strength, ksi (MPa)	18.3 (126)	11.9 (82)	15.5 (107)	10.7 (74)
Failure strain (%)	1.34	0.91	0.91	0.97
Iosipescu shear				
Strength, ksi (MPa)	10.6 (73)	8.7 (60)	6.5 (45)	5.3 (37)
Failure strain (%)	1.15	0.81	0.52	0.35
Interlaminar toughness				
Mode I (J m^{-2})	434	303	472	247
Mode II (J m^{-2})	666	527	464	402
$(0/90)_{8s}$ Laminate				
Notched compression strength, ksi (MPa)	42.8 (295)	42.3 (292)	45.5 (314)	45.3 (312)
Unnotched compression strength, ksi (MPa)	72.6 (501)	71.8 (495)	87.6 (604)	87.1 (601)

two different epoxy matrix systems, it is clear that the PVP interphase produced a 'better' material. These data will complement the conclusions developed in the fatigue durability study following.

As with the work of Swain [7] and Subramanian [9], the *S–N* curve for the 8 series shows substantial differences between the performance of the A and O fiber sized composites. Reviewing the *S–N* curve for the 8 series compression–compression, $R = 10$ fatigue, a best fit line through each of the data sets assists in delineation of the difference in performance of the A and O interphases, as shown in Fig. 11.1. The applied cyclic stress levels are presented as a percentage of their ultimate notched compression strength, as reported in Table 11.1. The 810-A composite exhibited a slightly lower rate of life reduction with increasing load level. However, all lifetimes

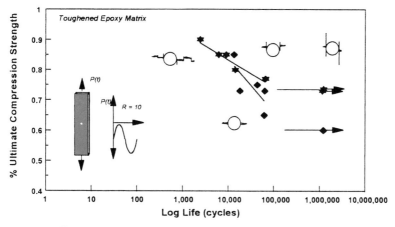

Figure 11.1 Applied gross compression stress (as a percentage of the ultimate notched $[0/90]_{8s}$ compression strength) versus the cycles (log of *n*), for the 810-A and ○ materials under *R* = 10 fatigue. ◆, epoxy interphase; ✷, PVP interphase

recorded for 810-A, with the exception of that at 85% of the ultimate compression strength (UCS) of the notched laminate, were less than those measured for the PVP sizing. The 810-O composite showed relatively good agreement to the fit revealing a greater sensitivity to load level. At lower load levels it appeared to perform better than its epoxy sized counterpart. A fatigue limit, at approximately 73% UCS, was recorded for the 810-O. By the limited data set of the 810-A, a best possible fatigue limit was potentially expected at a load level no greater than at 65% UCS and positively identified at 60% UCS. Thus, a considerable difference in fatigue limit between the 810-A and 810-O seems to emerge.

As with the results of the 8 series composites, substantial differences in fatigue lifetime are shown in the *S–N* curve of the 3 series composites between the PVP interphase materials and the epoxy interphase materials (Fig. 11.2). An assumed linear fit to the low cycle fatigue data, from specimens which were not interrupted, appears to indicate that the PVP interphase provided an order of magnitude improvement in life. The significance of this difference between the linear fits to this data are proven to a 90% level of confidence. This difference was still maintained in the light of the similarity in the 310-A and 310-O unidirectional compression strengths as well as notched compression strengths. Thus, the inherent compression strength (axial performance) might be considered in conjunction with the off-axis characteristics of the material to understand the difference in fatigue performance between the A and the O interphase composites. Recall that essentially all of the inelastic off-axis characteristics of the materials

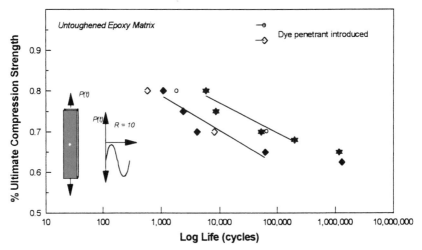

Figure 11.2 Applied gross compression stress (as a percentage of the ultimate notched [0/90]$_{8s}$ compression strength) versus the cycles (log of *n*), for the 310-A and O materials under *R* = 10 fatigue. ◆, epoxy interphase; ✱, PVP interphase

with the O interphase were better than those of the materials with the A interphase for both the 8 and 3 series composites (Table 11.1). The data might also be viewed as remarkable in light of the fact that the quasistatic compression strength of these two laminates is nearly identical.

An apparent fatigue limit was recorded for the 310-O at 65% of UCS, which is below that measured for the 810-O (at 74% UCS). This was not entirely unexpected, because of the difference in the matrix toughness between the 8 and 3 series composites. Once again, the transition from low cycle fatigue to high cycle fatigue characteristics appears to take place over a small window of load level as was observed for the 8 series. At 68% of UCS the 310-O failed at 10^5 cycles. Radiographs of the 310-O at the 65% load level revealed little evidence of damage which might be considered detrimental to the life of the laminate (this is discussed in the following section). Considering the 310-A at the 65% of UCS load level, there is at the very least, an order of magnitude difference in lifetime.

Cycling the 310-A composite at 62.5% of UCS did not reveal a fatigue limit. Thus, it is possible that there is a difference in fatigue limit between the two interphases, as was noted in the 8 series composites. According to Kortschot and Beaumont [15], the strength of a notched composite is controlled either by the ability of the laminate to relieve stress concentration by the growth of longitudinal splits (LS) or by the intrinsic strength of the material. An investigation was made of the growth rate of the longitudinal 0° ply splits which extended at a tangent to the hole's edge. This was

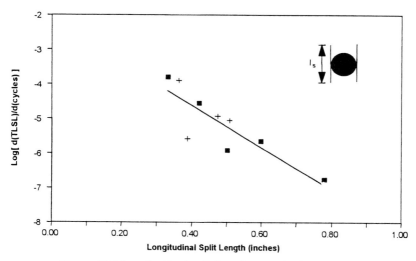

Figure 11.3 Longitudinal split length growth characteristics for the 3 series materials from X-ray radiography data. ■, 310-O; +, 310-A

accomplished by interrupting fatigue tests at particular intervals and assessing the split length by X-ray radiography. Compiling this data for the different interphases in the 3 series composites, it appears that there was no effect of the interphase on the LS growth rate, as demonstrated in Fig. 11.3. Thus, the second feature which could control the strength of a notched laminate must now be examined: the intrinsic strength of the material. It is evident from the review of quasistatic properties that the compression strengths of the unidirectional 310-A and 310-O composite materials were nearly identical.

This was also true for the notched [0/90]$_{8s}$ laminates (see Table 11.1). However, we also recognize that the failure strain and the interlaminar toughness of the PVP interphase materials were consistently higher for almost all modes of deformation. Thus, the authors suggest that it is not necessarily the inherent strength of the composites which controls the laminate strength, but the intrinsic toughness of the PVP interphase material which enhances the life of the 810-O and 310-O notched laminates. This inference is supported when considering the dominant failure mode, the growth of ply buckling from the hole's edge perpendicular to the loading axis, which appears to control the residual strength and life of these laminates. Surface ply buckling (SPB) was recorded by video and revealed distinguishing attributes for each of the two interphases. For example, it was observed at a load level of 70% of the ultimate compression strength that the 310-O material initiates SPB at approximately 10 000 cycles, compared with 3000 cycles for the 310-A material as shown in Fig. 11.4.

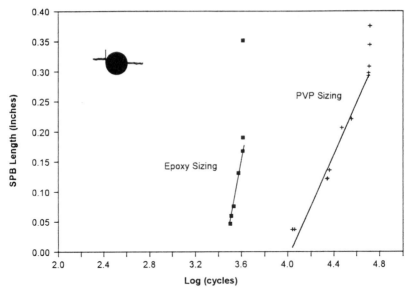

Figure 11.4 Surface ply buckling growth characteristics for the 3 series composites fatigued at 70% of UCS

The PVP interphase material possessed a particular advantage over the conventional epoxy sized material in the reduced rate of SPB growth. In addition, the data also suggested that the critical length at which unstable growth of ply buckling occurred was greater for the PVP interphase than for the conventional epoxy interphase. This may suggest that the 810-O and 310-O materials possessed a greater apparent critical K (stress intensity factor).

In summary, the authors conclude that it is not the LS splitting which controls the difference in lives between the A and O interphase materials, but the inherent characteristics of the material which dominate their fatigue response. In addition the authors are of the opinion that it is not merely the inherent strength of the material that influenced the durability of these varied interphase systems but, more exactly, the toughness or damage tolerance of the material.

11.2.2 Fully reversed ($R = -1$) fatigue on composites: effect of two controlled thermoplastic sizings

More recently, tests have been performed on composites made from carbon fiber (Hercules AS4) with two different sizings, namely polyvinyl pyrrolidone (PVP K-17) and polyhydroxyether (phenoxy) with a vinyl ester (Derakane 441-400) matrix. The phenoxy polyhydroxyether (PKHW 35)

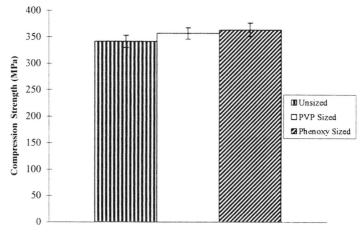

Figure 11.5 Quasistatic notched compressive strength for various sizing materials

sizing material was obtained from Phenoxy Associates, Rock Hill, SC. Data obtained from the company indicates that the polymer had a tensile ultimate strength of 55.16 MPa and a strain to failure of 40–100%. This polymer was therefore a tough polymer. This material was obtained as a 35 wt% dispersion of approximately 1 μm diameter particles in water. The number average molar mass, $\overline{M}_n$ of the phenoxy was 19 000 g mol^{-1} (by gas phase chromatography) and it had a T_g of 97°C (by differential scanning calorimetry). The Luviskol K-17 polyvinylpyrrolidone sizing material was obtained from the BASF Corporation. Although the properties of K-17 are not known, K-90, a higher molecular weight version, had an ultimate tensile strength of 62.06 MPa and a strain to failure of only 0.9%. This indicates that it was an extremely brittle polymer.

In addition to unidirectional panels, four sets of $[0/90]_{7s}$ laminates were prepared using resin film infusion (RFI) molding. In spite of a statistically indistinguishable difference in notched quasistatic compression strength (Fig. 11.5), substantial differences in the fatigue limit were noted between the polyhydroxyether sized material, the polyvinylpyrrolidone sized material and the unsized materials.

As seen in Fig. 11.6, the fatigue limit for the polyhydroxyether sized material was about 210 MPa, that is a 60% improvement in the fatigue limit compared with that for the unsized material, 130 MPa. The fatigue limit for the polyvinylpyrrolidone sized material was about 160 MPa, about 20% higher than the limit for the unsized material. At a given stress level, the type of sizing material had a significant effect on the lifetime of the composites. For example, at a 207 MPa stress level, the composite sized with

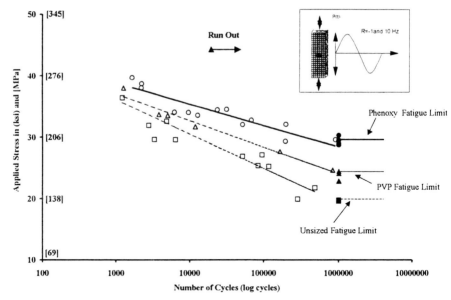

Figure 11.6 Fatigue limit *S–N* curve for various sizing material. The open symbols indicate specimens that had a finite life, whereas the corresponding dark symbols are for samples that exhibited run-out. ○, phenoxy sizing; △, PVP (K-17) sizing; □, unsized

polyhydroxyether lasted for 200 000 loading cycles, the composite sized with polyvinylpyrrolidone lasted for 50 000 loading cycles and the unsized material lasted for only 9000 loading cycles.

11.3 Conclusions and remarks

This discussion demonstrates the influence the interphase and its microstructure has on the compression fatigue performance of continuous fiber composites. However, much work remains to be done in attempting to develop the fundamental description of how specific interphase properties and characteristics control the interaction of the fiber and matrix to affect the mechanics of composite strength and life. In order to understand the interaction mechanics of the fiber, matrix and the interphase, an understanding of the size of the interphase region is critical. Estimates of both the morphology and the mechanical properties are also important across this region. Both the elastic, and more importantly the inelastic, properties across the interphase are needed, in order to describe both damage initiation and propagation in composites with different interphases.

Recently, extensive work has been on going in the area of atomic force microscopy (AFM) where tests such as nanoindentation are being used to characterize the hardness across the interphase region. Much work [16] is also ongoing in the area of micro-Raman spectroscopy to measure strains in single fiber model composites.

Information from these small scale tests will be extremely important in understanding the exact nature of the interactions and in helping to understand the role of the interphase in polymer matrix composites better.

Acknowledgments

The authors would like to thank the Office of Naval Research (ONR) for their financial support (grant number N00014-95-1-0340). We would also like to thank the members of the Designed Interphase Group (DIG) for their support, in particular Dr J S Riffle in Chemistry, Dr R Davis and Norman Broyles in Chemical Engineering and Dr S Davis who is now at Kodak.

References

1 L H Sharpe, 'The interphase in adhesion', *J Adhesion* 1972 **4**(1) 51–64.
2 L H Sharpe, 'Some thoughts about the mechanical response of composites', *J Adhesion* 1974 **6**(1) 15–21.
3 M S Madhukar and L T Drzal, 'Effect of fiber-matrix adhesion on the longitudinal compressive properties of graphite/epoxy composites', *Proceedings Fifth Technical Conference American Society for Composites*, 1990, pp 849–858.
4 M S Madhukar and L T Drzal, 'Fiber-matrix adhesion and its effect on composite mechanical properties. I. In-plane and interlaminar shear behavior of graphite/epoxy composites', *J Compos Mater* 1991 **25** 932–957.
5 M S Madhukar and L T Drzal, 'Fiber-matrix adhesion and its effects on composite material properties. II. Tensile and flexural behavior of graphite/epoxy composites', *J Compos Mater* 1991 **25** 958–991.
6 M S Madhukar and L T Drzal, 'Fiber-matrix adhesion and its effects on composite mechanical properties: IV. Mode I and mode II fracture toughness of graphite/epoxy composites', *J Compos Mater* 1992 **26**(7) 936–968.
7 R E Swain III, '*The role of the fiber/matrix interphase in the static and fatigue behavior of polymeric composite laminates*', Dissertation, Department of Engineering Science and Mechanics, Virginia Polytechnic Institute & State University, February, 1992.
8 Y S Chang, J J Lesko, S W Case, D A Dillard and K L Reifsnider, 'Mechanical properties of thermoplastic composites: the interphase effect', *Proceedings Seventh Technical Conference American Society for Composites*, 1992, pp 871–826.
9 S Subramanian, '*Effect of fiber/matrix interphase on the long term behavior of cross-ply laminates*', Dissertation, Department of Engineering Science and Mechanics, Virginia Polytechnic Institute & State University, January, 1994.
10 J J Lesko, R E Swain, J M Cartwright, J W Chin, K L Reifsnider, D A Dillard and J P Wightman, 'Interphase developed from sizings and their chemical-

structural relationship to composite performance', *J. Adhesion* 1994 **45** 43–57.

11 J J Lesko, J S Elmore, S W Case, R E Swain, K L Reifsnider and D A Dillard, 'A global and local investigation of compressive strength to determine the influence of the fiber/matrix interphase', *Compression Response of Composite Structures*, ASTM STP 1185, 228–240, Philadelphia, American Society for Testing and Materials, 1994.

12 J J Lesko, S Subramanian, D A Dillard and K L Reifsnider, 'The influence of fiber size developed interphases on interlaminar fracture toughness', *Proceedings of 39th International Symposium*, Anaheim, CA; Covina, CA SAMPE, April 1994.

13 J J Lesko, '*Effect of fiber/matrix interphase on the long term behavior of cross-ply laminates*', Dissertation, Department of Engineering Science and Mechanics, Virginia Polytechnic Institute & State University, January, 1994.

14 H Oyama, J J Lesko and J P Wightman, 'Inter-diffusion at the interface between poly(vinylpyrrolidone) and epoxy', *J Polym Sci* 1997 **35** 331–346.

15 M T Kortschot and P W R Beaumont, 'Damage mechanics of composite materials: II – A damage-based notched strength model', *Compos Sci Technol* 1990 **39** 303–326.

16 L S Schadler, N Melanitis, J Figueroa, C Laird and C Galiotis, *J Mater Sci* 1992 **19** 3640–3648.

12

Computer models for predicting durability

SAMIT ROY

12.1 Introduction

A comprehensive analytical model for predicting the long term durability of polymers and fiber reinforced plastics (FRP) should, in general, take into account at least three kinds of change in the materials. First there is the characteristic time-dependent mechanical behaviour of polymers, mentioned briefly in Chapter 1 (involving viscoelastic/viscoplastic creep); second is the combined influence of moisture and temperature, that is hygrothermal effects, mentioned in Chapter 3, and third are the effects of physical and chemical aging (physical aging is discussed in Chapter 4). These effects, in turn, are influenced by a multitude of factors such as polymer morphology, service temperature, ambient relative humidity, internal moisture concentrations, air pressure, fiber volume fraction, fiber architecture, applied stress level, degree of damage and aging time.

The purpose of this chapter is to describe an analytical tool which, when coupled with accelerated material characterization, is capable of predicting the long term durability of reinforced plastics for applications in hostile environments. Throughout the chapter, the words 'resin' and 'polymer' are used interchangeably, and the resin can be thermoplastic or thermosetting, but it must be remembered that thermosetting resins are much less prone to creep and other viscoelastic effects than are thermoplastics.

The analytical tool consists of a specialized test-bed finite element code, NOVA-3D, that can be used for the solution of complex stress analysis problems including interactions between non-linear material constitutive behavior and environmental effects [1].

Some of the key durability modeling issues are introduced in Sections 12.2 to 12.4 of this chapter. The modeling approaches are described in detail in Section 12.5. Model verification examples and applications are presented in Section 12.6.

12.2   Viscoelastic creep in a polymer

Viscoelastic creep manifests itself in the time-dependent deformation of a material. Experimental data obtained from a laboratory creep test under constant applied stress for a viscoelastic solid is shown in Fig. 12.1. Traditionally, a creep curve consists of three stages. In the first stage, also known as primary creep, the creep strain rate decreases with time until it reaches a constant value. The second stage, known as steady state creep, is defined as the region where the slope of the creep strain is a constant with respect to time. In the third and final stage, termed tertiary creep, the creep strain rate increases with time through progressive failure and terminates with the rupture of the specimen.

It is now well established that the thermomechanical response of glassy polymers and their composites is viscoelastic at temperatures near to, and above the glass transition temperature. Therefore, an accurate long term durability model at elevated temperatures for resins and reinforced plastics (FRP) must necessarily include viscoelastic behavior. This is especially true

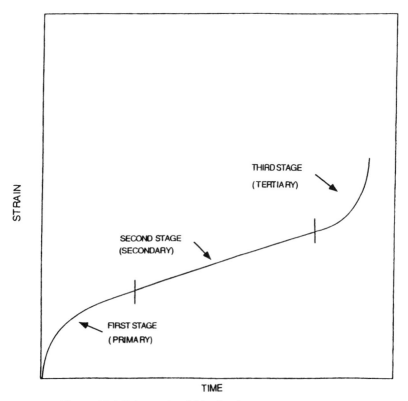

Figure 12.1 Schematic of idealized creep curve

for matrix-dominated properties which, in addition to being time dependent, are also very sensitive to environmental conditions such as temperature and moisture. Clearly, accelerated material characterization techniques such as the time–temperature superposition principle are needed in order to obtain the short term material characterization data that are required as input for the long term predictive models.

At first sight, it may seem questionable whether service temperatures ever exceed the matrix glass transition temperature, T_g, since it is not normally recommended to use resins within about 20°C of their T_g. Absorbed moisture however can drastically lower the T_g of many resins. A strong direct correlation exists between the absorbed moisture level and the reduction in the T_g of the matrix, both for thermosetting resins and for thermoplastics. The topic is discussed elsewhere in this book. As a result, it is possible that the service temperature of the composite might approach, or even exceed the T_g of the wet matrix, especially in the case of polar thermosetting resins such as epoxy and polyester resins, as well as polyamide thermoplastics. All of these materials can absorb more than 3% moisture, sometimes more. The resulting plasticization of the matrix due to moisture absorption lowers the yield strength, increases the fracture toughness and increases the magnitude of the viscoelastic response. Postcuring for the purpose of raising the dry T_g of a thermosetting epoxy resin will also tend to increase the equilibrium moisture absorption level, so the final outcome is only a minimal increase in the T_g of the material after reaching equilibrium moisture absorption.

12.3   Diffusion in a polymer

Most polymers and FRPs absorb moisture and are therefore hygroscopic in nature. In general, the moisture absorption process in a polymer has been modeled using Fick's law. Fick's law states that the rate of transfer of a diffusing substance through unit area of a section is proportional to the concentration gradient measured normal to the section, that is,

$$F = -D\frac{\partial C}{\partial x} \tag{12.1}$$

where F is the moisture flux of the diffusing substance per unit area of the solid, C is the concentration of the diffusing substance, x is the spatial coordinate measured normal to the section and D is the diffusion coefficient, also referred to as the diffusivity of the solvent, in the given solid. It should be noted that the form of the above equation is valid for one-dimensional diffusion in an isotropic medium, although Fick's law can easily be extended to enable it to model three-dimensional diffusion in anisotropic media.

The fundamental governing equation of one-dimensional diffusion in an isotropic medium can be derived as follows. Conservation of mass of diffusing substance within a volume element of the isotropic medium yields:

$$\frac{\partial C}{\partial t} = -\frac{\partial F}{\partial x}$$

[12.2]

Combining equations [12.1] and [12.2] yields the familiar governing equation for one-dimensional isotropic Fickian diffusion, also known as Fick's Law, and given by:

$$\frac{\partial C}{\partial t} = D\frac{\partial^2 C}{\partial x^2}$$

[12.3]

subject to the initial condition $C(x,0) = C_i$, and boundary conditions $C(x,t) = C_m$ when $x = a$, and $x = -a$. In deriving equation [12.3], it was assumed that the diffusivity, D, is a constant with respect to time, and that it is also independent of the spatial coordinate. The solution to equation [12.3] is given by:

$$\frac{C - C_i}{C_m - C_i} = 1 - \frac{4}{\pi}\sum_{j=0}^{\infty}\frac{1}{(2j + 1)}\sin\frac{(2j + 1)\pi x}{h}\exp\left[\frac{-(2j + 1)^2\pi^2 Dt}{h^2}\right]$$

[12.4]

The total mass of moisture in the polymer or resin, m, is obtained by integrating equation [12.4] over the specimen thickness, h, giving:

$$G = \frac{m - m_i}{m_m - m_i} = 1 - \frac{8}{\pi^2}\sum_{j=0}^{\infty}\frac{1}{(2j + 1)^2}\exp\left[\frac{-(2j + 1)^2\pi^2 Dt}{h^2}\right]$$

[12.5]

A typical Fickian curve, consisting of a plot of mass uptake versus square root of time, is depicted in Fig. 12.2. The curve consists of three main segments: an initial linear portion, followed by a non-linear 'knee' region and finally a flat plateau denoting complete saturation. The diffusivity, D, can be calculated from the slope of the initial linear segment, as shown in Fig. 12.2, while the flat plateau provides the maximum saturation level, m_m.

12.4   Physical aging in resins and FRP

In addition to long term influences due to creep and diffusion, at temperatures slightly below glass transition, an aging phenomenon occurs in a polymer which is evidenced by an increase in stiffness with elapsed time. This phenomenon, known as physical aging, can cause a polymer to become

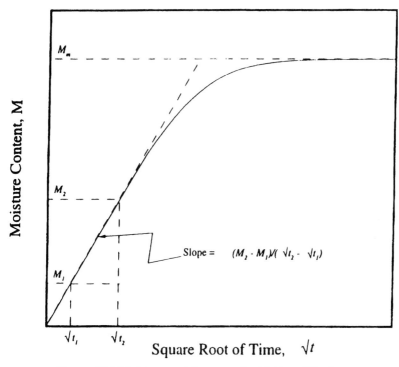

Figure 12.2 Moisture uptake curve for Fickian diffusion

less compliant and more brittle with age, as mentioned in Chapter 4, thereby increasing the likelihood of failure initiation. Therefore, it is important to take into account effects of aging when developing a long term durability predictive model for FRP.

Several investigators in the field of rheology have suggested that free volume is a good unifying parameter to describe changes in the timescale of material response in polymers. Free volume is the portion of the specific volume of the material that is unoccupied by the molecules. Researchers have applied the concept of free volume to develop a non-linear viscoelastic constitutive relationship [2] as well as for modeling coupled diffusion in viscoelastic materials [1].

Struik [3] originally associated the concept of polymer free volume with polymer transport mobility and, hence, with physical aging. Figure 12.3 shows a representation of changes in the free volume of a polymer with temperature. From this figure it is apparent that the amount of free volume available for molecular motion is strongly related to temperature, especially above the glass transition temperature (T_g). While a sample cooled very slowly from above T_g will follow the thermodynamic equilibrium path,

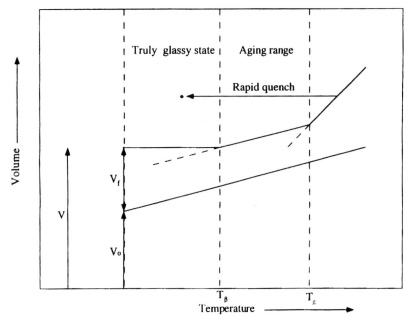

Figure 12.3 Change in polymer free volume with temperature

rapid quenching of a sample causes a specimen to deviate from this path, as shown in Fig. 12.3. In such an event, the sample attempts to return to its equilibrium state by a slow decay in its free volume with time, thereby giving rise to physical aging.

12.5   Durability models

12.5.1 Time-dependent non-linear constitutive behavior

Researchers have reported that composites exhibit non-linear viscoelastic behavior in tension as well as in shear [4–6]. Among the many single integral non-linearly viscoelastic constitutive laws that have been proposed, Schapery's law [6] was selected for implementation in the predictive model because (i) the non-linear parameters are relatively simple to characterize, and (ii) numerical implementation of the governing equations is straightforward and robust. In general, it has been observed that the creep and recovery behavior of a polymeric material can be described by a viscoelasticity constitutive law only when the material specimen is subjected to mechanical preconditioning. For an unpreconditioned material, the recovery behavior following creep cannot be adequately described by viscoelasticity alone. In such a situation, a combined viscoelastic/viscoplastic material model is necessary to characterize material behavior. One such model, proposed by Zapas and Crissman [7] employs a functional and accounts for the stress-history dependent non-linear viscoplastic strains. The combined Schapery–Zapas–Crissman model for uniaxial loading employed by Pasricha *et al.* [8] is given by:

$$\varepsilon^t = g_0^t \sigma^t D_0 + g_1^t \int_0^t \Delta D(\psi^t - \psi^s) \frac{d}{ds}\left[g_2^s \sigma^s\right] ds$$

$$+ \varphi\left[\int_0^t g_3^t(\sigma^s) ds\right] + \alpha \Delta T \qquad [12.6]$$

The first two terms on the right-hand side of equation [12.6] are viscoelastic terms proposed by Schapery, where ε^t represents uniaxial kinematic (or total) strain at time t, σ^t is the Cauchy stress at time t, D_0 is the instantaneous compliance and $\Delta D(\psi^t)$ is a transient creep compliance function. The factor g_0 defines stress and temperature effects on the instantaneous elastic compliance and is a measure of state dependent reduction (or increase) in stiffness. Transient compliance factor g_1 has a similar meaning, operating on the creep compliance component. The factor g_2 accounts for the influence of loading rate on creep. The function ψ^t represents a reduced timescale parameter defined by:

$$\psi^t = \int_0^t \left(\alpha_{\sigma_T}^S\right)^{-1} ds \qquad [12.7]$$

where α_{sT} is the stress-dependent timescale shift factor. In general, g_0, g_1, g_2 and α_{sT} are functions of stress level, temperature and moisture concentration. The third term on the right-hand side in equation [12.6] is the Zapas–Crissman functional and it accounts for the load history dependent viscoplastic strains. Finally, the fourth term on the right-hand side is the thermal strain. A benchmark verification of this model under thermomechanical cyclic loading is presented in Section 12.6.

12.5.2 Hygrothermal effects

A fiber reinforced composite material with a polymer matrix will typically absorb moisture in a humid environment and at elevated temperatures. Combined exposure to heat and moisture affects reinforced plastics in a variety of ways. First, the hygrothermal swelling causes a change in the residual stresses within the composite that could lead to microcrack formation. These microcracks in turn provide fast diffusion paths and thus alter the moisture absorption characteristics of the laminate. Second, heat and humidity may cause the matrix to become plasticized, thus causing an increase in the elongation to failure of the matrix.

Numerous diffusion models have been proposed over the years for modeling hygrothermal effects in FRP. The most frequently used model is the one-dimensional Fickian description. Unfortunately, this model tends to overestimate the moisture absorption in panels for short diffusion times, even when it is modified to take into account edge effects [9]. Some researchers have suggested that the deviation can be explained by a two-stage Fickian process [10,11]. Others claim that the diffusion process in FRP is really non-Fickian [12,13]. In reality, the nature of the diffusion process depends on the material and on the environmental conditions to which the material is exposed. For example, if the rate of viscoelastic relaxation in a resin is comparable to the rate of moisture diffusion, then the diffusion is likely to be non-Fickian. In addition, the presence of strong temperature and stress gradients have been known to engender non-Fickian driving forces. Hence, the applicability of Fick's Law for a given material system cannot be determined *a priori* but must be established through moisture uptake experiments under specified environmental conditions.

The concept of free volume was adapted by Roy *et al.* [14] to unify the effects of polymer viscoelasticity, temperature, volumetric change due to strain and penetrant concentration on the diffusion process within a polymer. The resulting non-linear diffusion coefficient is given by:

$$D = \frac{D_0}{T_0} T \exp\left[\frac{B}{f_0} \left\{ \frac{3(\alpha\Delta T + \gamma C^N) + \delta\left(\varepsilon_{kk} - \frac{1}{3} M_0 \sigma_{kk}\right)}{f_0 + 3(\alpha\Delta T + \gamma C^N) + \delta\left(\varepsilon_{kk} - \frac{1}{3} M_0 \sigma_{kk}\right)} \right\} \right]$$

[12.8]

where D is the non-linear diffusion coefficient at any material point at the current time, temperature, moisture concentration and dilatational strain, D_o is the diffusion coefficient at its reference state, T_o is the reference temperature, T is the current temperature, B is a material constant, f_o is the reference free volume fraction, α is the linear coefficient of thermal expansion, N is the exponent for the moisture concentration term in the definition of the non-linear diffusion coefficient (typically, $N = 1$), ΔT is the difference between the current temperature and the reference temperature, γ is the linear coefficient for change in free volume due to dilatational strain (typically $\delta = 1$), ε_{kk} is the dilatational strain, σ_{kk} is the dilatational stress, C is the moisture concentration at a material point and M_o is the instantaneous bulk compliance of the polymer.

Various types of coupled non-linear Fickian diffusion processes were numerically simulated using the free-volume approach given by equation [12.8], as well as non-Fickian transport. The non-Fickian transport was modeled as a stress-induced mass flux that typically occurs in the presence of non-uniform stress fields normally present in complex structures. The coupled diffusion and viscoelasticity boundary value problems were solved numerically using the finite element code NOVA-3D. Details of the non-linear and non-Fickian diffusion model have been described elsewhere [14]. A benchmark verification of the linear Fickian diffusion model defined by equations [12.3]–[12.5] under a complex hygrothermal loading is presented in Section 12.6.

12.5.3 Aging effects

Polymer-based composites exhibit time dependent properties and sub-T_g aging of these materials results in a combination of physical and chemical aging phenomena. Therefore, any durability study of polymer composites for elevated use-temperature applications should address both the physical and chemical aspects of aging. In general, aging causes polymers to become stiffer and more brittle with age, increasing the likelihood of more rapid progression of various damage states.

It is well known that a resin is generally not in thermodynamic equilibrium below its glass transition temperature (T_g). Below the glass transition, the volume, enthalpy and entropy of a glassy polymer changes with time, and it does that at an ever decreasing rate. This process of slow evolution toward an equilibrium state is what is known as 'physical aging' and it manifests itself in the mechanical properties of the polymer as well as in the thermodynamic quantities mentioned above. Physical aging has been correlated to changes in the free volume of a polymer with time after a sample has been quenched below its T_g. Hence, the mechanical behavior of glassy polymers and their composites depends on their thermal history. Physical aging results in a slow loss of polymer free volume which, in turn, results in a decrease in polymer chain segment mobility. The mobility of chain segments is inversely related to the relaxation time of the polymer chains. All effects of physical aging can be erased by heating a polymer to a temperature above its T_g, that is, physical aging is thermoreversible.

As stated in a preceding section, physical aging starts when a polymer is quenched to a temperature below its T_g, irrespective of the mechanical loading on the material. Therefore, when a sub-T_g polymer is subjected to mechanical load, two processes occur simultaneously: (1) physical aging and (2) stress relaxation or creep associated with mechanical load or deformation. Each of these processes has its own unique time and temperature dependence and hence, both effects must be accounted for in the long term viscoelastic/viscoplastic characterization of a polymer and its composites.

The two sequential processes that lead to the chemical degradation of a polymer due to oxidation are: (1) oxygen diffusion and (2) chemical reaction [15]. Researchers [16] have used the so-called 'unreacted core' model to characterize and predict thermo-oxidative degradation in a composite laminate. According to this model, the composite weight loss due to oxidation, q_c, can be expressed as a power law function of time:

$$q_c = D_E t^n \qquad [12.9]$$

where D_E and n are material constants. For oxygen diffusion controlled composite degradation, $n = 0.5$, and for a chemical reaction controlled degradation process, $n = 1$.

Struik [3] originally proposed a method to model physical aging through the use of a momentary creep master curve obtained from a series of short term creep tests performed at various aging times. The momentary creep master curve was then used in conjunction with the 'effective time' theory to predict long term creep in a polymer in the presence of physical aging.

Little data exists in the literature about the influence of physical aging on the long term viscoelastic properties of polymer composites. However, it is evident that for FRP, the composite properties are influenced by the viscoelastic behavior of its constituent polymer matrix. Struik demonstrated that the creep properties of glassy polymers are profoundly influenced by the physical aging process. Also, physical aging is a process that can persist for long periods of time and can influence toughness, yielding and non-linear creep. The time–temperature superposition principle that has been successfully applied to characterize polymer behavior above its T_g is generally not applicable to the long term viscoelastic behavior of polymers below their T_g. In such situations, the momentary creep master curve for the FRP must be obtained by performing creep tests where the test duration is much less than the time the sample was aged prior to loading. Sullivan [17] demonstrated that theoretical predictions of long term creep of FRP from momentary creep master curves using an 'effective time' theory were within 5% of experimental results.

The effective time, λ', is related to real time t by an integral expression given by:

$$\lambda' = \int_0^t \left(\frac{t_a}{t_a + \xi} \right)^\mu \, d\xi \qquad [12.10]$$

where t_a is the aging time, and μ is the aging shift rate.

The concept of effective time was employed by Roy et al. [14] to predict the effect of physical aging on the diffusion coefficient of a stretched polystyrene film. It was found that in the case of polystyrene, a faster rate of physical aging caused the diffusion coefficient to decay more slowly.

12.6 Model verification examples

12.6.1 Hygrothermal cycling of FRP laminates

In this model verification case, the moisture uptake history for a five-harness satin-weave [0/90/0/90]$_s$ graphite/epoxy laminate predicted by the model was compared with actual hygrothermal test data. To simulate in-service conditions, laminated panels were exposed continuously at 85% relative humidity at 85°F (29°C), interrupted daily by one simple simulated flight cycle conducted in an air-circulating oven. During the simulated flight cycle, laminates were heated from 85°F to the designated maximum temperature of 250°F (121°C) in 10 min and then held at that temperature for 80 min followed by a cool down to 85°F over 20 min. Five simulated 'flights' were conducted from Monday to Friday (one cycle daily) with no oven exposure on the weekend. One week downtimes from simulated flight cycles were introduced at the 4th and 12th weeks to simulate airline sched-uled short term maintenance. Two week downtimes were incorporated into the hygrothermal load history at the 7–8th and 15–16th weeks to simulate scheduled long term maintenance. During these weeks the laminate was stored in the humidity chamber. A total of 51 flight cycles were completed over the 4 month test period.

Since little damage was observed within the laminate in the course of the experiment, it was assumed that the diffusion process in the laminate essen-tially obeyed Fick's law given by equation [12.3]. A two-dimensional plane strain finite element model of the laminate was generated and the NOVA-3D finite element program was used to solve for the moisture uptake and stresses within the laminate. Based on characterization test data, the tem-perature dependent through-thickness diffusivity definition used in this analysis is given by:

$$D_z(T) = D_0 e^{A_1\left(\frac{1}{T_{REF}}-\frac{1}{T}\right)+A_2\left(\frac{1}{T_{REF}}-\frac{1}{T}\right)^2} \qquad [12.11]$$

with the in-plane diffusivity $D_x = 2.58\, D_z$. The material constants given in equation [12.11] are defined in Table 12.1.

Figure 12.4 depicts the comparison of the measured moisture uptake history for the laminate with NOVA-3D predictions. In general, the model prediction agrees reasonably well with test data over the 112 day period used for this comparison, especially during the absorption cycles. Figure 12.5 shows the predicted evolution of the in-plane stress at the exposed laminate surface with hygrothermal cycling. The increase in in-plane tensile

Table 12.1 Material constants for temperature
dependent diffusion

D_0 (cm^2 s^{-1})	2.40×10^{-9}
A_1	7.7023488×10^3
A_2	-3.5539691×10^6
α_1 (K^{-1})	9×10^{-6}
α_3 (K^{-1})	2.75×10^{-5}
γ_1 (cm^3 g^{-1})	1.28×10^{-3}
γ_3 (cm^3 g^{-1})	3.92×10^{-3}
T_{ref} (K)	302

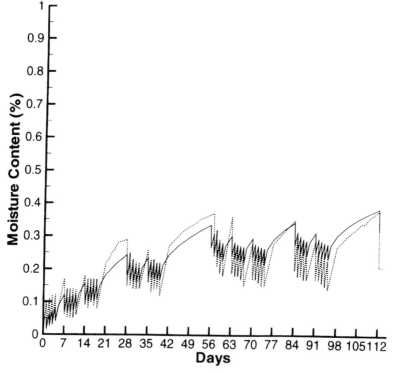

Figure 12.4 Cyclic moisture uptake in a woven graphite/epoxy
laminate. ——, NOVA-3D; ...., experimental

stress at the laminate surface during a dry cycle immediately following an
extended absorption cycle is due to the increased moisture induced expansion in the inner layers of the laminate. The progressive increase in peak
tensile stress could ultimately lead to damage initiation in the form of
matrix cracking at the exposed surface of the laminate.

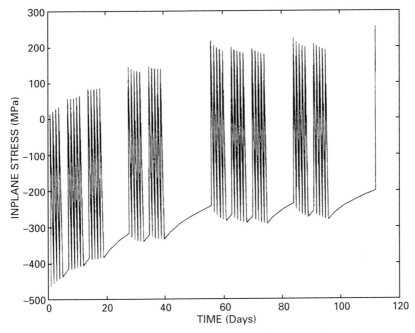

Figure 12.5 Variation in in-plane stress due to cyclic moisture uptake in a woven graphite/epoxy laminate

12.6.2 Non-linear creep and recovery of FRP laminates

In order to verify some of the predictive capabilities of the finite element model described in the previous sections, the transverse creep response of a IM7/5260 composite investigated in a earlier study [8] was used as a benchmark case. Two separate load histories were considered: (1) transverse creep and recovery of a $[90]_{16}$ specimen under isothermal conditions, and (2) transverse creep of a $[90]_{16}$ specimen subjected to cyclic thermomechanical loading for extended periods of time.

The Schapery–Zapas–Crissman (SZC) model that was described in an earlier section was used to model the constitutive behavior of a $[90]_{16}$ specimen. Under conditions of uniaxial loading transverse to the fiber direction at constant temperature, the SZC constitutive model takes the form:

$$\varepsilon_{22}^{t} = g_{0}^{t}\sigma_{22}^{t}D_{22} + g_{1}^{t}\int_{0}^{t}\Delta D_{22}(\psi^{t} - \psi^{s})\frac{\mathrm{d}}{\mathrm{d}s}\left[g_{2}^{s}\sigma_{22}^{s}\right]\mathrm{d}s \qquad [12.12]$$

$$+ \varphi\left[\int_{0}^{t}g_{3}^{s}(\sigma_{22}^{s})\mathrm{d}s\right] + \alpha_{22}\Delta T$$

Table 12.2 Material properties for the Schapery–Zapas–Crissman constitutive model

Viscoelastic coefficients:

Prony terms	1	2	3	4	5
D_r (MPa^{-1})	1.147×10^{-4}	1.200×10^{-6}	1.795×10^{-6}	8.823×10^{-7}	1.954×10^{-7}
λ_r (min^{-1})	—	0.001	0.01	0.1	1.0

g_0	g_{0T} for $\tau_{oct} < 9.65$ MPa $g_{0T} + 2.306 \times 10^{-3} (\tau_{oct} - 9.65)$ for $\tau_{oct} > 9.65$ MPa
g_1	1.0
g_2	1.0
g_{0T}	$-0.000\,846\,(T_{ref} - T) + 1.056\,23$

Viscoplastic coefficients:

C_{vp}	$1.1714 \times 10^{-6} \times e^{(T_{ref} - T)/71.925}$
N	6.25
n	0.14

The creep compliance function $\Delta D_{22}(t)$ is modeled by using a Prony series of the form,

$$\Delta D_{22}(t) = \sum_{r=1}^{n} D_r \left(1 - e^{-\lambda_r \psi}\right) \qquad [12.13]$$

$$g_3^t(\sigma_{22}^t) = C_{VP}(\sigma_{22})^N \qquad [12.14]$$

Expressing the viscoplasticity function $g_3(s_{22})$ as a power law in stress [7], the resulting viscoplasticity functional is given by:

$$\phi\left[\int_0^t g_3^s(\sigma_{22}^s)ds\right] = \left[\int_0^t C_{VP}(\sigma_{22})^N ds\right]^n \qquad [12.15]$$

where C_{VP}, N and n are temperature-dependent viscoplasticity material constants.

All material parameters used in this benchmark verification were obtained from material characterization data for a IM7/5260 composite published by another group [8] and are given in Table 12.2. The authors of that study performed creep and recovery tests to characterize the non-linear creep behavior of the composite.

In this section, NOVA-3D predictions are verified against the exact solution for non-linear creep and recovery of an IM7/5260 $[90]_{16}$ specimen for two different stress levels. Figure 12.6 shows a comparison of the creep and recovery of an IM7/5260 $[90]_{16}$ specimen predicted by NOVA-3D with exact solution. The constitutive law expressed in equation [12.12] was used for NOVA-3D predictions. The exact solution for this case is given by:

$$\varepsilon_{22_c}^t = \left[g_0^t D_{22} + C g_1^t g_2^t \left(\frac{t}{a_\sigma} \right)^p \right] \sigma_{22} + C_{VP}^n \sigma_{22}^{Nn} t^n \qquad [12.16]$$

and

$$\varepsilon_{22_R}^t = \frac{C g_1^t g_2^t}{g_1^t} \left(\frac{t_1}{a_\sigma} \right) \left[(1 + a_\sigma \lambda)^p + (a_\sigma \lambda)^p \right] \sigma_{22} \qquad [12.17]$$

for the creep and recovery strains, respectively. In the above expressions $\lambda = (t - t_1)/t_1$ and a power law function of the form $\Delta D(t) = Ct^p$ was used to define the tensile creep compliance. Values of $C = 4.8761 \times 10^{-7}$ and $p = 0.325$ were obtained through a least-squares curve fit and used for this comparison. Excellent agreement is obtained for all stress levels, as can be seen in Fig. 12.6.

12.6.3 Creep of a laminate under cyclic thermomechanical loading

12.6.3.1 Case 1: without physical aging

In order to verify some of the long term predictive capabilities of the finite element model described in the previous sections, the transverse creep of a IM7/5260 $[90]_{16}$ specimen subjected to cyclic thermomechanical loading investigated in an earlier study [8] was used as a benchmark. The authors of that study performed creep and recovery tests to characterize the non-linear creep behavior of the composite and then subjected the composite specimens to cyclic thermomechanical loading for up to 6 months.

The Schapery–Zapas–Crissman (SZC) model described in an earlier section was used to model the constitutive behavior of a $[90]_{16}$ specimen. All material parameters used in this benchmark verification are listed in Table 12.2.

Figure 12.7 depicts the thermomechanical loading cycle that was imposed on the 6 month test. The IM7/5260 $[90]_{16}$ specimen was subjected to a

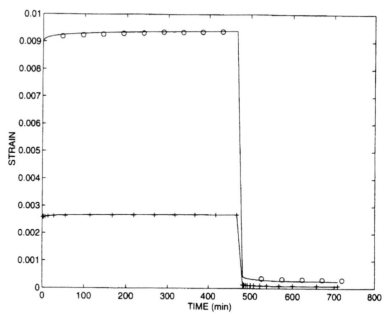

Figure 12.6 Comparison of NOVA-3D predictions with exact solution for transverse creep and recovery of an IM7/5260 [90]$_{16}$ specimen. ——, NOVA-3D; +, exact 21 MPa; ○, exact 70 MPa

periodic stress of 38.5 MPa. The thermal cycle consisted of maintaining a temperature of 121°C for 4 h in phase with the mechanical load, followed by a gradual reduction to 79°C in approximately 30 min and then held there for 1 h. Finally, the temperature was gradually increased back up to 121°C in 30 min and held there for 1 h, thus completing the 7 h cycle.

Figure 12.8 depicts a comparison of NOVA-3D predictions for thermomechanical cycling of the IM7/5260 [90]$_{16}$ specimen with test data. To avoid making the figure unnecessarily cluttered, predictions for cycles 1, 96, 192, 288 and 384, representing one month intervals, are presented. Except for a 2% discrepancy in the creep strain at the start of each cycle, good agreement is observed between the unaged NOVA-3D predictions indicated by the broken line and test data.

12.6.3.2 Case 2: with physical aging

The instantaneous modulus from every cycle can be determined by dividing the applied stress by the instantaneous elastic strain at the beginning of each cycle. A comparison of the instantaneous strains at the beginning of each of the cycles from 96 to 384 reveals a steady decrease in the strain magnitude. The resulting steady increase in the measured instantaneous modulus indi-

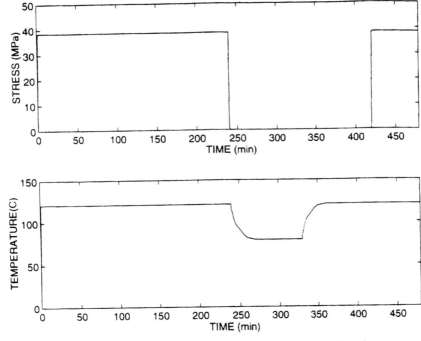

Figure 12.7 Thermomechanical loading cycle imposed on long term creep test

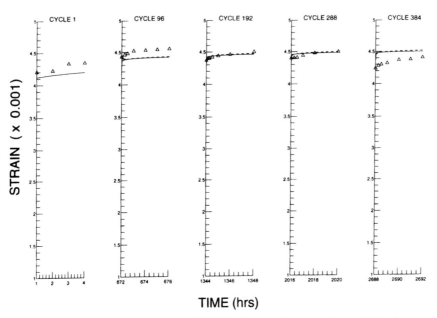

Figure 12.8 Comparison of NOVA-3D predictions with test data for thermomechanical cycling of an IM7/5260 [90]$_{16}$ specimen. ------, without aging; ——, with aging; △, test data

cates an aging phenomena that causes the instantaneous modulus of the specimen to become stiffer with time. In order to investigate this behavior, physical aging was introduced in the NOVA-3D governing equation by the use of an effective time parameter defined by equation [12.10]. Due to the unavailability of aging parameters (t_a and μ) as well as the momentary creep master curve for the IM7/5260 composite, generic values were assigned to these parameters. The value of aging time (t_a) was assumed to be 20000 min which is about 12% of the test duration, and the aging shift rate (μ) was conservatively assumed to be equal to 1.0 for this study. In addition, the momentary creep master curve was assumed to be identical to the linear creep compliance curve obtained using the time–temperature superposition principle.

While the creep strain curves shown by the solid line in Fig. 12.8 exhibit a slight stiffening effect due to physical aging, it is apparent that the progressive stiffening of the instantaneous modulus is not a viscoelastic phenomenon and therefore it cannot be accounted for by a physical aging model alone. It is likely that the stiffening of the instantaneous modulus is due to chemical changes in the resin, engendered by postcuring of the resin during thermal cycling.

12.7   Summary and conclusions

A comprehensive analytical model for predicting long term durability of resins and of fibre reinforced plastics (FRP) taking into account viscoelastic/viscoplastic creep, hygrothermal effects and the effects of physical and chemical aging on polymer response has been presented. An analytical tool consisting of a specialized test-bed finite element code, NOVA-3D, was used for the solution of complex stress analysis problems, including interactions between non-linear material constitutive behavior and environmental effects.

Computer simulations of the hygrothermal cycling of a woven graphite/epoxy laminate and transverse creep response of a unidirectional IM7/5260 composite were presented as benchmark problems to demonstrate some of the predictive capabilities of the proposed durability prediction model. Good agreement was obtained between the proposed model (NOVA-3D) predictions, exact solutions and published results for cases where physical aging was not considered. When physical aging was included in the proposed model using an effective-time parameter, the predicted increase in the rate of stiffening was found to be small when compared with actual test data. It is therefore likely that in addition to physical aging, there are other aging processes that may be causing the IM7/5260 specimen to become increasingly 'stiff' with continued thermomechanical cycling. However, the accuracy of the aging predictions can be completely verified only after the momentary creep properties for the IM7/5260 composite are determined from short time tests.

Acknowledgments

The author would like to acknowledge the contribution of Mr. Richard Cornelia for providing the hygrothermal test data and Mr. Vikas Gupta for his assistance in preparing the manuscript.

References

1 S Roy and J N Reddy, 'A finite element analysis of adhesively bonded composite joints with moisture diffusion and delayed failure', *Computers & Structures* 1988 **29**(6) 1011–1031.

2 W G Knauss and J J Emri, 'Nonlinear viscoelasticity based on free volume considerations', *Compos Struct* 1981 **13** 123–128.

3 L C E Struik, *Physical Ageing in Amorphous Polymers and Other Materials*, Elsevier, Amsterdam, 1978.

4 R Mohan and D F Adams, 'Nonlinear creep-recovery response of a polymer matrix and its composites', *Exper Mech* 1985 **25** 262–271.

5 M E Tuttle and H F Brinson, 'Prediction of the long-term creep compliance of general composite laminates', *Exper Mech* 1986 **26** 89–102.

6 R A Schapery, 'A theory of non-linear thermo-viscoelasticity based on irreversible thermodynamics', *Proceedings 5th US National Congress Applied Mechanics*, ASME, 1966, p 511.

7 L J Zapas and J M Crissman, 'Creep and recovery behavior of ultra-high molecular weight polyethylene in the region of small uniaxial deformations', *Polymer* 1984 **25** 57–62.

8 A Pasricha, M Tuttle and A Emery, 'Time-dependent response of IM7/5260 composites subjected to cyclic thermo-mechanical loading', *Compos Sci Technol* 1996 **56** 55–62.

9 C H Shen and G S Springer, 'Moisture absorption and desorption of composite materials', *J Compos Mater* 1976 **10** 2–20.

10 S Y Lo, H T Hahn and T T Chaio, 'Swelling of Kevlar 49/epoxy and S2-glass/epoxy composites', *Progress in Science and Engineering of Composites*, ICCM IV, Tokyo 1982, pp 987–1000.

11 M E Gurtin and C Yatomi, 'On a model for two phase diffusion in composite materials', *J Compos Mater* 1979 **13** 126–130.

12 C D Shirrell, 'Diffusion of water vapor in graphite/epoxy composites', *ASTM STP*, 1978 **658** 21–42.

13 J M Whitney and C E Browning, 'Some anomalies associated with moisture diffusion in epoxy matrix composite materials', *ASTM STP*, 1978 **658** 43–60.

14 S Roy, D R Lefebvre, D A Dillard and J N Reddy, 'A model for the diffusion of moisture in adhesive joints. Part III: Numerical simulations', *J Adhesion* 1989 **27** 41–62.

15 H Parvatareddy, J Z Wang, D A Dillard and T C Ward, 'Environmental aging of

high-performance polymeric composites: effects on durability', *Compos Sci Technol* 1995 **53** 399–409.

16 J D Nam and J C Seferis, 'Anisotropic thermo-oxidative stability of carbon fiber reinforced polymeric composites', *SAMPE Quart* October 1992, 10–18.

17 J L Sullivan, 'Creep and physical aging of composites', *Compos Sci Technol* 1990 **39** 207–232.

Index